Solid Base Catalysts

Solid Base Catalysts

Synthesis, Characterization, and Applications

Edited by *Ravi Tomar, K. K. Pant, and Ramesh Chandra*

WILEY-VCH

Editors

Dr. Ravi Tomar
Department of Chemistry
SRM Institute of Science & Technology
Delhi-NCR Campus
Modinagar, Ghaziabad
India, 201204

Prof. K. K. Pant
Indian Institute of Technology Delhi
New Delhi
India, 110016

Prof. Ramesh Chandra
Department of Chemistry
University of Delhi
New Delhi
IN, 110007

Cover: © Xander/Adobe Stock Photos, © ibreakstock/Shutterstock

Library of Congress Card No.: applied for

British Library Cataloguing-in-Publication Data: A catalogue record for this book is available from the British Library.

Bibliographic information published by the Deutsche Nationalbibliothek
The Deutsche Nationalbibliothek lists this publication in the Deutsche Nationalbibliografie; detailed bibliographic data are available on the Internet at <http://dnb.d-nb.de>.

Print ISBN: 978-3-527-35376-7
ePDF ISBN: 978-3-527-84670-2
ePub ISBN: 978-3-527-84669-6
oBook ISBN: 978-3-527-84671-9

Typesetting Straive, Chennai, India

Printed and bound by CPI Group (UK) Ltd, Croydon, CR0 4YY

C9783527353767_011024

Contents

Preface

In 1958, Pines and Haag conducted a series of investigations on solid base catalysts. They found that alkali metal-supported alumina can catalyze double-bond isomerization of alkenes via carbanions as intermediates. Subsequently, in the 1970s, other metal oxides were also discovered to have catalytic activity for base-catalyzed reactions. The advantages of using solid base catalysts over liquid base catalysts in industrial processes are numerous, such as not requiring neutralization of the reaction mixture, allowing high-temperature reactions, employing flow reactor systems for highly productive processes, and facilitating easy separation of catalysts from products. The rapid development of material chemistry and catalysis chemistry has ignited significant interest in the heterogeneous catalysis field, with sophisticated arrangements and the production of solid basic catalysts. In contrast, homogeneous catalysts have several drawbacks, such as being nonrecyclable, producing low-quality side products, and generating a large amount of waste that leads to energy and chemical waste. Solid basic catalysts play a crucial role in many base-catalyzed industrial procedures, offering opportunities for aiding separation, minimizing corrosion, and resolving pollution issues. This book provides a comprehensive overview of the latest advanced studies in the synthesis and characterization of solid base catalysts, along with their applications in various fields such as industrial, catalytic, biomedical, among others. The book focuses on how advanced techniques can be applied to characterize various types of solid base catalysts. It can serve as a valuable educational resource for students and professionals in the field of material and organic synthesis, providing insights and knowledge from experts in the industry.

Dr. Ravi Tomar
Department of Chemistry
SRM Institute of Science & Technology
Delhi-NCR Campus, Modinagar, Ghaziabad
India, 201204

Prof. K. K. Pant
Department of Chemical Engineering
Indian Institute of Technology Delhi
Hauz Khas, New Delhi, India, 110016

Prof. Ramesh Chandra
Department of Chemistry, University of Delhi
Delhi, India, 110007

1

Introduction to Solid Base Catalyst

Indu Sindhu[1], *Ravi Tomar*[2], *and Anshul Singh*[1]

[1] *Baba Mastnath University, Department of Chemistry, Asthal Bohar, Rohtak, 124021, India*
[2] *SRM Institute of Science & Technology, Department of Chemistry, Delhi-NCR Campus, Modinagar, Ghaziabad 201204, India*

1.1 Introduction

With the expeditious growth of material chemistry and catalysis chemistry, the sophisticated arrangement and manufacture of solid base catalysts have sparked stupendous interest in the heterogeneous catalysis field. Moreover, there is an array of shortcomings for the homogeneous catalysts: they are nonrecyclable, produce low-quality side products, and generate a large amount of waste that leads to energy and chemical waste that still needs to be addressed. The principal supremacy of heterogeneous catalysts over homogeneous catalysts is their harmless nature and highly basic nature. In the context of catalyst regeneration and the potential to be reused in ongoing processes, heterogeneous catalysts reduce the drawbacks of homogeneous catalysis. The heterogeneous catalytic process has been reported to have greater economic potential than the homogeneous one, as reported in the literature. For example, the solvent role is very different in solid base catalysis in comparison to homogeneous catalysis because in solid catalysis the reaction can be executed without the use of a solvent [1]. As a consequence, when the reaction occurs entirely, there is no incentive to separate the solvents. Evidently, no solvent waste is emitted and solid catalysts could potentially be used in additional reactions [2, 3]. For instance, heterogeneous catalysts have 4–20% lower refinery expenses as compared with homogeneous catalysts [4, 5]. Overall, solid base catalysts have a number of incentives encompassing effectiveness in separating their components from the reaction mixture, the potential to be reutilized, stability in challenging reaction conditions and the ability to proceed in a shorter time [6–8]. They are often involved in a variety of industrial processes, notably hydrogenation, aldol condensation, transesterification, Henry reaction, Knoevenagel, Wittig reaction, Michael addition and Cannizzaro reaction. These catalysts facilitate the reaction pathways by having basic sites on their surfaces that can receive protons (H^+) or donate lone pairs of electrons to reactants [9]. Solid base catalysts are present in a different phase, generally a solid, while the reactants are in a different phase, typically a

Solid Base Catalysts: Synthesis, Characterization, and Applications, First Edition.
Edited by Ravi Tomar, K.K. Pant, and Ramesh Chandra.

liquid or a gas. This is in contrast to conventional homogeneous catalysts, which are present in the similar phase as the reactants [10]. In essence, solid base catalysts ought to replace the role of homogeneous base catalysts in industrial processes for the purpose of streamlining procedures and sustaining the environment. For the purpose of elucidating their catalytic properties in solid base catalysts, their basic properties must be scrutinized. It depends on the locality of basic sites, the definite count of basic sites and their basic dominance factors.

The fundamental constituents of solid base catalysts are the poisoning of the active site by acidic molecules such as HCl, CO_2 and water, which persuades the firmness of the basic sites. Characterization methods and several kinds of other techniques, such as the variation in color of acid–base indicators and the adsorption of acidic molecules, signify the presence of basic sites on surfaces [11]. The reaction pathway includes the inclusion of anionic intermediates, as evidenced by spectroscopic analyses and the reaction proceedings are extremely comparable to base-catalyzed reactions that are prominent in homogeneous systems. Since the beginning of the twentieth century, it has been discovered that plenty of reactions are possible using various solid base catalysts. Numerous studies on solid base catalysts suggest that their catalytic behavior was principally determined by structure, property and the existence of active basic sites on surface [12–14].

1.2 History and Main Facts on Solid Base Catalysts

Pines and Haag conducted the foremost work on the heterogeneous base catalyst, demonstrating that sodium metal dissolved in alumina acted as a potent catalyst for the double-bond isomerization of alkenes [15]. "Solid Acids and Bases" written by Professor Kozo Tanabe was an innovative work that popularized the concept of "solid base" in the catalysis world in 1970 [2]. The work addressed significant research on solid acid and base catalysis carried out throughout the 1950s and 1960s. As a sequel to "Solid Acids and Bases," Tanabe, Misono, and other authors published "New Solid Acids and Bases" in 1989, providing an overview of the advances in the area during the 1970s and 1980s [16]. It described contemporary advances in the field throughout the 1970s and 1980s. Both solid acids and solid bases were covered in the two publications, but the focus was primarily on the former, which was typical for those decades. Significant strides have been made in solid base-catalyzed processes and catalytic materials since the early 1990s. This is because solid base catalysts has been proven to be environmentally benign. Consequently, there is an expanding utilization of solid base catalysts in organic reactions.

Until that time, research on solid bases has advanced significantly in terms of catalyst materials and catalytic processes, but with greater slack than research on solid acids. When the catalysts were prepared under vacuum, Tanabe and colleagues reported in 1972 that magnesium oxide and calcium oxide demonstrated extraordinarily high catalytic activity for 1-butene isomerization [4, 17]. This research elucidated the significance of the base catalyst pretreatment and preparation techniques.

Tanabe and his colleagues conducted a thorough investigation of the fundamental characteristics of different metal oxides, combined metal oxides, and the wide range of processes that these substances catalyzed in the 1970s. A family of solid bases for new materials entered the market in the interim period. Different approaches to identifying the reaction intermediates and describing the fundamental characteristics of solid surfaces were also developed [18].

1.3 Literary Perspective of Solid Base Catalyst

The coprecipitation approach was adopted to fabricate heterogeneous solid catalysts made of mixed oxides of CaO and ZrO_2, containing various Ca-to-Zr ratios. Temperature-programmed desorption, Brunauer–Emmett–Teller (BET), X-ray diffraction (XRD), and X-ray photoelectron spectroscopy(XPS) were employed for the characterization by Dehkordi and Ghasemi. The increase in catalyst activity with an uptick in the Ca-to-Zr ratio is corroborated by the results of the experiment. These chemically synthesized mixed oxides were tested as catalysts in the transesterification of residual cooking oil and methanol at 65°C to yield biodiesel fuel [19].

Wang et al. designed Mg–Al oxides, which were utilized as catalysts for the transesterification of ethanol and dimethyl carbonate. Mg–Al-layered double hydroxide was crushed without the assistance of a solvent. Fourier transform infrared (FTIR), scanning electron microscopy (SEM), XPS, and XRD were employed to validate the surface properties. With moderate basic sites and high BET surface areas, a molar ratio of 2.0 was shown to yield satisfactory catalytic activity [20].

The solid base catalysts was fabricated by Wei et al. by heating a porous hydroxyapatite doped with $Sr(NO_3)_2$ at 873 K. The thermogravimetric analysis (TGA), XRD, FTIR, SEM, BET, and indicator techniques were implemented for the characterization. With an 85% conversion rate, these innovative solid base catalysts demonstrate use in the transesterification of soybean oil [21].

Magnesium oxide/carbon mesoporous composites were fabricated utilizing potassium chloride as the salt template and alkali lignin as the carbon source, resulting in a range of Mg-doping ratios by Wang and research team. For characterization, the BET, FTIR, SEM, and transmission electron microscopy (TEM) were employed. To optimize the fructose yield during glucose isomerization, these chemicals were employed as base catalysts [22].

Na-modified graphite carbon nitrides were synthesized by Kim et al. and used to transesterify soybean oil and methanol. Melamine and sodium hydroxide were polymerized co-thermally to produce the catalyst. Carbon dioxide diffuse reflectance, SEM, FTIR, and XRD were employed for the characterization. The density functional theory (DFT) theory and CO_2 diffuse reflectance infrared Fourier transform spectroscopy verified the basicity produced by the electron transfer from the sodium to nitrogen atoms [23].

Lithium-doped Li/NaY zeolite materials were designed using microemulsion-assisted impregnation with varying Li_2CO_3 to NaY molar ratios by Li and research group. Several alkali species were added to the zeolites, which increased the strength and catalytic property of Li/NaY catalysts in the ethanol-castor oil transesterification reactions [24].

The TiO_2-based Na–SiO_2 heterogeneous base catalysts were synthesized and characterized with SEM, XPS, FTIR, and BET measurements by Naeem et al. This catalyst was successfully implemented in production of biodiesel from waste cooking oil with 98% yield. Another important potential of these catalysts is their ability to be reused after five successive runs in reactions [25].

Ye et al. synthesized Mg–Al hydrotalcite catalyst using bottom-up approach. Characterization was performed via DFT and FTIR studies. These were applicable in glucose to fructose isomerization, and experimental data confirm conversion through base-catalyzed route. The interaction between catalyst and glucose rises with increase in Mg–Al molar ratio [26].

A two-dimensional (2D) nanoporous carbonaceous solid base catalyst was synthesized by one-step carbonization method utilizing chitosan by Kong and his team. They were tested for their applicability in transesterification to generate biodiesel and Knoevenagel condensation reactions. The result suggests their adequate potency in these reactions. The literature also claims their use in biodiesel production [27].

Mg–Al-based hydrotalcites were fabricated by Park and his research team employing various approaches including coprecipitation, simple, and urea methods. The results exemplified that synthesis method strongly affects catalytic properties of catalyst. Coprecipitation method-based catalysts were best ones that are utilized for glucose isomerization due to their small crystalline size [28].

1.4 Solid Basic Sites

Solid basic sites are some discrete sites present on surface of solid bases, which can act as a base. These basic sites can extract protons from reactant species (RH) and lead to formation of carbanions (R^-).

$$RH + B \rightarrow R^- + BH \tag{1.1}$$

In this reaction (1.1), the basic center on the surface of solid behaves as Brønsted base. In some cases, such as in carbonyl groups (aldehyde and ketones), these basic sites can act as Lewis base (1.2), as observed in aldol condensation and Knoevenagel reactions. As observed in many reactions, these basic sites can simultaneously act as Lewis base (e-donor) and Brønsted base (proton acceptor).

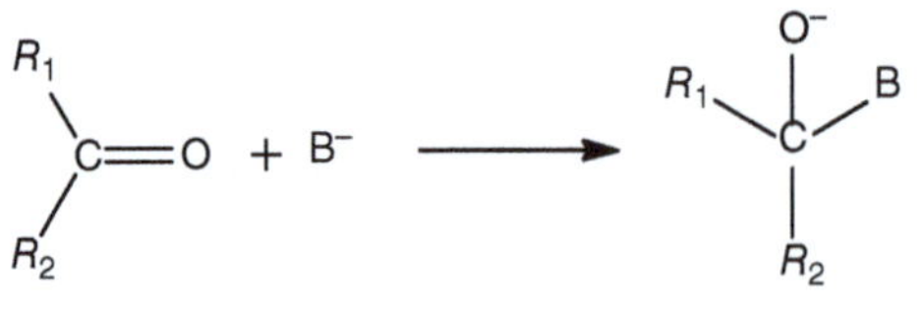

(1.2)

For most of the basic catalysts, banishing oxygen and water molecules over the surface and pretreating them at extremely high temperatures is the most crucial step to exhibit the ability to illustrate basic characteristics that foster base-catalyzed reactions. The temperature required for pretreatment mainly depends on decomposition temperature of adsorbed oxides and hydroxide molecules. The molecules that are attached frailly are detached at lower temperatures and the sturdy molecules are sequestrated at high-temperature conditions. In infrequent instances, with increasing pretreatment temperature repositioning of bulk and surface particles eventuates, which expedites the unusual change in nature of basic center on surface, and this alteration reflects in their catalytic power on variation with temperature [17].

1.5 Types of Solid Base Catalysts

To remove acidic impurities adsorbed on surfaces of basic sites pretreatment is essential and this treatment generally follows volcanic-type variation as a function of temperature. The left side in Figure 1.1 points toward emergence of basic site and right side focuses on deliberation of basic sites by surface realignment of atoms [29].

1.5.1 Metal Oxides

1.5.1.1 Alkaline Earth Oxides

With some anomalies, this type of catalyst catalyzed almost every base-catalyzed reaction. MgO, CaO, BaO, and SrO are examples of alkaline earth metal oxides but among all, MgO is the most extensively explored owing to its definite structure with large surface area and synthesizes effortlessly as compared to CaO, SrO, and BaO preparation. Because of structural similarity of CaO, SrO, and BaO with magnesium oxide, their catalytic properties can be clarified with the study of MgO catalytic activity. The basicity trend followed by alkaline oxides is BaO > SrO > CaO > MgO. CaO and MgO can be prepared smoothly from thermal decomposition of their hydroxide salts. SrO and BaO active sites are synthesized from their carbonate salts, as their

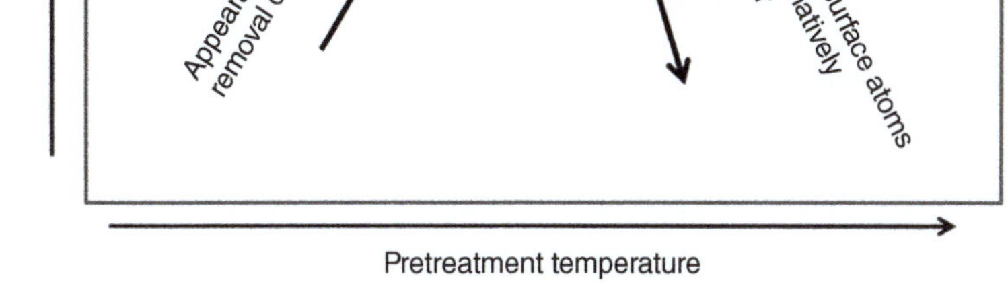

Figure 1.1 Volcano shape in catalysis as a function of temperature in solid base catalysis. Reprinted with permission Ref. [29] © Elsevier, 2015.

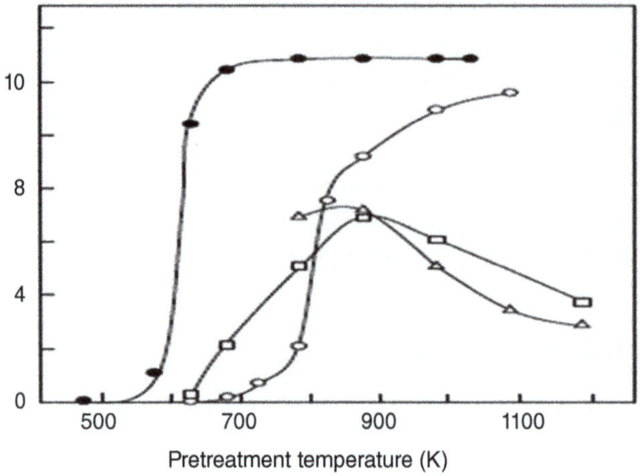

Figure 1.2 Variations in (•) amount of water evolved. (Δ) surface area/100 $m^2 g^{-1}$. (□) Activity of 1-butene isomerization/a.u. as a function of pretreatment temperature of CaO. (○) Amount of CO_2 evolved/20 mmol g^{-1}. Reprinted with permission Ref. [18] © Springer Nature, 2011.

hydroxide salt melts at a certain temperature range. Figure 1.2 represents that with deaeration temperature the active sites of CaO are produced by detachment of water and CO_2 gases [30, 31].

To determine how effortless it is to eradicate protons from the reactant under the reaction conditions deployed, differing basic sites need to be met for base-catalyzed reactions. When a proton is easily extracted from a reactant, the reaction is feasible even on a weak basic site, whereas if a proton is laborious to remove, the reaction will only occur on a strongly basic site. The exact kind of reaction and the particular form of alkaline earth oxide ascertain the pinnacle of pretreatment temperature. Figure 1.3 emphasizes how the activity of MgO produced from hydroxide by decomposition in a vacuum fluctuates based on the pretreatment temperature for different processes [32]. Depending on the type of preparedness, various temperatures are recommended for pretreatment. The temperatures for MgO demonstrating the most stringent instances of reactivity are 800 K just for 1-butene isomerization, 973 K for methane–D_2 exchange, and 1300 K for 1,3-butadiene hydrogenation. The dominance of the basic site needed to trigger the reaction is depicted in the variance in the optimum pretreatment temperature. Proceeding from 1-butene isomerization toward 1,3-butadiene hydrogenation necessitates more formidable basic sites.

1.5.1.2 Zirconium Oxides

Although ZrO_2 catalyzes numerous base-catalyzed reactions, still its functions are highly different from other alkali-based oxides. The acid–base bifunctional characteristics of ZrO_2 are extensively utilized for deciphering the catalytic characteristics that distinguish it from other solid base catalysts. Some industrial reactions in which zirconium oxides are utilized as catalysts are:

- Hydrogenation of carboxylic acids (aromatic) to aromatic aldehydes [33]
- Conversion of isobutyraldehyde into diisopropyl ketone [34]

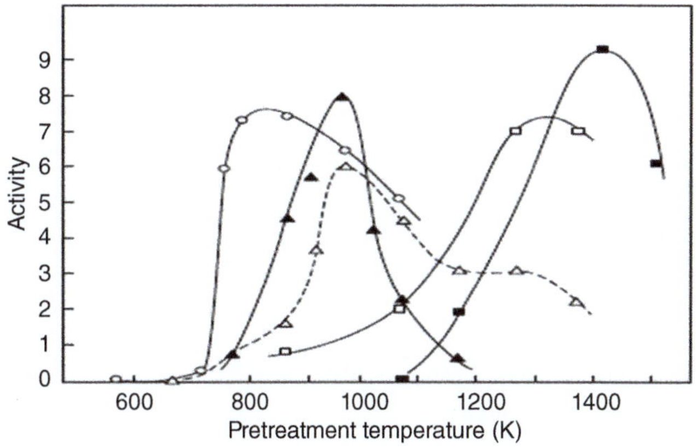

Figure 1.3 Variations in MgO activities with various reactions versus pretreatment temperature as a function. (o) 1-Butene isomerization at 303 K, (□) 1,3-butadiene hydrogenation at 273 K, (Δ) 1,3-butadiene amination at 273 K, (▲) methane-D_2 exchange at 673 K, (■) ethylene hydrogenation at 523 K. Reprinted with permission Ref. [18] © Springer Nature, 2011.

- Dehydration reaction of 2-hydroxypropylamine [35]
- Production of cyclohexylethylene from 1-cyclohexylethanol dehydration [36, 37].

The temperature at which ZrO_2 is produced from hydrous ZrO_2 influences its catalytic activity. The sort of reaction dictates the temperature at which the highest activity emerges. Figure 1.4 provides examples of the hydrogenation of 1,3-butadiene and the H_2–D_2 equilibration, both demonstrate optimal activity at 873 K for pretreatment and the activities vanish at 1073 K for pretreatment. However, the activity maxima for isomerization of 1-butene and transfer hydrogenation of 1,3-butadiene with cyclohexadiene occur at a temperature of 1073 K during the pretreatment phase. As the pretreatment temperature rises, the surface area reduces, and the fraction of monoclinic phase is enhanced. Tanabe and Yamaguchi hypothesize that the bond gap between Zr and O varies with pretreatment temperature, which, in consequence, creates fluctuations in the activity [38].

1.5.1.3 Rare Earth Oxides

Rare earth oxides have not been frequently used in base-catalyzed processes despite having strong basic characteristics. The oxides are also endowed with an innate capacity to oxidize. Rare earth oxides are created from the hydroxides either through air calcination or high-temperature vacuum decomposition. By hydrolyzing aqueous nitrates with aqueous ammonia, hydroxides can be generated [37].

The conditions under which hydroxides decompose, including temperature and environment, affect the surface areas of the rare earth oxides that are produced. The oxides produced by hydroxide breakdown at 873 K in a vacuum have certain surface areas between 10 and 50 $m^2\,g^{-1}$. The synthesis of oxides from hydroxides typically transpires through three different stages. Figure 1.5 displays the thermogram of $La(OH)_3$ to La_2O_3 in a vacuum [37]. The initial step of $La(OH)_3$ decomposition was obtained in the temperature range of 523–623 K and led to the generation of a distinct

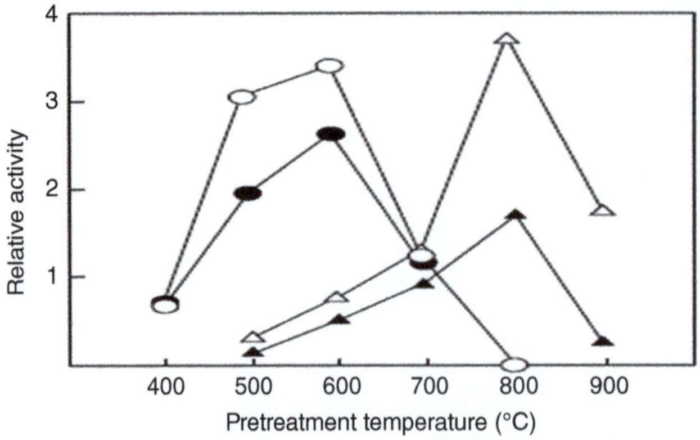

Figure 1.4 Catalytic activity of ZrO_2 at various pretreatment temperatures. (Δ) Isomerization of 1-butene; (o) 1,3-butadiene hydrogenation with H_2; (•) equilibration of H_2–D_2; (▲) 1,3-butadiene hydrogenation with cyclohexadiene. Reprinted with permission Ref. [38] © Elsevier, 1994.

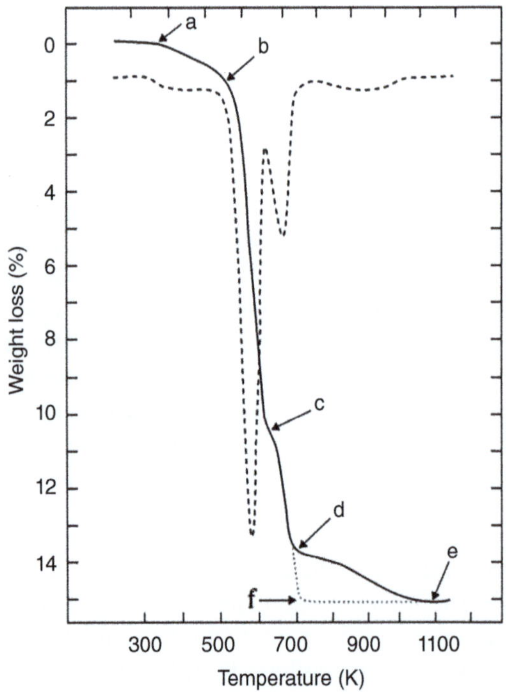

Figure 1.5 TG curve of $La(OH)_3$ procured at 2 K min^{-1} in vacuum. Dotted line (…..) path followed by rehydrated La_2O_3 sample throughout 2nd step of dehydration; broken line (- - -) represents time/temperature derivative; solid line (-) represents weight loss curve. Reprinted with permission Ref. [37] © Elsevier, 1977.

hexagonal LaO(OH) intermediate, delineated by the point c break in the integral thermogravimetric curve. This is after a primary small weight loss at 373–473 K (a–b) attributed to the eradication of adsorbed water and crystallization water. At 523–693 K (c–d), oxyhydroxide undergoes subsequent dehydration to La_2O_3, which is finished at that temperature. When the highly basic trihydroxide precursor interacts with the atmospheric CO_2 during formation and handling, carbonate species that always exist in the oxide undergo decomposition, which results in the eventual weight loss that happens in the temperature range of 723–1073 K (d–e) and results in generation of active site.

1.5.1.4 Titanium Oxides

Titanium oxides exist in three stable polymorphforms, i.e. rutile, brookite, and anatase. Among them, rutile and anatase are most frequently utilized in catalyzing reactions as revealed by literature study.

$$\text{Anatase } (TiO_2) \rightarrow \text{Rutile } (TiO_2) \text{ at } 1073\text{ K temperature range}$$

Both the anatase and the rutile polymorphs of titanium dioxide exhibit acidic and basic characteristics on their surfaces and they catalyze methods in which acidic and/or basic sites are significant. The basic character of ZrO_2 is higher than TiO_2's basic nature [39].

Inevitably, titanium oxides can be generated by hydrolyzing and then calcining titanium chlorides or alkoxide salts. Some examples where $TiO_2$is used as catalyst:

- Dehydration of 2-propanol to propene and finally to acetone by dehydrogenation reaction [40]
- Acetaldehyde aldolization to form crotonaldehyde occurs faster on rutile than anatase polymorph of TiO_2[41]
- Isomerization of 1-butene [42].

Butene isomerization occurs at 473 K over TiO_2. In a vacuum, the pretreatment temperature impacted both the activity of TiO_2 and the reaction process. The activity began to become apparent at 473 K and accelerated as the pretreatment temperature rose, eventually culminating at 673 K. The activity dropped as the temperature rose over 673 K, while the cis/trans ratio of the 2-butenes that were generated increased. Ti^{3+} also increased as the pretreatment temperature climbed [42].

1.5.1.5 Zinc Oxide

Despite it being apparent that acidic sites have a stake in the reactions, ZnO serves as a catalyst that fosters base-catalyzed reactions. Experimental evidence bears out the prevalence of both basic and acidic properties in zinc oxide, thereby hypothesizing it to be amphoteric. By introducing ammonium hydroxide to a simple aqueous solution of zinc nitrate and scorching the resultant mixture to 353 K until almost completely precipitated, zinc oxide can typically appear [43]. The precipitate gets filtered, rinsed with water, patted dry, and calcined in air between 673 and 773 K. ZnO acts as solid base catalysts in:

- Acetone formation from dehydrogenation of 2-propanol [44]
- Formation of *cis-* and *trans*-butene from 1-butene isomerization [45, 46]
- Cis–trans isomerization of 1,3-pentadiene [47].

By altering ZnO by alkylsilylation, the isomerization of 1-butene, 1-pentene, and 1,3-pentadiene was increased. Specifically, the addition of triethylsilane increased the activity of 1,3-pentadiene by a ratio of 89. The improvement was attributed to the reactant's irreversible adsorption being averted with the goal of preventing self-poisoning of the active sites [47].

1.5.1.6 Alumina

In spite of its tremendous mechanical strength and beneficial liaisons with metals and metal oxides, which foster widespread dispersion across the supported compounds, Al_2O_3 is frequently employed as a support for commercial catalysts. By dehydrating at high temperatures, oxyhydroxide (AlO(OH)) and hydroxide ($Al(OH)_3$) are converted to alumina. While referring to surface characteristics, alumina is usually considered acidic rather than basic. However, the relevance of fundamental sites in alumina's catalytic activity is undoubtedly exaggerated. Alumina (Al_2O_3) is implemented as a support for metals and metal oxides in addition to serving as a catalyst for an assortment of processes. Al_2O_3 serves as a catalyst in an ordinary reaction

- wherein alcohols are dehydrated to yield alkenes
- Al_2O_3-specific operation called the H–D exchange between CH_4 and D_2 transpires even at ambient temperature [48].

This is the mechanism through which alcohol is activated [49, 50]. It is thought that proton oscillations between the surface enable the activation and the adsorbed alcohol molecules, which can cause the molecule to polarize. It is proposed that the alcohol molecule itself has some rotational or vibrational freedom in comparison to the surface, allowing the β-H to go closer to a basic O^{2-} ion when the alcohol is in its antiperiplanar conformation. The major isotope effect is noticed when D substitutes for the H linked to β–C is employed for estimating the removal of H connected to β–C by the basic site on Al_2O_3 (O^{2-}), which is the step that establishes the rate [51].

1.5.1.7 Mixed Oxides

Two distinct categories of metal oxides are amalgamated to generate mixed oxides. Alkali metal oxide containing mixed oxides and the Al–Mg oxides generated through calcination of hydrotalcite are the most extensively researched mixed oxides possessing basic characteristics. Certainly, one of the component oxides has basic characteristics, especially when it pertains to mixed oxides that reflect these characteristics. In most instances, the addition of the second component oxide modifies the basicity of the first component oxide to certain extent. Different techniques, including coprecipitation, sol–gel, quick combustion, kneading, impregnation, and molten mixture are employed for manufacturing mixed oxides. The coprecipitation approach is the most prevalent [18].

ZnO can dehydrogenate cyclohexanol and transfer hydrogenate cyclohexanone with 2-propanol. When CeO_2 was incorporated, the activity of both reactions increased. When ZnO was blended with CeO_2, its strong basicity was reduced and its acidity was increased, which increased the activity of the two processes [52].

1.5.1.8 Alkali Metal-Loaded Metal Oxides

Alkali metal compounds can change the surface properties for a sort of reason, and it can be exigent to pinpoint the exact cause of the increased activity in each situation. Some reasons are:

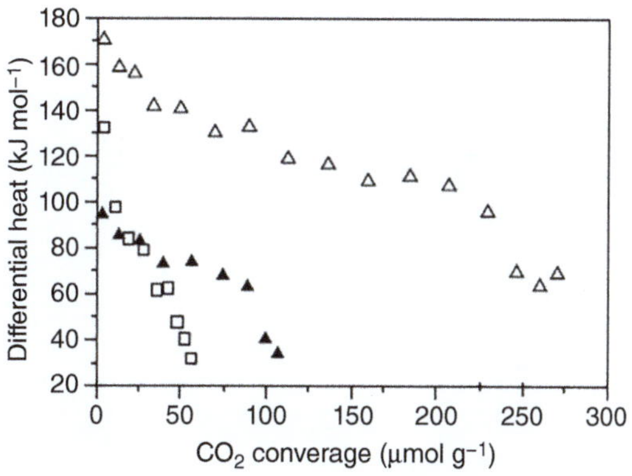

Figure 1.6 Differential heat versus adsorbate coverage for absorption of CO_2 on gamma-Al_2O_3 (□) at 873 K (Δ) and 1273 K (▲). Reprinted with permission Ref [53] © Elsevier, 2003.

- Alkali metal reduces the reaction ability of acidic sites by neutralizing the acidic sites present on surface.
- These alkali metal compounds are converted into its oxides, which display basic characteristics.
- The alkali can react with surface compounds and assist the inception of active sites that can act as catalysts.

The catalytic activity of the alkali metal-loaded metal oxides varies with variation in metal compounds. Incorporation of alkali metals on different metal oxides enhances activity, which was reported in these examples:

(a) Addition of metal oxide in alumina [53]
(b) 1-Butene isomerization in which addition of alkali metal on MgO [54]

Figure 1.6 illustrates the increase in the heat of CO_2 adsorption, which happens when alkali compounds are introduced to Al_2O_3. The incorporation of K^+ ions into γ-Al_2O_3 and subsequent calcination at 873 K resulted in the production of extreme basic sites. For γ-Al_2O_3 loaded with K^+ ions, the heat of adsorption was 170 kJ mol^{-1}, whereas it was 100 kJ mol^{-1} for pure γ-Al_2O_3 [55] reflects the categorization of the various types of metal oxides (Figure 1.7).

1.5.2 Zeolites

Zeolites are aluminosilicates fabricated from SiO_4 and AlO_4 tetrahedra that share an apical oxygen atom with their neighboring tetrahedral. These are frequently implemented as solid acid catalysts, as is well established, but in the early decades, they were employed as base catalysts in their ion-exchanged forms [56, 57]. Zeolites that have been subjected to alkali cations exchanged with them result in weak bases; however, zeolites that offer alkali cations in excess of their ion exchange capacity, rendering them stronger bases. Zeolites become more basic when alkali metals are infused into them, as contrasted with zeolites that are solely loaded with alkali metal oxide. Many research investigations have assessed the catalytic characteristics of these zeolites incorporating alkali metals in numerous ways [58–63].

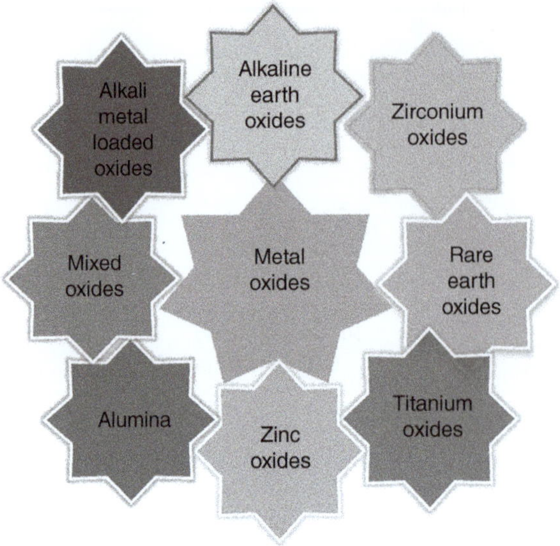

Figure 1.7 Metal oxides categorization.

According to assessments, X- and Y-type zeolites, additionally recognized as faujasites, exchange with alkali cations and facilitate side-chain alkylation reactions. For instance, styrene and ethylbenzene are the products of the reaction of toluene with formaldehyde in the occupancy of alkali cations exchanged from zeolites. As a whole, basic category of catalysts contributes to the side-chain alkylation reaction, while acidic-type catalysts induce ring alkylation. Until then, the fundamental and catalytic properties of alkali cation-exchanged zeolites, particularly those of the faujasite type, have been thoroughly investigated [64]. The delicate basic nature and narrow pore structure are the distinctive properties of the catalysis by alkali cation-exchanged zeolites. These types of exchanged zeolites obsess weak basic sites as compared with alkaline earth oxides and alkali loaded (excess) ion-exchanged zeolites.

Strong solid base catalysts derived from different basic zeolites with caged alkali metals have been explored. Ionic or neutral metal clusters can be generated via imbibing zeolite with NaN_3 or CsN_3, subsequently permitting the alkali azide to decompose under controlled ambient temperatures. First, occluded metallic sodium zeolites were developed by Martens and colleagues by thermally decomposing sodium azide that had been adsorbed on the zeolite. These catalysts were active in the hydrogenation of acetylene, benzene, and *cis*-2-butene as well as the isomerization of *cis*-2-butene [65–67]. Some reactions that were reported to be catalyzed by basic zeolites are:

- Alkylation of toluene with methanol [68]
- Double bond isomerization [69]
- Aldol condensation [70]
- Cycloaddition reactions [71]
- Alcohol dehydrogenation [72]
- Knoevenagel condensation reactions [73].

In the reaction 1-butene isomerization, the catalytic activity clearly demonstrates the production of extreme base sites on alkali metal oxide-loaded zeolites as contrasted to those on alkali cation-exchanged zeolites. Strong active sites for 1-butene isomerization allow CO_2 to be retained as bidentate carbonate at temperatures as high as 573 K. After CO_2 poisoning the active sites, the activity starts to recuperate by deaeration at 573 K and at 673 K, the previous activity is fully reinstituted [69].

1.5.3 Mesoporous Materials

Zeolites having basic nature have been utilized in a wide range of processes; however, in some instances, their employment in the production of fine compounds is restricted due to their narrow pore orifices, which prohibit large molecules from accessing the active sites in zeolite cages. Due to their tunable bigger pore sizes, novel categories of mesoporous silicas, such as MCM-41 [74] and SBA-15 [75], provide fresh avenues for support. The following methods can be exploited for fabricating basic mesoporous materials:

(a) Cation exchange with alkali metal ions (such as Na^+, K^+, and Cs^+);
(b) Impregnation with alkali salt solution and calcination; and
(c) Organic group functionalization.

The first two types of mesoporous materials can be prepared in a manner that is comparable to that used for basic zeolites. Mesoporous catalysts are frequently employed in green catalytic chemistry, owing to its primacy of anticorrosive nature, facile separation, and rapid mass transport. Despite this, these catalysts are also utilized in fabrication of organics and fine chemicals via catalytic ways and engaged in environmental catalysis field. These mesoporous solids have made significant strides during the past several decades.

An abundance of emphasis has been devoted to developing basic sites on mesoporous materials. These mesoporous hosts rarely exhibit basic features, so it is necessary to get acquainted with basic guests. Because the surfactants have obscured their pores, the as-synthesized samples have a low surface area. Typically, these samples are extracted or calcined to detach the surfactants, facilitating guest species to be introduced to the supports with open mesopores. Remarkably for the Knoevenagel condensation reaction, the as-synthesized mesoporous silica MCM-41 was discovered to be an effective basic catalyst [76].

1.5.4 Clay Minerals (Hydrotalcite)

Catalysts and catalyst precursors crafted from hydrotalcite and other comparable elements are crucial. Numerous publications have reviewed the creation, composition and catalytic employability of hydrotalcite [77, 78]. A naturally occurring mineral referred to as hydrotalcite has the approximated chemical formula $Mg_6Al_2(OH)_{16}CO_3{\cdot}4H_2O$ and the structure of hydrotalcite is dimorphous, positively charged, comprised up of brucite-like hydroxide layers surrounding negatively charged interlayers, possessing anion exchange capabilities. In brucite, hydroxyl groups are positioned octahedrally around the magnesium cation, forming sheets of unlimited thickness from the resulting octahedral sharp edges. Al^{3+} ions are isomorphously substituted for some of the magnesium ions in the brucite layer of hydrotalcite. The interlayers of hydrotalcite naturally embrace CO_3^{2-} ions. These materials are frequently utilized as base catalysts for various organic reactions [79–83]. The base catalysis originating from hydrotalcite is categorized into subcategories: first one is, the hydrotalcite layer structure was retained and interlayer anions are changed and second one, by calcination of materials with hydrotalcite structure produced mixed oxides catalysts. Hydrotalcite catalyst is active against these reactions:

- Michael's addition of 2-acetylcyclopentanone
- 2-Methylcyclo-hexane-1,3-dione to methyl vinyl ketone [84]

The Mg/Al ratio influences the catalytic activity and basic characteristics of $MgO–Al_2O_3$. According to a titration procedure, the number of basic sites reaches its maximum when Mg/Al = 3. The highest percentage was detected at a calcination temperature of around 773 K. The basicity is equally dependent on the calcination temperature [18].

1.5.5 Oxynitride

Lednor and Ruiter first synthesized silicon oxynitride, Si_2N_2O, through the execution of a gas–solid reaction of amorphous silica with NH_3 at 1373 K and noticed that the obtained substance showcased a solid base catalysis in the Knoevenagel condensation reaction [85–88]. Following that, aluminophosphateoxynitrides (AlPON), generated by nitriding high-surface-area amorphous aluminophosphate precursors under NH_3 at about 1073 K, were identified as a novel class of solid base catalysts [89, 90]. These catalysts communicated highly intriguing activities in the methyl isobutyl ketone synthesis and in Knoevenagel condensation reactions. The surface acid–base characteristics of the precursors can be tweaked through the inclusion of nitrogen in the aluminophosphate anionic framework, especially by minimizing the quantity of acid sites and raising the proportion of basic sites. Recognition of oxynitrides' basic sites is quite difficult due to presence of various species on surface that can behave as basic sites. The catalytic characteristics of oxynitrides are more commonly investigated for Knoevenagel condensation reactions.

1.5.6 Calcined Metal Phosphates

Metal phosphates that were reported most frequently are K_3PO_4, $Ca_3(PO_4)_2$, and Na_3PO_4 and among them potassium phosphate possesses great base strength. Calcium phosphate is most commonly employed in methylation reaction of phenol to *o*-cresol and 2,6-dimethylphenol [91]. The natural phosphates are widely investigated for numerous base catalysis reactions such as:

- Knoevenagel reaction and
- Michael addition reaction as reported [102, 103]. Numerous reaction has been catalyzed by various kinds of catalyst and some of them are reported here (Table 1.1).

Table 1.1 Different types of catalyst are used in different reactions as base catalysts.

Types of catalysts	Reaction catalyzed	Temperature condition (K)	Ref.
Alkaline earth oxide	(a) Alkene hydrogenation	—	[30]
	(b) Tischenko reaction	—	[92]
Rare earth oxide	(a) Dehydration of alcohol	623–700;	[93]
	(b) Hydrogenation of ethylene	195	[94]
ZrO_2	(a) Isomerization of 1-butene	373	[95]
	(b) Dehydration of 2-butanol	473	[96]
	(c) Conjugated dienes dehydrogenation	323	[96]
	(d) Alkylamine decomposition to nitriles	673	[97]
	(e) Formation of diisopropyl ketone from isobutyraldehyde	733	
	(f) Meerwein–Ponndorf–Verley reduction of ketones, aldehydes, carboxylic acids, nitriles, anhydrides and lactones	573	
TiO_2	(a) Dehydration of 2-propanol to yield propene (anatase)	423–523	[40]
	(b) Aldolization of acetaldehyde to form crotonaldehyde	313 (rutile)	[41]
	(c) Alkylation of phenol	753 (rutile)	
	(d) Isomerization of 1-butene	473	[48]
	(e) Dehydration of glucose to produce 1,6-anhydroglucose	473 (anatase)	
	(f) Isomerization of glucose to form fructose and further dehydration occurs	473 (anatase)	
ZnO	(a) Dehydrogenation of 2-propanol into acetone	483	[44]
	(b) Production of acetone, acetylene (1:1) from 2-methyl-3-butyn-2-ol (MBOH)	453	
	(c) Double bond isomerization of 1-butene to form *cis*/*trans*-2-butene	Room temp.	[48]
	(d) Isomerization of *cis*-1,3-pentadiene into *trans* form	293	
Alumina	(a) Co-isomerization of 1-butene and *cis*-2-butene d_0/d_8	300	[48]
Mixed oxides with MgO	(a) Ortho selective alkylation of phenol over $MgO\text{-}CeO_2$	723–823	[98]
	(b) Conversion of 1-propanol into 3-pentanone	723	[99]
KF/Al_2O_3	(a) Isomerization of 1-pentene	473	[100]
	(b) Tischenko reaction of benzaldehyde	473	[101]

(Continued)

Table 1.1 (Continued)

Types of catalysts	Reaction catalyzed	Temperature condition (K)	Ref.
KNH_2/Al_2O_3	(a) Double bond isomerization of alkenes (b) Production of enamines from unsaturated amine	— —	[48]
Hydrotalcite	(a) Aldol condensation of benzaldehyde and acetone	723	[70]
Zeolites	(a) Meerwein–Ponndorf–Verley reduction of aldehydes with 2-propanol (b) Nitroaldol condensation of aromatic aldehydes with nitroalkanes	373 and 453 413	[73]
ZrPON (zirconium phosphate oxynitride)	(a) Knoevenagel condensation of benzaldehyde with malononitrile	323	[48]
VAlPON (alumino-vanadate oxynitride)	(a) Knoevenagel condensation of benzaldehyde with malononitrile	323	[48]

1.6 Why Solid Base Catalysts Have Fascinated the Scientific Community?

Base-catalyzed reactions, a significant area of synthetic chemistry with numerous industrial applications, have garnered a lot of fascination since they yield a wide range of compounds with added value. Currently, liquid alkali (such as aqueous NaOH, KOH, and triethylamine) has been broadly implemented as a homogeneous catalyst in the fabrication of fine compounds because of its high activity and affordability. However, a number of issues are yet unresolved, including the production of copious amounts of waste liquid, atmospheric contamination, and appliance erosion. In addition, the process of isolating the product from the liquid alkali catalyst is exceptionally challenging and frequently leads to further side reactions. Solid base catalysts are therefore important since they can help with separation, prevent corrosion, and reduce pollution in profuse base-catalyzed industrial procedures [104–107]. Additionally, in terms of the recycling of catalysts, solid base catalysts expedite the easy separation and retrieval of products from reaction systems [108–111]. Basic oxide catalysts with substantial specific surface area and a high concentration of accessible basic sites on their surface are considered to have significant catalytic activity. But solid base catalysts are able to achieve the equivalent yield in less time and at lower temperatures. Hence, for a variety of base-catalyzed processes, solid base catalysts provide an affordable and environmentally safe approach.

Considering several examples of industrial environmental effects, one of the main disruptions to the aquatic ecosystem is wastewater from different industrial activities that contains organic contaminants. The need for heterogeneous solid base catalysts to be used in the treatment of such wastewater is growing along with their growth. Numerous catalysts, including those based on copper, aid in the remediation of solid waste as well as the breakdown of organic contaminants [112]. In order to remove dyes from contaminated water in the textile industry, a number of metal oxides, including solid base catalysts such as lanthanum oxide and calcium oxide, are also employed [113]. However, in spite of this, solid base heterogeneous catalysts recognizes application in an assortment of other processes, including amination, polymerization, alkylation and isomerization.

1.7 Advantages and Disadvantages of Solid Base Catalysts Over Inorganic/Organic Bases

- **Advantages**
  - **Minimize wastes**: The utilization of hazardous chemicals such as methyl iodide and dimethyl sulfate was reduced in case of solid base catalysis turning the reaction more efficient and productive.
  - **Solvent selection**: Solid bases have a significant solvent advantage over homogeneous bases. Both the reactants and the base catalyst need to dissolve in the solvent during homogeneous phase reactions. This significantly limits the solvent selection. Hazardous solvents such as methyl chloride and dimethyl sulfoxide (DMSO) are therefore frequently used. Otherwise, a phase-transfer system must be employed, just like in the reaction example (1.3). This could lead to tedious complications with workup procedures such as separation. Solvents in solid-base-catalyzed processes are only required to dissolve reactants and products. This greatly broadens the choice of solvent [15, 114]. For example, water and hydrocarbons such as pentane perform as alternate solvents in the Knoevenagel condensation of 2-furaldehyde with ethyl cyanoacetate on calcium silicate [104]. Moreover, there is no need for a solvent to conduct the reactions [16]. The reactions can even proceed in the vapor phase, as demonstrated by the case above. This makes the process less complicated, considering there is obviously no requirement for separating the solvents when the reactions have concluded. There are no solvent-based wastes generated, such as organic solvent-containing alkaline water or organic solvent-containing water.

$$PhCH_2CN + NaOH + CH_3I \rightarrow PhCH(CH_3)CN + H_2O + NaI \quad (1.3)$$

  - **Corrosion resistance**: The reactor and disposal device must be made of corrosion-resistant materials since solutions containing bases, such as sodium hydroxide are corrosive in the case of acid–base catalysts. Solid base catalyst reaction systems rarely experience this kind of issue.
  - **Reaction smoothness**: Furthermore, solid base catalysts can occasionally be the only ones that can accomplish eminent activity and selectivity for an assortment of reactions. Many reactions can occur without mishap when acid

and basic sites work together. The major issue in acid-catalyzed reactions is coking, which basic catalysts typically do not encounter. When compared to acidic sites, functionalized organic molecules such as amines have less interaction with basic sites. As a result, these molecules desorb from the surface more readily and do not become trapped there. Consequently, over solid bases, reactions involving these molecules proceed more easily. Examples of this type of reaction include aldol condensation, hydrogenation, and transfer hydrogenation.

- **Catalytic efficiency**: Furthermore, the use of organometallic reagents must be stoichiometric. Solid bases provide more convenient and environmentally friendly paths for the same products. Both of these reactions are catalytic and occur well in the existence of KF/Al_2O_3 and KNH_2/Al_2O_3, respectively [115, 116].

$$H_2SiEt_2 + tBuC \equiv CH \rightarrow tBuC \equiv C - SiEt_2H + H_2 \tag{1.4}$$

$$H_2SiEt_2 + C_6H_5CH_3 \rightarrow C_6H_5CH_2 - SiEt_2H + H_2 \tag{1.5}$$

- **Reusability**: These solid base catalysts are reusable and can be regenerated easily after completion of a reaction.
- **Bifunctional characteristics**: Having dual-purpose catalytic properties that permit basic and acidic sites to coexist on the surface of solid base catalysts, which, in the case of liquid basic catalysts, is not attainable.
- **Energy efficiency**: The reaction with solid base catalysts can occur even at mild conditions. These catalysts are less energy-intensive as reported.

• **Disadvantages**
  - **Poisoning**: When exposed to external surrounding or when coming in contact with air particles these catalysts turn into poison.
  - **Sensitivity**: These catalysts are sensitive to free fatty acid (FFA) content present in the oils owing to their basic nature.
  - **Leaching**: Leaching of catalyst active sites can occur and lead to contamination in the products.
  - **Decreased yields**: If FFA content in feedstocks exceed 2 wt% then it leads to excess soap formation in reaction thus decreasing the yield of the reaction.

1.8 Role of Solid Base Catalysts in Green Chemistry

Green chemistry's primary objective is to synthesize organic molecules more effectively and in an environmentally friendly way. Synthesizing solid base catalysts that enable cleaner, more effective chemical reactions with the goal of producing zero waste in industrial processes is one of the significant challenges confronting academia nowadays.

Biodiesel production: Biodiesel is a green fuel made by transesterification of oil and alcohol with a catalyst, which renders it cost-effective and environmentally friendly. It is a potential diesel fuel extender that is speculated to comprise fatty acid methyl esters extracted from vegetable oil. Triglycerides are transesterified with methanol with the assistance of an acid or base catalyst to yield methyl esters of fatty acids. This transesterification method is preferred owing to its mild conditions for reaction and fewer byproducts compared to biodiesel production. The characteristic of the feedstock and the kind of reactions has a major impact on the transesterification process's efficiency and the catalyst's effectiveness is crucial to the formulation of biodiesel [117–119]. For the production of biodiesel, a variety of heterogeneous base catalysts have been employed extensively, comprising base zeolites, hydrotalcite, base-doped alumina, and base earth metal oxides [120]. Owing to its many benefits and higher biodiesel output, heterogeneous catalyst use in catalyzed transesterification reactions is becoming more and more prevalent. Furthermore, by calcining the ashes from these residues, which are regarded as wastes, heterogeneous catalysts have been produced from a variety of agricultural residues. Consequently, because of its renewable, recoverable and reusable character, producing biodiesel is somewhat less expensive than using a homogeneous catalyst [121, 122].

Dimethyl carbonate (DMC) synthesis: DMC has long been fabricated by reacting methanol with phosgene. An enormous amount of study has been attempted in pursuit of non-phosgene, environmentally conscious methods to synthesize DMC. Utilizing a Cu or Pd catalyst DMC manufactured from methanol, CO, and oxygen has been industrialized. Dimethyl carbonate is a significant methylation and carbonylation agent in addition to being an essential step for polycarbonate resins. Due to DMC's minimal toxicity, it presents a viable alternative to the hazardous and caustic phosgene dimethyl sulfate or methyl iodide [45, 123, 124]. It might potentially be used as an octane enhancer for petrol. Solid base catalysts, on the other hand, have drawn a lot of interest because homogeneous basic catalysts, including trialkylamine and alkali alkoxides, produce complications with separation of the product and catalyst reuse.

Isoeugenol synthesis: Eugenol was utilized to produce isoeugenol, which is employed in medications and a wide range of floral compositions in perfumes. The isomerization reaction generally precedes as-synthesized hydrotalcite materials. The higher effectiveness was reported by a combination of Mg–Al and Ni–Al hydrotalcite [125].

Anethole synthesis: An effective way to fabricate an anethole is by isomerizing estragole, which is used in mouthwash and alcoholic beverages. It was revealed that compounds such as hydrotalcite that had been synthesized using combinations of Mg–Al, Ni–Al, and Mg–Ni–Al were also effective in isomerizing estragole to anethole. With a cis/trans ratio of 15/85, Mg–Ni–Al hydrotalcite yielded a 99% conversion of estragole [126].

Flavanone synthesis: A common intermediary for many medicinal and fine chemical products is flavanone. A typical pathway leading to flavonone involves the Claisen–Schmidt condensation process. However, research revealed that this solvent-free condensation reaction can occur in the presence of $MgO–Al_2O_3$ mixed oxides generated from calcining hydrotalcite [127].

Synthesis of phytosterol esters: The food and cosmetics sectors can benefit from phytosterol esters, which can be produced through esterification or transesterification reactions. Solid bases such as MgO, ZnO, and La_2O_3 facilitate the transesterification of β-sitosterol and methyl dodecanoate, which produce phytosterol ester [128, 129].

1.9 Future Prospects for Solid Base Catalysts

The following significant solid base catalysts challenges require deeper study.

(i) **Clarification of the fundamental sites**: The majority of solid-base-catalyzed reactions begin with the removal of a proton from the reactant, which results in the constitution of an anionic species. It is necessary to stabilize the anionic species somewhere, presumably on the conjugate acid sites. It is a crucial step to dictate the location of the conjugate acid sites.

(ii) **Shape-selective catalysts**: The fabrication of shape-selective solid catalysts is a crucial step in the catalysis of solid acids, but no shape-selective solid base catalysis has been discovered yet. Materials with both fundamental and shape-selective characteristics are readily synthesized. These substances may pave the way for new developments in solid base catalysis.

(iii) **Characterization of basic sites**: In order to get more accurate information, more techniques should be used to analyze basic sites. To enhance characterization, new techniques for characterization should be created.

(iv) **Theoretical computation**: In comparison to acid-catalyzed reactions, the theoretical computation for solid-base-catalyzed reactions is inadequate. Solid base catalysts would advance significantly if theoretical calculations could describe the parameters controlling the basic site's strength and the reactant's interaction with the surface to start the base-catalyzed reaction.

(v) **Clarification of the unique characteristics of alumina-supported alkalis**: High activity is exhibited by alumina-supported alkalis in a variety of processes. Even yet, nearly five decades ago, Pines et al. introduced Na metal supported by alumina as the first solid base catalyst. It is yet uncertain whether alumina contributes to the creation of the fundamental sites.

1.10 Conclusion

Overall, solid base catalysts are a promising area of research in the field of catalysis that has numerous avenues for use. The technology's significance stems from its knack for furnishing effective and sustainable substitutes for existing liquid-base catalysts. Several prominent movements and opportunities will shape the direction of solid base catalysts. Alkali metal and alkaline earth oxides, aside from other salts, either in their pure form or on an adequate carrier, are the heterogeneous base catalysts that have been researched most thoroughly. They provide a reasonably quick reaction rate and are not prohibitively costly. The specific surface area and distribution of pore size serve an essential role in defining the activity of these catalysts since the active sites on the surface ultimately determine their activity. Thus, preparation and pretreatment procedures are extremely important, as the catalyst is poisoned by the existence of water, carbon dioxide and free fatty acids. Zeolites are covered with alkali and alkaline earth metals since they are ineffective as catalysts on their own. As catalysts, clays and organic solids have also been produced, but all of these are still in the laboratory. In its entirety, the foreseeable future of solid base catalysts will ultimately be governed by a dynamic interaction of basic research, technological advancements and practical implementations. The field is intended to advance environmentally friendly and efficient chemical processes across several kinds of areas as it keeps expanding.

References

1 Laszlo, P. (1986). *Accounts of Chemical Research* 19: 121. https://doi.org/10.1021/ar00124a004.

2 Wang, X., Tseng, Y.H., Chan, J.C.C., and Cheng, S. (2005). *Journal of Catalysis* 233: 266.

3 Climent, M.J., Corma, A., Iborra, S., and Velty, A. (2004). *Journal of Catalysis* 221: 474.

4 Boro, J., Deka, D., and Thakur, A.J. (2012). *Renewable and Sustainable Energy Reviews* 16: 904.

5 Dube, M.A., Tremblay, A.Y., and Liu, J. (2007). *Bioresource Technology* 98 (3): 639.

6 Chen, X., Qian, W.W., Lu, X.P., and Han, P.F. (2012). *Natural Product Research* 26 (13): 1249.

7 Li, E. and Rudolph, V. (2008). *Energy & Fuels* 22: 145.

8 Liu, X., Piao, X., Wang, Y., and Zhu, S. (2008). *Energy & Fuels* 22: 1313.

9 Bing, W. and Wei, M. (2019). *Journal of Solid State Chemistry* 269: 184. https://doi.org/10.1016/j.jssc.2018.09.023.

10 Sharma, Y.C., Singh, B., and Korstad, J. (2010). *Fuel* 90 (4): 1309.

11 Liu, Z., Li, W., Pan, C. et al. (2011). *Catalysis Communications* 15: 82.

12 Ge, J., Zeng, Z., Liao, F. et al. (2013). *Green Chemistry* 15: 2064.

13 Lei, X., Lü, Z., Guo, X., and Zhang, F. (2012). *Industrial and Engineering Chemistry Research* 51: 1275.

14 Frey, A.M., Bitter, J.H., and de Jong, K.P. (2011). *Chemical Catalysis Chemistry* 3: 1193.

15 Pines, H. and Haag, W. (1958). *The Journal of Organic Chemistry* 23: 328.

16 Tanabe, K., Misono, M., Ono, Y., and Hattori, H. (1989). *New Solid Acids and Bases*, 51, 0-444-98800-9. Kodansha-Elsevier.

17 Che, M. and Bond, G.C. (1985). *Adsorption and Catalysis on Oxide Surface*, 319, 9780080960555. Elsevier.

18 Ono, Y. and Hattori, H. (2011). *Solid Base Catalysis*. Springer Series in Chemical Physics https://doi.org/10.1007/978-3-642-18339-3.

19 Dehkordi, A.M. and Ghasemi, M. (2012). *Fuel Processing Technology* 97: 45.

20 Wang, H., Liu, W., Wang, Y. et al. (2020). *Industrial and Engineering Chemistry Research* 59: 5591.

21 Wei, C., Zhiliang, H., Yu, L., and Qianjun, H. (2008). *Catalysis Communications* 9: 4.

22 Wang, S., Guo, D., Kang, R. et al. (2023). *International Journal of Biological Macromolecules* 232: 123341.

23 Kim, S.E., Kim, J.H., Kim, D.K. et al. (2023). *Fuel* 341: 127548.

24 Li, Z., Ding, S., Chen, C. et al. (2019). *Energy Conversion and Management* 192: 335.

25 Naeem, A., Khan, I.W., Farooq, M. et al. (2021). *Bioresource Technology* 328: 124831.

26 Ye, X., Shi, X., Li, J. et al. (2022). *Chemical Engineering Journal* 440: 135844.

27 Kong, W., Liu, F., and Liu, Y. (2020). *Green Chemistry* 3: 903.

28 Park, S., Kwon, D., Kang, J.Y., and Jung, J.C. (2019). *Green Energy & Environment* 4 (3): 287.

29 Hattori, H. (2015). Solid base catalysts: fundamentals and their applications in organic reactions. *Applied Catalysis A: General* 504: 103–109. https://doi.org/10.1016/j.apcata.2014.10.060

30 Hattori, H., Yoshii, N., Tanabe, K. (1972). Proceedings of the 5th International Congress on Catalysis, Miami Beach, 233.

31 Utiyama, M., Hattori, H., and Tanabe, K. (1978). *Journal of Catalysis* 53: 237.

32 Yokoyama, T., Setoyama, T., Fujita, N. et al. (1992). *Applied Catalysis A* 88: 149.

33 M. Fukui, S.Hayashi, K. Okamoto, I. Koga, T. Inoi, US Pat. 3966822 (1976).

34 K. KagakuEur. Pat. 0433959 (1991).

35 M. Araki, K.Takahashi, T.Hibi, Eur. Pat. 0150832 B1 (1985).

36 Takahashi, K., Hibi, T., Higashio, Y. et al. (1993). *Catalysts* 35: 12.

37 Rosynek, M.P. and Magnuson, D.T. (1977). *Journal of Catalysis* 46: 402.

38 Tanabe, K. and Yamaguchi, T. (1994). *Catalysis Today* 20: 185.

39 Busca, G. (1999). *Physical Chemistry Chemical Physics* 1: 723.

40 Haffad, D., Chambellan, A., and Lavalley, J.C. (2001). *Journal of Molecular Catalysis A* 168: 153.

41 Rekoske, J.E. and Barteau, M.A. (1999). *Langmuir* 15: 2061.

42 Hattori, H., Ito, M., and Tanabe, K. (1975). *Journal of Catalysis* 38: 172.

43 Huffman, J.R. and Dodge, B.F. (1929). *Industrial and Engineering Chemistry* 21: 1056.

44 Lahousse, C., Bachelier, J., and Lavalley, J.C. (1994). *Journal of Molecular Catalysis* 87: 329.

45 Chang, C.C., Conner, W.C., and Kokes, R.J. (1973). *The Journal of Physical Chemistry* 77: 1957.

46 Dent, A.L. and Kokes, R. (1971). *The Journal of Physical Chemistry* 75: 487.

47 Imizu, Y., Narita, T., Fujito, Y., and Yamada, H. (2000). *Studies in Surface Science and Catalysis* 130: 2429.

48 Saunders, P.C. and Hightower, J.W. (1970). *The Journal of Physical Chemistry* 74: 4323.

49 Knozinger, H., Buehl, H., and Kochloefl, K. (1972). *Journal of Catalysis* 24: 57.

50 Knozinger, H. and Scheglila, A. (1970). *Journal of Catalysis* 17: 252.

51 Kraus, M. (1997). *Handbook of Heterogeneous Catalysis* (ed. G. Ertl, H. Knozinger, and J. Weitkamp), 1051. Wiley-VCH.

52 Mishra, B.G. and Rao, G.R. (2006). *Journal of Molecular Catalysis A* 243: 204.

53 Zou, H., Ge, X., and Shen, J. (2003). *Thermochemica Acta* 397: 81.

54 Matsuda, T., Sasaki, Y., Miura, H., and Sugiyama, K. (1985). *Bulletin of Chemical Society Japan* 58: 1041.

55 Berteau, P., Kellens, M.-A., and Delmon, B. (1991). *Journal of the Chemical Society, Faraday Transactions* 87: 1425.

56 Weitkamp, J., Hunger, M., and Rymsa, U. (2001). *Microporous and Mesoporous Materials* 48: 255.

57 Corma, A. (1995). *Chemical Reviews* 95: 559.

58 Pines, H., Vesely, J.A., and Ipatieff, V.N. (1955). *Journal of the American Chemical Society* 77: 347.

59 Barthomeuf, D. (1996). *Catalysis Reviews: Science and Engineering* 38: 521.

60 Dartt, C.B. and Davis, M.E. (1994). *Catalysis Today* 19: 151.

61 Ono, Y. (1997). *Catalysis Today* 38: 321.

62 Weitkamp, J. and Hunger, M. (2007). *Studies in Surface Science and Catalysis* 168: 787.

63 Davis, R.J. (2003). *Journal of Catalysis* 216: 396.

64 Yashima, T., Sato, K., Hayashi, T., and Hara, N. (1972). *Journal of Catalysis* 26: 303.

65 Xu, B. and Kevan, L. (1991). *Journal of the Chemical Society, Faraday Transactions* 87: 2843.

66 Xu, B. and Kevan, L. (1992). *The Journal of Physical Chemistry* 96: 2642.

67 Martens, L.R.M., Grobet, P.J., Vermeiren, W.J.M., and Jacobs, P.A. (1986). *Studies in Surface Science and Catalysis* 28: 935.

68 Engelhardt, J., Szanyi, J., and Valyon, J. (1987). *Journal of Catalysis* 107: 296.

69 Yagi, F., Tshji, H., and Hattori, H. (1997). *Microporous Materials* 9: 237.

70 Ballini, R., Bigi, F., Gogni, E. et al. (2000). *J. Catal.* 191: 348.

71 Tu, M. and Davis, R.J. (2001). *Journal of Catalysis* 199: 85.

72 Hathaway, P.E. and Davis, M.E. (1989). *Journal of Catalysis* 116: 263.

73 Corma, A., Martin-Aranda, R.M., and Sanchez, F. (1991). *Studies in Surface Science and Catalysis* 59: 503.

74 Kresge, C.T., Leonowicz, M.E., Roth, W.J. et al. (1992). *Nature* 359: 710.

75 Zhao, D., Feng, J., Huo, Q. et al. (1998). *Science* 279: 548.

76 Kubota, Y., Nishizaki, Y., Ikeya, H. et al. (2004). *Microporous and Mesoporous Materials* 70: 135.

77 Cavani, F. and Trifiro, T. (1991). *Catalysis Today* 11: 173.

78 Vaccari, A. (1998). *Catalysis Today* 41: 53.

79 Tomar, R., Singh, N., Rathee, G. et al. (2017). Synthesis and Characterization of Hybrid $Mg(OH)_2$ and $CeCO_3OH$ Composite with Improved Activity Towards

Henry Reaction. *Asian Journal of Organic Chemistry* 6 (12): 1728–1732. Portico. https://doi.org/10.1002/ajoc.201700485.

80 Tomar, R., Ebitani, K., and Chandra, R. (2019). Hydrotalcite-Supported Ceria Nanoparticles as a Heterogeneous Catalyst for One-Pot Synthesis of Imines under Atmospheric Air. *ChemistrySelect* 4 (12): 3577–3581. https://doi.org/10.1002/slct.201900750.

81 Tomar, R., Singh, N., Kumar, N. et al. (2019). Base-Free Suzuki–Miyaura Coupling Reaction Using Palladium(II) Supported Catalyst in Water. *Catalysis Letters* 149 (6): 1589–1594. https://doi.org/10.1007/s10562-019-02723-9.

82 Tomar, R., Rathee, G., Chandra, I. et al. (2018). Synthesis and Characterization of Magnesium Hydroxide & Cerium Oxide Composite: Application in Organic Transformation. *ChemistrySelect* 3 (6): 1645–1649. Portico. https://doi.org/10.1002/slct.201702947.

83 Kumar, L., Verma, N., Sehrawat, H. et al. (2022). *New Journal of Chemistry* 46 (7): 3472–3481. https://doi.org/10.1039/d1nj05690e.

84 Prescott, H.A., Li, Z.J., Trunschke, A. et al. (2005). *Journal of Catalysis* 234: 119.

85 Lednor, P.W. and de Ruiter, R. (1989). *Journal of the Chemical Society, Chemical Communications* 5: 320.

86 Lednor, P.W. and de Ruiter, R. (1991). *Journal of the Chemical Society, Chemical Communications* 22: 1625.

87 Lednor, P.W. (1992). *Catalysis Today* 15: 243.

88 Benitez, J.J., Odriozola, J.A., Marchand, R. et al. (1995). *Journal of the Chemical Society, Faraday Transactions* 91: 4477.

89 Massinon, A., Odriozola, J.A., Bastians, P. et al. (1996). *Applied Catalysis A* 137: 9.

90 Nozaki, F. and Kimura, I. (1977). *Bulletin of Chemical Society Japan* 50: 614.

91 Sebti, S., Solhy, A., Tahir, R. et al. (2003). *Journal of Catalysis* 213: 1.

92 Lundeen, A.J. and van Hoozen, R. (1967). *The Journal of Organic Chemistry* 32: 3386.

93 Minachev, K.M., Khdakov, Y.S., and Nakhshunov, V.S. (1977). *Journal of Catalysis* 49: 207.

94 Yamaguchi, T., Sasaki, H., and Tanabe, K. (1973). *Chemistry Letters* 2: 1017.

95 Xu, B.-Q., Yamaguchi, T., and Tanabe, K. (1991). *Applied Catalysis* 75: 75.

96 Takahashi, K., Shibagaki, M., and Matsushita, H. (1992). *Bulletin of Chemical Society Japan* 65: 262.

97 Sato, S., Koizumi, K., and Nozaki, F. (1998). *Journal of Catalysis* 178: 264.

98 Kamimura, Y., Sato, S., Takahashi, R. et al. (2003). *Applied Catalysis A* 252: 399.

99 Tsuji, T., Kabashima, H., Kita, H., and Hattori, H. (1995). *Reaction Kinetics and Catalysis Letters* 56: 363.

100 Handa, H., Baba, T., Sugisawa, H., and Ono, Y. (1998). *Journal of Molecular Catalysis* 134: 171.

101 Kantam, M.L., Choudary, B.M., Reddy, C.V. et al. (1998). *Chemical Communications* 1033: 163.

102 Abouki, Y., Zahouly, M., Rayadh, A. et al. (2002). *Tetrahedron Letters* 43: 8951.

103 Seki, T., Kabashima, H., Akutsu, K. et al. (2001). *Journal of Catalysis* 204: 393.

104 Faba, L., Diaz, E., and Ordonez, S. (2013). *ChemSusChem* 6: 463.

105 Boukha, Z., Fitian, L., López-Haro, M. et al. (2010). *Journal of Catalysis* 272: 121.

106 Keller, T.C., Isabettini, S., Verboekend, D. et al. (2014). *Chemical Science* 5: 677.

107 Shen, W., Tompsett, G.A., Hammond, K.D. et al. (2011). *Applied Catalysis A* 392: 57.

108 Kandel, K., Althaus, S.M., Peeraphatdit, C. et al. (2012). *Journal of Catalysis* 291: 63.

109 Bai, R., Liu, P., Yang, J. et al. (2015). *ACS Sustainable Chemistry & Engineering* 3: 1292.

110 Bai, R., Zhang, H., Mei, F. et al. (2013). *Green Chemistry* 15: 2929.

111 Radhakrishan, R., Do, D.M., Jaenicke, S. et al. (2011). *ACS Catalysis* 1: 1631.

112 Basumatary, S., Nath, B., and Kalita, P. (2018). *Journal of Renewable and Sustainable Energy* 10 (4): 043105.

113 Mogha, N.K., Gosain, S., and Masram, D.T. (2020). *Arabian Journal of Chemistry* 13 (1): 1367.

114 Tanabe, K. (1970). *Solid Acids and Bases*. Kodansha/Academic Press.

115 Baba, T., Kato, A., Yuasa, H. et al. (1998). *Catalysis Today* 44: 271.

116 Baba, T., Yuasa, H., Handa, H., and Ono, Y. (1998). *Catalysis Letters* 50: 83.

117 Costantini, A. and Califano, V. (2021). *Catalysts* 11: 629.

118 Nguyen, H.C., Wang, F.M., Dinh, K.K. et al. (2020). *Energies* 13: 2167.

119 Nguyen, H.C., Liang, S.H., Li, S.Y. et al. (2018). *Journal of the Taiwan Institute of Chemical Engineers* 85: 165.

120 Borges, M.E. and Díaz, L. (2012). *Renewable and Sustainable Energy Reviews* 16: 2839.

121 Olatundun, E.A., Borokini, O.O., and Betiku, E. (2020). *Renewable Energy* 166: 163.

122 Dhawane, S.H., Kumar, T., and Halder, G. (2016). *Renewable Energy* 89: 506.

123 Ono, Y. (1996). *Pure and Applied Chemistry* 68: 367.

124 Tundo, P. (2001). *Pure and Applied Chemistry* 73: 1117.

125 Kishore, D. and Kannan, S. (2004). *Applied Catalysis A* 270: 227.

126 Kishore, D. and Kannan, S. (2006). *Journal of Molecular Catalysis A* 244: 83.

127 Climent, M.J., Corma, A., Iborra, S., and Primo, J. (1995). *Journal of Catalysis* 151: 60.

128 Pouilloux, Y., Courtols, G., Boisseau, M. et al. (2003). *Green Chemistry* 5: 89.

129 Vallange, S., Beauchaud, A., Barrault, J. et al. (2007). *Journal of Catalysis* 251: 113.

2

Synthesis of Solid Base Catalysts

Chetna Kumari, Nishu Dhanda, Nirmala Kumari Jangid, and Sudesh Kumar

Banasthali Vidyapith, Department of Chemistry, Banasthali, Rajasthan 304022, India

2.1 Introduction

Base catalysis is a crucial component of organic synthesis and has major commercial applications, especially in the petrochemical, fine chemical, and pharmaceutical sectors. Homogeneous bases such as $Ca(OH)_2$, NaOH, and KOH are commonly used in commercial base-catalyzed processes [1, 2]. Significant efforts were made to develop processes mediated by heterogeneous catalysts because these bases are hazardous, require more than stoichiometric amounts, have high operating costs, and pose serious environmental risks related to base neutralization, corrosion, waste generation, and product separation and purification [2–10]. Because the type of basicity of the solid base can be designed in accordance with the requirements of the reactions, solid bases have been found to be the best alternative for solving the previously mentioned homogeneous base problems as well as helping to make the reaction more selective [2, 11]. Due to their low-temperature functioning, environmental friendliness, and improved selectivity, when necessary, along with simplicity of product isolation, solid bases are also of industrial interest.

2.2 K_2O/Al_2O_3–CaO

A solid base catalyst consisting of potassium coated with mixed oxides (K_2O/Al_2O_3–CaO) was synthesized. It effectively facilitated the synthesis of pyrano[2,3-*d*]pyrimidinone derivatives, as well as derivatives of benzylidenemalononitrile and benzylidene barbituric acid. One of the most important components of green chemistry is catalysis, which is important for the creation of sustainable chemical manufacturing methods as well as cleaner technologies [12, 13]. Catalysis is modified to become nanocatalysts by the need to eliminate the use of harmful substances and produce minimal or no waste [14, 15]. Properties of both heterogeneous and homogeneous catalysts are included in nanocatalysts [16–18]. Nevertheless, during the reaction, they might clump together [19]. Immobilizing or coating them with

Solid Base Catalysts: Synthesis, Characterization, and Applications, First Edition.
Edited by Ravi Tomar, K.K. Pant, and Ramesh Chandra.

strong supports overcomes this disadvantage [20–22]. To form carbon–carbon bond reactions, the catalytic activity of produced alkali metal coated with mixed oxide support solid base catalyst is examined.

2.2.1 Preparation of K_2O/Al_2O_3–CaO

2.2.1.1 Preparation of Al_2O_3–CaO Mixed Oxides Basic Support

Alumina–calcium oxide (Al_2O_3–CaO), the active oxides basic support, was produced by mechanical grinding and calcination at various temperatures. In a pre-weighed agate mortar, one gram of basic alumina (Al_2O_3) and 1 g of calcium carbonate ($CaCO_3$) were added. The mixture was then mechanically ground for the required amount of time at normal temperature to produce a free-flowing powder of the combined oxides. Following a three-hour calcination at 500 °C, it was allowed to cool to ambient temperature, forming the alumina–calcium oxide (Al_2O_3–CaO) basic support [23].

2.2.1.2 Potassium Nitrate Loading with Calcined Mixed Oxides Basic Support

Using the initial wetness impregnation approach, the KNO_3 was loaded over the basic support. In a round-bottom flask, the combination of calcined mixed oxides basic support (Al_2O_3–CaO, 1 g) and potassium nitrate (KNO_3, 0.5 g) was stirred constantly for 24 hours at room temperature. Subsequently, the solid support loaded with alkali metal was separated through filtration and continuously washed with distilled water to eliminate any surplus unloaded metal ions from the support surface. The KNO_3/Al_2O_3–CaO mixture was filtered, dried in an oven for two hours, and then calcined for three hours at 500 °C to produce a highly effective K_2O/Al_2O_3–CaO base catalyst (Scheme 2.1) [23].

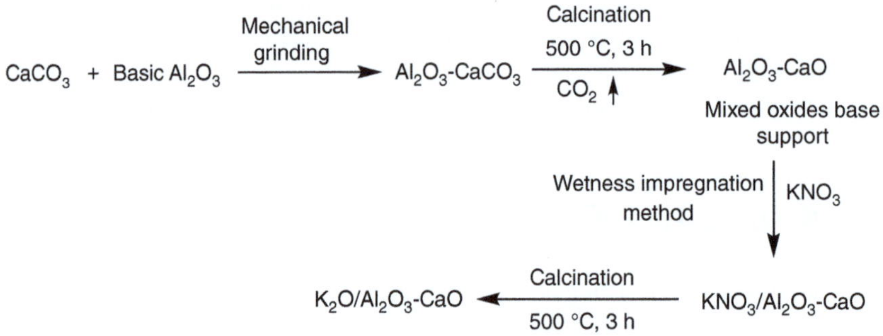

Scheme 2.1 Synthesis of K_2O/Al_2O_3–CaO.

2.2.2 Catalytic Activity of K_2O/Al_2O_3–CaO in the Knoevenagel Condensation Process for the Preparation of Benzylidene Barbituric and Benzylidenemalononitrile Derivatives

The preparation of benzylidene barbituric and benzylidenemalononitrile derivatives by Knoevenagel condensation was used as a test for the catalytic activity of K_2O/Al_2O_3–CaO. p-Nitrobenzaldehyde (0.5 mmol), barbituric acid (0.5 mmol) malononitrile, and m-nitrobenzaldehyde (0.5 mmol) were chosen as test substrates to maximize the reaction conditions. They used various catalysts, various kinds of solvents, and varied temperature ranges for their reactions. Initially, a variety of catalyst types and amounts were used to try the reaction. It was discovered that potassium-impregnated mixed oxides (K_2O/Al_2O_3–CaO, 0.025 g) worked well as a solid base catalyst, effectively catalyzing the reaction and resulting in a good product yield. Additionally, polar aprotic, polar protic, and nonpolar solvents such as acetonitrile, water, toluene, and ethanol were used to carry out the reaction between the test's substrates. Additionally, the reaction was tested without the use of solvents. We found that ethanol produced the greatest results in terms of reaction time, selectivity, and yield after watching the reaction under these circumstances.

To determine the temperature that most effectively catalyzed the reaction under mild conditions, the reaction was further conducted at several temperature ranges. The reaction was reported to proceed more smoothly and provide a higher product yield with high selectivity at 80 °C. Once the reaction circumstances were optimized, the protocol's range of use for different aromatic aldehydes was determined. When reactive methylene compounds such as barbituric acid and malononitrile are substituted with different aromatic aldehydes that donate or withdraw electrons, the reaction results in an outstanding yield of products. Aromatic aldehydes substituted with electron-withdrawing groups react more quickly than heteroaromatic aldehydes and aromatic aldehydes substituted with electron-donating groups. Since the reaction happened under mild circumstances, a three-hour microwave reaction involving malononitrile and 3,4-dimethoxybenzaldehyde was also attempted, but only 50% of the reaction was successful (Scheme 2.2) [23].

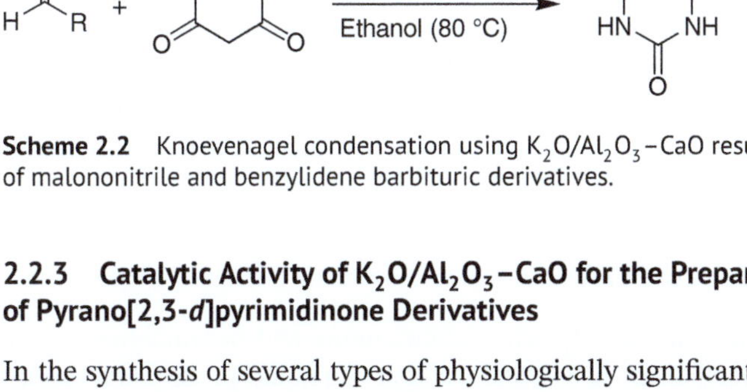

Scheme 2.2 Knoevenagel condensation using K_2O/Al_2O_3–CaO results in the preparation of malononitrile and benzylidene barbituric derivatives.

2.2.3 Catalytic Activity of K_2O/Al_2O_3–CaO for the Preparation of Pyrano[2,3-*d*]pyrimidinone Derivatives

In the synthesis of several types of physiologically significant pyrano[2,3-*d*]pyrimidine derivatives, the catalytic efficacy of K_2O/Al_2O_3–CaO was also investigated. As model substrates, the reaction between p-nitrobenzaldehyde (0.5 mmol), barbituric acid (0.5 mmol), and malononitrile (0.5 mmol) was chosen. Preliminary research was done to identify the ideal reaction conditions. To determine the most effective and selective catalyst, the reaction was first tested with various types of catalysts in varying concentrations. It was found that the reaction worked best when K_2O/Al_2O_3–CaO (0.025 g) was used as the solid base catalyst (Scheme 2.3) [23].

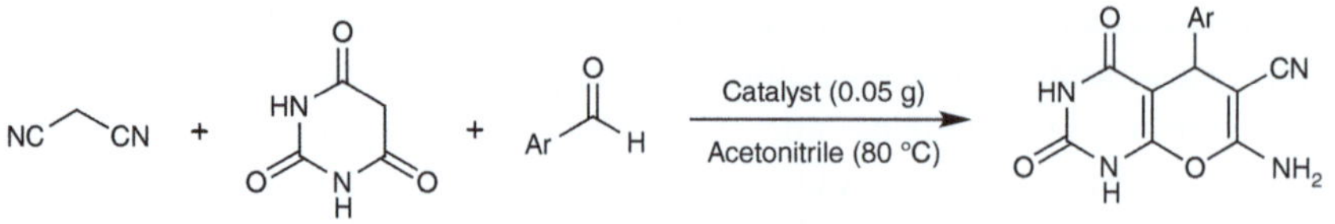

Scheme 2.3 Preparation of pyrano[2,3-*d*]pyrimidinone derivatives.

2.3 Solid Base Fly Ash

Fly ash has been utilized as a catalyst in a variety of processes, including solid plastic pyrolysis, hydrocarbon oxidation, gas phase oxidation of volatile organic compounds, solvent-free organic synthesis, H_2 generation, deSOx, deNOx, and hydrocracking [24]. In this work, fly ash of the F-type, which is high in alumina and silica content, is chosen to transform it into a solid base. This is in contrast to zeolites, which possess enough —OH groups to catalyze organic conversions. As per recent research from our laboratory, fly ash can be utilized as an economical, reusable, and active solid acid catalyst for the production of aspirin and wintergreen oil [25]. This work's solid base, made of fly ash, is a nanocrystalline, recyclable catalyst with enough active basic sites for cyclohexanone and benzaldehyde solvent-free condensation to produce α,α′-dibenzylidene cyclohexanone, which finds extensive use in the fine chemical industries.

2.3.1 Synthesis

The fly ash was chemically activated with NaOH to create the solid base fly ash (SBFA) catalyst . Fly ash and NaOH mixtures were used in a stirred reactor for the chemical activation process. The mixtures were heated to 110 °C while being stirred and the temperature was maintained for two days. Following the aging process, surplus NaOH and leached chemicals were eliminated from the pulp by washing it with distilled water. The resulting pulp was dried for 24 hours at 110 °C. After being calcined at 450 °C for four hours, the chemically activated fly ash was thermally stabilized [26].

2.3.2 Catalytic Activity of SBFA

Claisen–Schmidt condensation between cyclohexanone and benzaldehyde yields α,α′-dibenzylidene cyclohexanone as a test reaction in a liquid-phase batch reactor, which is used to assess the catalytic performances of the nanocrystalline SBFA (Scheme 2.4) [26].

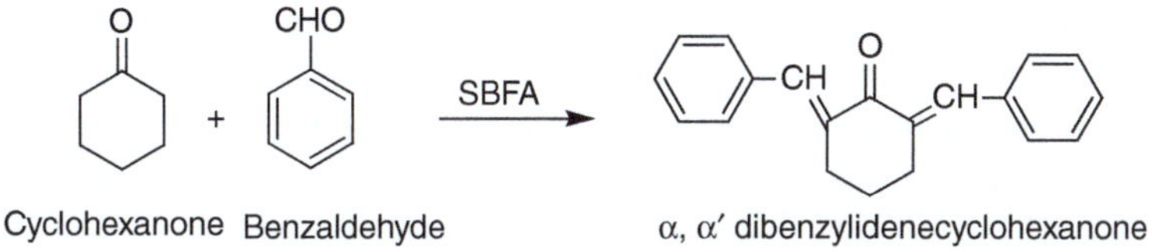

Scheme 2.4 SBFA catalyst to facilitate the Claisen–Schmidt condensation.

2.3.3 Condensation Between Benzaldehyde and Cyclohexanone

In a liquid-phase batch reactor, benzaldehyde and cyclohexanone were condensed using SBFA. During the process, a round-bottom flask (50 ml) with a magnetic stirrer and condenser was filled with cyclohexanone and benzaldehyde (molar ratio: 1 : 2) and immersed in an oil bath maintained at a constant temperature. After being activated at 450 °C for two hours in static air before the reaction, the catalyst (weight ratio = 10:1) was added to the reaction mixture. The reaction mixture was heated for 30 minutes to four hours at atmospheric pressure, with the necessary reaction temperature varying between 90 and 140 °C. After the reaction was finished, the mixture was cooled, filtered to remove the catalyst, and then examined using a gas chromatograph with a flame ionization detector, an HP-5 capillary column, a programmed oven temperature of between 50 and 280 °C, and N_2 (1.5 ml/min) as the carrier gas [26].

2.3.4 Catalyst Regeneration

Once the catalyst was used, it was filtered out of the reaction mixture and regenerated for another use. Before the procedure, the recovered catalyst was completely cleaned with acetone, dried in an oven at 110 °C for 12 hours, and then activated for two hours at 450 °C in a static environment. Consequently, regenerated catalyst was employed in further reaction cycles with comparable reaction conditions [26].

2.4 Calcined Water Sludge

Waste materials including marble slurry, shells, red mud, and bones have all been studied as potential catalysts [27–30]. The finest heterogeneous base catalysts for the production of biodiesel were discovered to be these waste products, which serve as the primary source of CaO. Therefore, it is highly desirable to create waste-derived catalysts with excellent catalytic activity for the low-cost generation of biodiesel. Water treatment precipitate, sometimes referred to as water sludge, is produced as solid waste following the Nalgonda method of defluoridating drinking water. Coagulation is used in the well-known Nalgonda defluoridation process. Here, alum is introduced in the form of alumina-fluoro complexes along with $CaCO_3$ as a pH controller. The disposal of waste sludge leads to major complications, including environmental concerns and health risks. $CaCO_3$ and alumino-fluoro complexes make up the majority of water sludge, suggesting that it can be a viable feedstock for the creation of efficient catalysts for the generation of biodiesel. Consequently, drinking water defluoridation sludge was modified to create a solid base catalyst, which was then investigated for use in the production of biodiesel in this work. The purpose of this work was to prepare a very simple heterogeneous catalyst for the generation of biodiesel from soybean oil using drinking water treatment sludge [31].

2.4.1 Catalyst Preparation

Water sludge was produced by the water precipitate that was left over after the defluoridation process, in which alum was introduced together with lime. After repeatedly rinsing the water sludge with distilled water to get rid of contaminants and unwanted material, it was dried in a hot air oven set at 105 °C for roughly twenty-four hours. Then, the dried water sludge was ground into a fine powder known as uncalcined water sludge, and it was calcined for three hours at 950 °C in a muffle furnace to create the catalyst known as calcined water sludge (CWS) [31].

2.5 Oxides of Rare Earth

Rare earth oxides exhibit strong basic characteristics, but their application in base-catalyzed processes has been limited. The oxides exhibit the potential to oxidize as well [32].

2.5.1 Preparation

The hydroxides are produced into rare earth oxides by either air-calcination or high-temperature vacuum decomposition. By hydrolyzing aqueous nitrates with aqueous ammonia, the hydroxides are produced. Rare earth oxides generated in this manner are stable in sesquioxide M_2O_3 stoichiometry, except for three elements (Ce, Pr, and Tb). In the nominal compositions of Tb_4O_7, CeO_2, and Pr_6O_{11}, the oxides of the three exceptions are stable. Normal process flow for the synthesis of oxides from hydroxides consists of three stages [33]. Infra-red measurement during $La(OH)_3$ heat treatment shows that after outgassing at 773 K, CO_2 is substantially adsorbed and maintained on the surface. After 923 K, the IR bands corresponding to carbonates nearly completely disappeared. At the 923 K outgassing temperature, the number of strong basic sites appears to be at its maximum. The hydroxide breakdown process yields rare earth oxides, whose surface areas are dependent on temperature, environment, and other factors. The oxides generated by hydroxide breakdown in a vacuum at 873 K. Rare earth oxides undergo structural changes when stored in air at ambient temperature unless particular care is taken because of their strong interactions with CO_2 and H_2O. Freshly synthesized La_2O_3 underwent bulk carbonation and hydration, forming the amorphous hydroxycarbonate $La_2(OH)_4CO_3{\cdot}nH_2O$ phase and hexagonal $La(OH)_3$, when it was held for five years under normal conditions [34]. To renew the oxide phase, thermal treatment at a temperature of roughly 1100 K is required.

2.6 Titanium Dioxide

Three stable polymorphs of titanium dioxide exist at atmospheric pressure: brookite, rutile, and anatase. Anatase and rutile are often utilized as supports and catalysts. When heated in air to a temperature of around 1073 K or higher, anatase transforms into rutile. Both the anatase and rutile polymorphs of titanium dioxide exhibit basic and acidic surface characteristics, and they both catalyze processes where acidic or basic sites are significant. By using infrared spectroscopy of adsorbed probe molecules, researchers have experimentally demonstrated that the acidic characteristics of TiO_2, ZrO_2, and CeO_2 are arranged in the following order: $TiO_2 > ZrO_2 > CeO_2$ [35]. The basic character is arranged in reverse [36].

2.6.1 Preparation

Usually, TiO_2 is made by hydrolyzing and then calcining titanium alkoxides or $TiCl_4$. Even when the precipitate is thoroughly washed, $SO4^{2-}$ ions are included in the final TiO_2 when $TiOSO_4$ or $Ti(SO_4)_2$ is used as the raw material. Because of the $SO4^{2-}$ ions present, the titanium oxide containing sulfates have a large surface area. The following is an illustration of how TiO_2 is made from $TiCl_4$. To prevent the solution's temperature from rising, $TiCl_4$ is put dropwise into iced water. The mother solution's pH is raised to 7.0 by adding ammonia water to the mixture.

Deionized water is used to wash the precipitate until no Cl^- ions are found using a silver nitrate aqueous solution. After being allowed to dry at room temperature, the precipitate is calcined at the targeted extreme temperature. On the other hand, the hydrolysis of the previously mentioned titanium tetrachloride and water solution. To create a precipitate, the TiO_2 aqueous solution is heated for seven hours. As before, deionized water is used to wash the precipitate. The following is an illustration of how titanium alkoxides are used to prepare TiO_2. Recrystallization from 2-propanol is used to purify titanium tetraisopropoxide. Three hundred grams of 2-propanol dissolve 300 g titanium isopropoxide. Five liters of water is mixed with the isopropanol solution at a rate of 30 ml min^{-1}. The precipitate is dried at 393 K after being rinsed with water using draining until no chloride ions are left behind. A higher temperature is used to calcine the dried sample. Calcining below 573 K results in amorphous crystalline structures, calcining between 623 and 773 K results in anatase crystalline structures, and calcining over 873 K produces rutile crystalline structures. Many techniques have been proposed for a high surface area TiO_2. They are all based on a sol–gel technique that uses titanium alkoxides as the raw ingredient. Here are descriptions of two techniques of the former kind and numerous instances of the latter kind.

Tanaka et al. used a salt catalytic sol–gel method using diluted titanium *n*-butoxide and water solution without the use of a template molecule to create a titanium oxide with a microporous structure [37, 38]. Five mL of a butanol solution with titanium *n*-butoxide (6.25 mmol) and a different butanol solution (7.5 ml) with ammonium acetate (6.25 mmol) and water (6.25 mmol) were made under a nitrogen environment. When 12.5 ml of the two butanol solutions were combined, a sol–gel reaction began. Twenty-four hours at 298 K and one week at 338 K were spent keeping the reaction temperature stable. After drying at 333 K for 90 minutes, the precipitate was calcined at a temperature between 523 and 773 K. The TiO_2 samples were calcined at 523, 573, 623, 673, 723, and 773 K. At 623 K and lower, the TiO_2 was amorphous, while at 673 K and higher, it was anatase. By using a similar process, Liu et al. produced mesoporous titanium oxide, but instead of using ammonium acetate as the hydrolysis catalyst, they employed nitric acid [39]. After being calcined at 723 K, the resultant TiO_2 surface area decreased in its prepared state.

Another approach makes use of active carbon [40]. After adding Ti $(OiPr)_4$ to the ethanol suspension and homogenizing it, the same volume of water was added. The active carbon had been introduced in a specific amount. Precipitation occurred when ammonium hydroxide was added until the pH reached 9.The precipitate was calcined for three hours at 723 K and dried at 383 K.

2.7 Zinc Oxide

Experimental evidence has demonstrated the presence of both acidity and basicity in ZnO, which is thought to be amphoteric. ZnO functions as a catalyst by facilitating base-catalyzed reactions, even if these processes require acidic sites.

2.7.1 Preparation

Normally, zinc oxide is made by adding NH_4OH to a water solution of zinc nitrate at 353 K, causing the hydroxide to precipitate until virtually all of it does [35]. The precipitate is filtered, cleaned with H_2O, dried, and heated to 673–773 K in the air for calcination. It was noted that the precipitating reagent affected the catalytic activity for the production of methanol; the catalyst made with sodium carbonate exhibited greater activity than the catalyst made with NH_4OH [36]. Zinc metal is also ignited to produce zinc oxide.

2.8 Alkaline Earth Oxides

Solid base catalysts such as MgO, CaO, SrO, and BaO are commonly employed in many base-catalyzed processes. The order of the surface basic strength is thought to be BaO > SrO > CaO > MgO. The reason magnesium oxide has been investigated the most out of all of them is likely because it is much easier to generate samples with a clear structure and a large surface area through thermal pretreatment than samples of BaO, SrO, and CaO. Because BaO, SrO, and CaO, have structures that are comparable to MgO's, it is possible to infer information about their surface and catalytic qualities from MgO's performance. Even though RaO and BeO are classified as alkaline earth oxides, their respective radioactivity and toxicity have prevented further research on them. The creation of basic sites, the description of active sites, and the manufacture of catalysts with a high surface area and distinct pore structure are primarily discussed for magnesium oxide in this part, with a smaller amount of information about the other alkaline earth oxides.

2.8.1 Preparation

Typically, carbonates are used to prepare SrO and BaO, while hydroxides are used to prepare MgO and CaO. In a vacuum or with inert gases such as Ar and N_2, the thermal breakdown is carried out. Because barium hydroxide and strontium hydroxide melt at lesser temperatures, handling samples can be challenging. However, BaO and SrO are also generated via thermal decomposition of strontium hydroxide and commercially available barium oxide. Magnesium hydroxide was converted to magnesium oxide, and the layered structure known as brucite was replaced with a cubic structure known as periclase [41].

2.8.1.1 Conventional Method for MgO Catalyst

The surface area of magnesium oxide that is sold commercially is small. From a commercially available magnesium oxide, high surface area magnesium oxide is manufactured using the following steps. Overnight, magnetic stirring is used to boil commercially available magnesium oxide in deionized water. The slurry forms $Mg(OH)_2$, which is subsequently filtered out. The filter cake is then dried in an oven at 393 K. The powder is dehydrated and then gradually heated ($<10\ K\ min^{-1}$) in the presence of nitrogen or vacuum to 773 K or higher. H_2O should be kept out of the heat treatment process at high temperatures to prevent sintered MgO and reduced surface area.

In a vacuum, the temperature range of 553–623 K is where brucite magnesium hydroxide transforms into periclase magnesium oxide [42]. It took five hours to heat up. As the temperature rises from 573 to 583 K, the modification of magnesium hydroxide to magnesium oxide becomes more noticeable. The temperature at which $Mg(OH)_2$ is heat treated as well as the length of heating time affect this transformation into MgO. A 15% MgO phase was visible after one hour of heat treatment at 623 K in a vacuum. After 10 hours of heating, the MgO phase percentage increased to 94%. At a temperature between 473 and 673 K, brucite turns into periclase when magnesium hydroxide is calcined in air [43].

2.8.1.2 Effects of Starting Magnesium Salt

The characteristics of the resultant MgO are influenced by the kind of magnesium salt that is used to hydrolyze $Mg(OH)_2$. Main characteristics are suppressed by residual anions. Chloride ions in particular have a significant impact on the fundamentals of the resultant MgO. By hydrolyzing magnesium compounds such as nitrate, sulfate, carbonate, chloride, acetate, and oxalate with aqueous NH_3, then washing and calcining at 823 K, Matsuda et al. produced MgO. Cl^- ion addition significantly reduced basicity and activity. Cl^- was proposed to be retained in magnesium oxide produced from $MgCl_2$ and to have an impact on MgO's surface characteristics [44]. By hydrolyzing magnesium salts to $Mg(OH)_2$, Choudhary et al. produced MgO, which was then calcined under various circumstances. They altered the precipitating agent, aging period, calcination temperature, precipitation pH and temperature, and kind and concentration of magnesium salt.

Every circumstance had an impact on the final MgO's surface characteristics. Low surface area has a noticeable impact on the MgO produced from $MgCl_2$ [45]. Using a variety of synthetic methods, such as the application of the sol–gel technique, precipitation with urea, and the use of different salts, Aramendia et al. produced MgO. The most active magnesium oxide was created by rehydrating and then calcining magnesium oxide that had earlier been made from commercially available magnesium hydroxide by calcining it for two hours at 873 K in air [46].

2.8.1.3 Preparation of MgO by Sol–Gel Method

The following process was used to manufacture sol–gel magnesium oxide [47]. To create gel, the refluxing mixture of magnesium ethoxide and ethanol was mixed with oxalic acid ($H_2C_2O_4$) to bring the pH down to 5. The product was dried in air at 343 K to produce white powder after the surplus ethanol was evaporated. This powder was then calcined in air at 673, 873, and 1073 K. When brucite magnesium hydroxide was calcined to become periclase magnesium oxide [48]. Magnesium ethoxide was combined with a homogenous reaction media to create brucite by the sol–gel method. A shift in morphology from a needle-shaped crystallite to a tiny crystallite with a cubic structure occurred when the calcination temperature was raised from 540 to 773 K. This resulted in the breakdown of brucite to create periclase. By using a modified sol–gel approach, Gulkova et al. produced high surface area magnesium oxide by treating MgO with water/methanol or water/ethanol and then simply drying the resultant alcogel of magnesium hydroxide. The alcogel was calcined at 623–663 K in flowing air to produce magnesium oxide [49]. Klabunde's and coworkers used the autoclave to create an aerogel method to produce a high surface area magnesium oxide, which they called AP–MgO [50–52]. Clean metal fragments were dissolved in methanol in an inert atmosphere to create metal methoxide. The hydroxide gel was then autoclave-dried to remove the solvent. After being heated for 24 hours under a dynamic vacuum (0.05 Torr), the hydroxide underwent partial transformation into the metal oxide with a nanoscale size. The magnesium oxide has a very large surface area because of the autoclave preparation. CaO preparation can alternatively use autoclave preparation [53].

2.8.1.4 Preparation of Mesoporous MgO

The surface area of the resultant MgO is increased by the addition of carboxylic acids to a water solution of Mg salts. The citric acid employed by Chen et al. [54]. Citric acid was added to a liquid magnesium nitrate solution. In order to create a viscous solution, the solution was heated to remove the water. To turn the viscous solution into a fluffy powder, it was placed in an oven and baked at 413–453 K for 30 minutes. An amorphous combination of carbon and MgO was obtained by heating the powder to 1073 K in a N_2 flow with O_2. To stop MgO particles from aggregating, carbon was applied to them. The carbon was then removed from the mixture by heating it to 673 K in air. With a mean pore size of 4.0 nm, the resultant MgO was obtained. Using the same technique as Marcilly et al. Nakayama et al. also synthesized magnesium oxide from a mixture of magnesium nitrate and citric acid [55, 56]. After calcining the MgO at 673 K in air, they obtained MgO. The $Mg(NO_3)_2$ aqueous solution was supplemented by Takenaka et al. with a sequence of carboxylic acids [57]. The alkyl-chain length of carboxylic acid in the range of 13–38 nm allowed for control over the pore size of the resultant MgO. Using mesoporous carbon (CMK)-type carbon as an exotemplate, an ordered mesoporous magnesium oxide was created by the exotemplating technique [58]. As an exotemplate, mesoporous silica is also used to create CMK-type carbon [59]. CMK-3 was dissolved in an aqueous magnesium nitrate solution and agitated for two hours to allow the $Mg(NO_3)_2$ to penetrate the mesopores. Following vacuum drying and filtration, the solution was boiled to 573 K at a steady pace of 2.5 K/min in an air environment to change the $Mg(NO_3)_2$ in the pores to MgO. Three times this process was carried out. In the end, the carbon was eliminated through heating to 1073 K.

2.8.1.5 Catalytic Activity for Claisen–Schmidt Reaction

In order to assess the catalytic activity of the calcined magnesium oxide (500 °C) morphologies and comprehend the connection between the reactivity and morphology, solvent-free Claisen condensation of acetophenone and benzaldehyde was carried out (Scheme 2.5). All morphologies of calcined magnesium oxide exhibit 100% selectivity for 1,3-diphenyl propenone, albeit at varying rates of acetophenone conversion. When compared to bulk magnesium oxide under the same reaction conditions, all of the calcined MgO (500 °C) samples with varying morphologies showed noticeably enhanced activity for the named reaction. After four hours, the acetophenone conversion using MgO with flakes-like morphologies was over 90%; specifically, the large structure yielded 99% and 97% conversion, respectively, while the bulk MgO only produced 22% conversion of acetophenone under the same conditions. The presence of low-coordinate oxide sites, step edges, step corners, and lattice defects (anion and cation vacancies) in large quantities is responsible for the high catalytic activity. Once more, increased activity was observed in the morphologies composed of nanoflakes, most likely as a result of the catalyst's large surface area. This is because the catalyst's surface area controls the amount of reactant adsorption on its active sites. Since reactants have complete access to the flake-like building blocks, the efficiency of organic species' diffusion through the thin flakes further promotes the reaction [60].

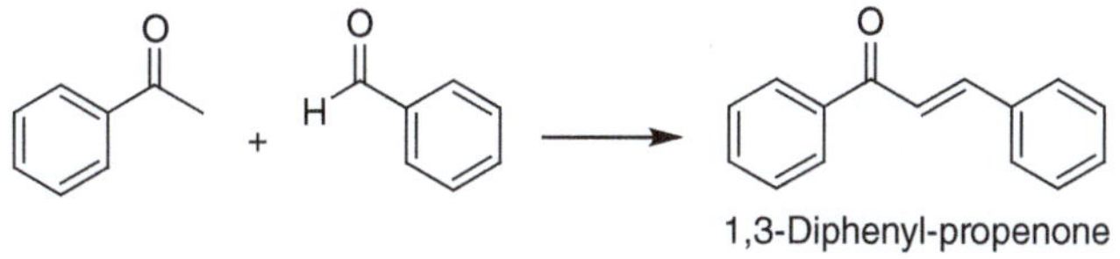

Scheme 2.5 Representation of Claisen condensation reaction.

2.9 Hydrotalcite

The approximate formula of hydrotalcite, a naturally occurring mineral, is $Mg_6Al_2(OH)_{16}CO_3 \cdot 4H_2O$. The hydroxide layers in hydrotalcite are negatively charged interlayers between positively charged, brucite-like layers. Hydroxyl groups encircle the magnesium cation in brucite, forming an octahedral share edge that forms infinite sheets. Al^{3+} ions isomorphously replace a portion of the Mg ions in the brucite layer of hydrotalcite. The interlayers of natural hydrotalcite contain CO^{3-} ions. Regarding the base catalysts, the most typically utilized starting materials are hydrotalcite containing Mg, Al, and CO_3^{2-} ions. Significantly significant catalysts and precursors for catalysts are hydrotalcite and related substances. Several writers have examined the synthesis, structure, and catalytic use of hydrotalcite [61–63]. An evaluation of hydrotalcite's catalytic properties has also been conducted [64, 65].

2.9.1 Synthesis of Hydrotalcite

The several methods for creating materials that resemble hydrotalcite have been examined [62, 66] coprecipitation is the most popular method, while hydrothermal treatment comes in second. It is also suggested to use urea hydrolysis as a homogenous precipitation technique to produce well-defined particles with regulated sizes [67–72]. Abelló and Pérez-Ramírez demonstrated the continuous manner of hydrotalcite generation (in-line dispersion–precipitation process) [73]. The sol–gel method of preparation is another extremely helpful technique [74–77]. Greenwell et al. described a comparative analysis of the crystal size, morphology, and catalytic activity for acetone aldol condensation using the coprecipitation method, sol–gel method, and urea method [72]. Below are some typical illustrations of the preparatory techniques.

2.9.1.1 Coprecipitation Method

The following aqueous solutions were gradually combined with vigorous stirring: $MgCl_2 \cdot 6H_2O$ (0.15 mol), NaOH (0.4 mol), $AlCl_3 \cdot 6H_2O$ (0.05 mol), and Na_2CO_3 (0.01 mol). The pH was kept between 8 and 10 at 343 ± 5 K. After the precipitate was aged for 15 hours at this temperature while being stirred in the mother liquor, it was repeatedly washed until the solution was clear of Cl^- ion. At 313 K, the products were dry. They have CO_3^{2-} and Cl^- as charge-compensating ions. To exchange chloride anions, a Na_2CO_3 solution was applied to the hydrotalcite; the dehydrated sample was mixed with 1.5×10^{-3} M sodium carbonate and left to stir at 343 K for two hours. The solids were filtered, then cleaned and dried at 313 K [78]. A subsequent exchange was conducted to reach a greater degree of exchange. Enhancing the crystallinity is frequently accomplished through aging or hydrothermal treatment [62, 79]. Microwave radiation can significantly reduce the amount of time needed for aging [80–82]. Textual properties are also altered by microwave radiation, resulting in a greater surface area [82]. To improve solid base catalysts, it is required to exchange Cl^- ions in the precipitate with carbonate. This is because the Cl^- ions in the calcined samples significantly decrease the quantity of carbon dioxide that can be adsorbed and the catalytic activity level for base-catalyzed processes [74, 78, 83]. The presence of chloride in the initial hydrotalcite also inhibits the catalytic activity of the rehydrated hydrotalcite [74]. As a result, metal nitrates rather than chlorides are commonly utilized.

2.9.1.2 Sol–Gel Method

Using HCl or HNO_3, an acid hydrolysis method is used to dissolve a sample of magnesium ethoxide (0.075 mol) in C_2H_5OH. With continual stirring, the solution refluxes at 353 K. A second solution is gradually added, and the pH is raised to roughly 10 with NH_4OH. This solution contains an appropriate amount of aluminum $C_5H_8O_2$ in a mixture of $(CH_3)_2CO/C_2H_5OH$ (1:1) to produce Mg^{2+}/Al^{3+}. Then, water is gradually added in molar amounts of Mg^{2+}. Up to 17 hours, the solution is refluxed at 353 K resulting in gel formation. The gel is agitated in an aqueous solution of Na_2CO_3 (10^{-1} M) to exchange chloride ions from hydrochloride. After being regularly cleaned in deionized water, the samples are dried at 353 K for an entire night [74].

2.9.1.3 Michael Addition

The Michael addition is a base-catalyzed C—C bond-forming process that involves the nucleophilic addition of a carbanion to α,β-unsaturated carbonyl compounds. For the production of monomers, coatings, corrosion inhibitors, and medicines, this reaction has found numerous industrial applications. Liquid bases such as NaH, alkoxides, and tertiary amines are employed as catalysts on an industrial scale [84–87]. KF supported by γ-Al_2O_3, α-Al_2O_3, phosphates, barium hydroxide, hydrotalcite, amines attached at surfaces, and organic resins are a few other solid bases that have been characterized [88–96]. One possible sequential reaction of the base-catalyzed Knoevenagel condensation is this addition, which requires less basic strength than the Michael addition reaction (Scheme 2.6) [97].

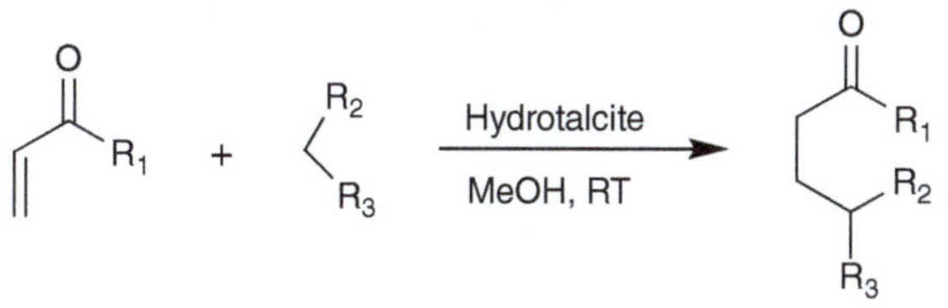

Scheme 2.6 Michael's addition of an unsaturated ketone.

2.10 Comparison of Different Solid Base Catalysts

See Table 2.1.

Table 2.1 Comparison of different base catalysts advantages, disadvantages, and applications.

Sr. no.	Catalysts	Advantages	Disadvantages	Applications	Ref.
1.	K_2O/Al_2O_3–CaO	High thermal stability	Production and maintenance of the K_2O/Al_2O_3–CaO catalyst can be relatively expensive	Petrochemical industry	[23]
		High surface area	It may require specific operating conditions for optimal performance, limiting its versatility in some applications	Environmental remediation	
		High activity		Biodiesel production	
		Excellent selectivity			
		It can be regenerated multiple times without significant loss of catalytic activity			
2.	Solid base fly ash	High surface area	Sensitivity to water	Environmental remediation	[26]
		Excellent basicity	Cost of production	Energy production	
		Environmental friendliness	Specific reaction limitations	Chemical synthesis	
3.	Calcined water sludge	Enhanced catalytic activity	High sensitivity to contaminants	Petrochemical industry	[31]
		Versatility in applications	Limited temperature tolerance	Environmental remediation	
		Exceptional reaction efficiency	Complex regeneration process	Wastewater treatment	
4.	Rare earth oxide	Chemical stability	Costly production	Environmental remediation	[61]
		Temperature resistance.	Careful monitoring and regeneration processes	Energy production	
		High surface area		Refining and petrochemicals	
		Environmental benefits enhanced reactivity			
5.	Titanium oxide	Chemical stability	High cost	Plastics and polymers	[61]
		UV light absorption	Environmental concerns	Cosmetics and sunscreen	
		High refractive index		Paints and coatings	

6.	ZnO	Low cost and abundance	Sensitivity to temperature and pressure	Chemical industry	[61]
		High catalytic activity			
		Antibacterial properties		Rubber industry	
		High thermal stability	Limited stability and durability	Cosmetics industry	
		Excellent UV protection		Food industry	
7.	Alkaline earth oxide	High reactivity	Limited to certain reactions	Petroleum refining	[61]
		Thermal stability	Sensitive to poisoning	Polymer production	
		Environmental enhanced selectivity	High activation energy	Environmental remediation	
		Friendliness abundant availability			
8.	Hydrotalcite	High selectivity	Limited applicability to certain reactions	Suitable for various organic transformations and environmental remediation processes	[98]
		High thermal stability			
		Selectivity in catalytic reactions			
		Regeneration capability			

2.11 Conclusion

With better basicity and higher product conversion values, the study offers a solid base catalyst made of fly ash that is effective and efficient. SBFA possesses enough basicity for catalytic activity and is nanocrystalline in nature. An effective way to produce biodiesel from soybean oil is to use a modified heterogeneous catalyst produced from water defluoridation sludge. Not only does this catalyst permit the utilization of hazardous waste, but it also holds great promise for the manufacture of biodiesel from an environmental and economic standpoint. Thus, the public's health and the environment are safeguarded, expenses are decreased, and resources are saved. A novel heterogeneous nanocatalyst K_2O/Al_2O_3–CaO solid base catalyst was created by loading K_2O onto an Al_2O_3–CaO mixed oxides basic support. The K_2O/Al_2O_3–CaO solid base catalyst had the benefit of being simple to prepare and handle, recyclable, and reusable, making it a green and sustainable catalytic method. As a result, it was easily recovered by filtration and reused for five successive reactions with a small decrease in activity.

Conflicts of Interest

The authors do not have any financial or technical conflicts.

Acknowledgment

We are indebted to Banasthali Vidyapith for providing the necessities to complete this work.

References

1 Corma, A., Abd Hamid, S.B., Iborra, S., and Velty, A. (2005). Lewis and Brönsted basic active sites on solid catalysts and their role in the synthesis of monoglycerides. *Journal of Catalysis* 234 (2): 340–347.

2 Abelló, S., Vijaya-Shankar, D., and Pérez-Ramírez, J. (2008). Stability, reutilization, and scalability of activated hydrotalcites in aldol condensation. *Applied Catalysis A: General* 342 (1–2): 119–125.

3 Tomar, R., Singh, N., Kumar, N. et al. (2019). Base-Free Suzuki–Miyaura Coupling Reaction Using Palladium(II) Supported Catalyst in Water. *Catalysis Letters* 149 (6): 1589–1594. https://doi.org/10.1007/s10562-019-02723-9.

4 Tomar, R., Ebitani, K., and Chandra, R. (2019). Hydrotalcite-Supported Ceria Nanoparticles as a Heterogeneous Catalyst for One-Pot Synthesis of Imines under Atmospheric Air. *ChemistrySelect* 4 (12): 3577–3581. https://doi.org/10.1002/slct.201900750.

5 Tomar, R., Kundra, P., Sharma, J. et al. (2024). An Overview: Synthesis of Menthol using Heterogeneous Catalysis. *Letters in Organic Chemistry* 21 (1): 16–28. https://doi.org/10.2174/1570178620666230623114308.

6 Kumar, L., Verma, N., Tomar, R. et al. (2023). Development of bioactive 2-substituted benzimidazole derivatives using an MnOx/HT nanocomposite catalyst. *Dalton Transactions* 52 (10): 3006–3015. https://doi.org/10.1039/d2dt02923e.

7 Kumar, L., Verma, N., Sehrawat, H. et al. (2022). Successive oxidation–condensation reactions using a multifunctional gold-supported nanocomposite (Au/MgCe–HDO). *New Journal of Chemistry* 46 (7): 3472–3481. https://doi.org/10.1039/d1nj05690e.

8 Tomar, R., Rathee, G., Chandra, I. et al. (2018). Synthesis and Characterization of Magnesium Hydroxide & Cerium Oxide Composite: Application in Organic Transformation. *ChemistrySelect* 3 (6): 1645–1649. https://doi.org/10.1002/slct.201702947.

9 Tomar, R., Singh, N., Rathee, G. et al. (2017). Synthesis and Characterization of Hybrid $Mg(OH)_2$ and $CeCO_3OH$ Composite with Improved Activity Towards Henry Reaction. *Asian Journal of Organic Chemistry* 6 (12): 1728–1732. https://doi.org/10.1002/ajoc.201700485.

10 Arya, K. and Tomar, R. (2013). Microporous zeolite catalyst system: an eco-approach for regioselective synthesis of pyrimidobenzimidazoles. *Research on Chemical Intermediates* 41 (6): 3389–3400. https://doi.org/10.1007/s11164-013-1441-4.

11 Climent, M.J., Corma, A., Iborra, S., and Velty, A. (2002). Designing the adequate base solid catalyst with Lewis or Bronsted basic sites or with acid–base pairs. *Journal of Molecular Catalysis A: Chemical* 182: 327–342.

12 Alinezhad, H., Tarahomi, M., Maleki, B., and Amiri, A. (2019). SO_3H-functionalized nano-MGO-D-NH_2: synthesis, characterization and application for one-pot synthesis of pyrano[2,3-d]pyrimidinone and tetrahydrobenzo [b] pyran derivatives in aqueous media. *Applied Organometallic Chemistry* 33 (3): e4661.

13 Clark, J.H. (1999). Green chemistry: challenges and opportunities. *Green Chemistry* 1 (1): 1–8.

14 Monireh, A., Javad, S., Maryam, T., and Ali, P. (2021). A facile green synthesis of $MgCoFe_2O_4$ nanomaterials with robust catalytic performance in the synthesis of pyrano[2,3-d]pyrimidinone and their bis-derivatives. *Molecular Diversity* 25 (4): 2183–2200.

15 Maleki, B., Nasiri, N., Tayebee, R. et al. (2016). Green synthesis of tetrahydrobenzo [b] pyrans, pyrano[2,3-c]pyrazoles and spiroindoline-3,4′-pyrano [2,3-c]pyrazoles catalyzed by nano-structured diphosphate in water. *RSC Advances* 6 (82): 79128–79134.

16 Shylesh, S., Schünemann, V., and Thiel, W.R. (2010). Magnetically separable nanocatalysts: bridges between homogeneous and heterogeneous catalysis. *Angewandte Chemie International Edition* 49 (20): 3428–3459.

17 Astruc, D., Lu, F., and Aranzaes, J.R. (2005). Nanoparticles as recyclable catalysts: the frontier between homogeneous and heterogeneous catalysis. *Angewandte Chemie International Edition* 44 (48): 7852–7872.

18 Karimi-Chayjani, R., Daneshvar, N., Shirini, F., and Tajik, H. (2019). New magnetic nanocatalyst containing a bis-dicationic ionic liquid framework for

Knoevenagel condensation reaction. *Research on Chemical Intermediates* 45: 2471–2488.

19 Argyle, M.D. and Bartholomew, C.H. (2015). Heterogeneous catalyst deactivation and regeneration: a review. *Catalysts* 5 (1): 145–269.

20 Shokouhimehr, M., Hong, K., Lee, T.H. et al. (2018). Magnetically retrievable nanocomposite adorned with Pd nanocatalysts: efficient reduction of nitroaromatics in aqueous media. *Green Chemistry* 20 (16): 3809–3817.

21 Sharma, P. and Gupta, M. (2015). Silica functionalized sulphonic acid coated with ionic liquid: an efficient and recyclable heterogeneous catalyst for the one-pot synthesis of 1,4-dihydropyridines under solvent-free conditions. *Green Chemistry* 17 (2): 1100–1106.

22 Kour, G., Gupta, M., Vishwanathan, B., and Thirunavukkarasu, K. (2015). Iron nanotube-silica composite (ZVI-S-PCAT modified silica composite) preparation, characterization and application as a recyclable catalytic system for 5-membered ring organic transformations. *Dalton Transactions* 44 (33): 14975–14990.

23 Vaid, R., Sharma, N., Gupta, M., and Chowhan, B. (2022). Nanosized calcined mixed oxides-supported alkali metal (K_2O/Al_2O_3–CaO) as solid base catalyst: preparation and investigation of its catalytic efficiency in the synthesis of benzylidene malononitriles/barbiturates and in pyrano[2,3-d]pyrimidines. *Journal of the Iranian Chemical Society* 19 (7): 2855–2870.

24 Wang, S. (2008). Application of solid ash-based catalysts in heterogeneous catalysis. *Environmental Science & Technology* 42 (19): 7055–7063.

25 Khatri, C. and Rani, A. (2008). Synthesis of a nano-crystalline solid acid catalyst from fly ash and its catalytic performance. *Fuel* 87 (13–14): 2886–2892.

26 Jain, D., Khatri, C., and Rani, A. (2011). Synthesis and characterization of novel solid base catalyst from fly ash. *Fuel* 90 (6): 2083–2088.

27 Liu, Q., Xin, R., Li, C. et al. (2013). Application of red mud as a basic catalyst for biodiesel production. *Journal of Environmental Sciences* 25 (4): 823–829.

28 Suryaputra, W., Winata, I., Indraswati, N., and Ismadji, S. (2013). Waste capiz (*Amusium cristatum*) shell as a new heterogeneous catalyst for biodiesel production. *Renewable Energy* 50: 795–799.

29 Boey, P.L., Maniam, G.P., Abd Hamid, S., and Ali, D.M.H. (2011). Utilization of waste cockle shell (*Anadara granosa*) in biodiesel production from palm olein: optimization using response surface methodology. *Fuel* 90 (7): 2353–2358.

30 Chakraborty, R., Bepari, S., and Banerjee, A. (2010). Transesterification of soybean oil catalyzed by fly ash and egg shell derived solid catalysts. *Chemical Engineering Journal* 165 (3): 798–805.

31 Gupta, J., Agarwal, M., and Dalai, A.K. (2017). Novel solid base catalyst derived from drinking water defluoridation for biodiesel synthesis. *Periodica Polytechnica Chemical Engineering* 61 (4): 288–294.

32 Rosynek, M.P. (1977). Catalytic properties of rare earth oxides. *Catalysis Reviews Science and Engineering* 16 (1): 111–154.

33 Rosynek, M.P. and Magnuson, D.T. (1977). Preparation and characterization of catalytic lanthanum oxide. *Journal of Catalysis* 46 (3): 402–413.

34 Bernal, S., Botana, F.J., Garcia, R., and Rodriguez-Izquierdo, J.M. (1983). Thermal evolution of a sample of La_2O_3 exposed to the atmosphere. *Thermochimica Acta* 66 (1–3): 139–145.

35 Busca, G. (1999). The surface acidity of solid oxides and its characterization by IR spectroscopic methods. An attempt at systematization. *Physical Chemistry Chemical Physics* 1 (5): 723–736.

36 Haffad, D., Chambellan, A., and Lavalley, J.C. (2001). Propan-2-ol transformation on simple metal oxides TiO_2, ZrO_2 and CeO_2. *Journal of Molecular Catalysis A: Chemical* 168 (1–2): 153–164.

37 Tanaka, K., Murakami, Y., Imai, T. et al. (2001). Microporous structure of titanium oxide prepared by a salt catalytic sol–gel process. *Chemistry Letters* 30 (12): 1280–1281.

38 Matsumoto, T., Murakami, Y., and Takasu, Y. (2000). Size control of titanium oxide sheets by regulating catalysis in a catalytic sol–gel process and their UV absorption properties. *The Journal of Physical Chemistry B* 104 (9): 1916–1920.

39 Liu, C., Fu, L., and Economy, J. (2004). A simple, template-free route for the synthesis of mesoporous titanium dioxide materials. *Journal of Materials Chemistry* 14 (7): 1187–1189.

40 Colón, G., Hidalgo, M.C., and Navıo, J.A. (2002). A novel preparation of high surface area TiO_2 nanoparticles from alkoxide precursor and using active carbon as additive. *Catalysis Today* 76 (2–4): 91–101.

41 Moodie, A.F. and Warble, C.E. (1986). MgO morphology and the thermal transformation of $Mg(OH)_2$. *Journal of Crystal Growth* 74 (1): 89–100.

42 Yoshida, T., Tanaka, T., Yoshida, H. et al. (1995). Study of dehydration of magnesium hydroxide. *The Journal of Physical Chemistry* 99 (27): 10890–10896.

43 Walling, S.A., Kinoshita, H., Bernal, S.A. et al. (2015). Structure and properties of binder gels formed in the system Mg $(OH)_2$–SiO_2–H_2O for immobilisation of Magnox sludge. *Dalton Transactions* 44 (17): 8126–8137.

44 Matsuda, T., Tanabe, J., Hayashi, N. et al. (1982). Properties of magnesium oxides prepared from various salts and their catalytic activity in 1-butene isomerization. *Bulletin of the Chemical Society of Japan* 55 (4): 990–994.

45 Choudhary, V.R. and Pandit, M.Y. (1991). Surface properties of magnesium oxide obtained from magnesium hydroxide: influence on preparation and calcination conditions of magnesium hydroxide. *Applied Catalysis* 71 (2): 265–274.

46 Aramendı′ a, M.A., Borau, V., Jiménez, C. et al. (2003). Influence of the preparation method on the structural and surface properties of various magnesium oxides and their catalytic activity in the Meerwein–Ponndorf–Verley reaction. *Applied Catalysis A: General* 244 (2): 207–215.

47 Wang, J.A., Bokhimi, X., Novaro, O. et al. (1999). Effects of the surface structure and experimental parameters on the isopropanol decomposition catalyzed with sol–gel MgO. *Journal of Molecular Catalysis A: Chemical* 145 (1–2): 291–300.

48 Wang, J.A., Novaro, O., Bokhimi, X. et al. (1998). Characterizations of the thermal decomposition of brucite prepared by sol–gel technique for synthesis of nanocrystalline MgO. *Materials Letters* 35 (5–6): 317–323.

49 Gulková, D., Šolcová, O., and Zdražil, M. (2004). Preparation of MgO catalytic support in shaped mesoporous high surface area form. *Microporous and Mesoporous Materials* 76 (1–3): 137–149.

50 Mel'gunov, M.S., Fenelonov, V.B., Mel'gunova, E.A. et al. (2003). Textural changes during topochemical decomposition of nanocrystalline $Mg(OH)_2$ to MgO. *The Journal of Physical Chemistry B* 107 (11): 2427–2434.

51 Utamapanya, S., Klabunde, K.J., and Schlup, J.R. (1991). Nanoscale metal oxide particles/clusters as chemical reagents. Synthesis and properties of ultrahigh surface area magnesium hydroxide and magnesium oxide. *Chemistry of Materials* 3 (1): 175–181.

52 Itoh, H., Utamapanya, S., Stark, J.V. et al. (1993). Nanoscale metal oxide particles as chemical reagents. Intrinsic effects of particle size on hydroxyl content and on reactivity and acid/base properties of ultrafine magnesium oxide. *Chemistry of Materials* 5 (1): 71–77.

53 Koper, O.B., Lagadic, I., Volodin, A., and Klabunde, K.J. (1997). Alkaline-earth oxide nanoparticles obtained by aerogel methods. Characterization and rational for unexpectedly high surface chemical reactivities. *Chemistry of Materials* 9 (11): 2468–2480.

54 Chen, L., Sun, X., Liu, Y., and Li, Y. (2004). Preparation and characterization of porous MgO and NiO/MgO nanocomposites. *Applied Catalysis A: General* 265 (1): 123–128.

55 Nakayama, T., Sato, S., and Nozaki, F. (1996). Physical and catalytic properties of MgO prepared using citric acid. *Bulletin of the Chemical Society of Japan* 69 (7): 2107–2110.

56 Marcilly, C., Courty, P., and Delmon, B. (1970). Preparation of highly dispersed mixed oxides and oxide solid solutions by pyrolysis of amorphous organic precursors. *Journal of the American Ceramic Society* 53 (1): 56–57.

57 Takenaka, S., Sato, S., Takahashi, R., and Sodesawa, T. (2003). Mesoporous MgO and Ni–MgO prepared by using carboxylic acids. *Physical Chemistry Chemical Physics* 5 (21): 4968–4973.

58 Roggenbuck, J. and Tiemann, M. (2005). Ordered mesoporous magnesium oxide with high thermal stability synthesized by exotemplating using CMK-3 carbon. *Journal of the American Chemical Society* 127 (4): 1096–1097.

59 Choi, M. and Ryoo, R. (2003). Ordered nanoporous polymer–carbon composites. *Nature Materials* 2 (7): 473–476.

60 Sutradhar, N., Sinhamahapatra, A., Pahari, S.K. et al. (2011). Controlled synthesis of different morphologies of MgO and their use as solid base catalysts. *The Journal of Physical Chemistry C* 115 (25): 12308–12316.

61 Ono, Y. and Hattori, H. (2011). Preparation and catalytic properties of solid base catalysts—II. Specific materials for solid bases. *Solid Base Catalysis* 21: 157–218.

62 Cavani, F., Trifiro, F., and Vaccari, A.J.C.T. (1991). Hydrotalcite-type anionic clays: preparation, properties and applications. *Catalysis Today* 11 (2): 173–301.

63 Vaccari, A. (1998). Preparation and catalytic properties of cationic and anionic clays. *Catalysis Today* 41 (1–3): 53–71.

64 Sels, B.F., De Vos, D.E., and Jacobs, P.A. (2001). Hydrotalcite-like anionic clays in catalytic organic reactions. *Catalysis Reviews* 43 (4): 443–488.

65 Tichit, D. and Coq, B. (2003). Catalysis by hydrotalcites and related materials. *Cattech* 7 (6): 206–217.

66 Reichle, W.T. (1986). Synthesis of anionic clay minerals (mixed metal hydroxides, hydrotalcite). *Solid State Ionics* 22 (1): 135–141.

67 Ogawa, M. and Kaiho, H. (2002). Homogeneous precipitation of uniform hydrotalcite particles. *Langmuir* 18 (11): 4240–4242.

68 Adachi-Pagano, M., Forano, C., and Besse, J.P. (2003). Synthesis of Al-rich hydrotalcite-like compounds by using the urea hydrolysis reaction—control of size and morphology. *Journal of Materials Chemistry* 13 (8): 1988–1993.

69 Shumaker, J.L., Crofcheck, C., Tackett, S.A. et al. (2007). Biodiesel production from soybean oil using calcined Li–Al layered double hydroxide catalysts. *Catalysis Letters* 115: 56–61.

70 Costantino, U., Curini, M., Montanari, F. et al. (2003). Hydrotalcite-like compounds as catalysts in liquid phase organic synthesis: I. Knoevenagel condensation promoted by $[Ni_{0.73}Al_{0.27}(OH)_2](CO_3)_{0.135}$. *Journal of Molecular Catalysis A: Chemical* 195 (1–2): 245–252.

71 Yang, P., Yu, J., Wang, Z. et al. (2004). Urea method for the synthesis of hydrotalcites. *Reaction Kinetics and Catalysis Letters* 83: 275–282.

72 Greenwell, H.C., Holliman, P.J., Jones, W., and Velasco, B.V. (2006). Studies of the effects of synthetic procedure on base catalysis using hydroxide-intercalated layer double hydroxides. *Catalysis today* 114 (4): 397–402.

73 Abelló, S. and Pérez-Ramírez, J. (2006). Tuning nanomaterials' characteristics by a miniaturized in-line dispersion–precipitation method: application to hydrotalcite synthesis. *Advanced Materials* 18 (18): 2436–2439.

74 Prinetto, F., Ghiotti, G., Durand, R., and Tichit, D. (2000). Investigation of acid–base properties of catalysts obtained from layered double hydroxides. *The Journal of Physical Chemistry B* 104 (47): 11117–11126.

75 Prinetto, F., Ghiotti, G., Graffin, P., and Tichit, D. (2000). Synthesis and characterization of sol–gel Mg/Al and Ni/Al layered double hydroxides and comparison with co-precipitated samples. *Microporous and Mesoporous Materials* 39 (1–2): 229–247.

76 Bolognini, M., Cavani, F., Scagliarini, D. et al. (2003). Mg/Al mixed oxides prepared by coprecipitation and sol–gel routes: a comparison of their physico-chemical features and performances in m-cresol methylation. *Microporous and Mesoporous Materials* 66 (1): 77–89.

77 Valente, J.S., Cantu, M.S., Cortez, J.G. et al. (2007). Preparation and characterization of sol–gel MgAl hydrotalcites with nanocapsular morphology. *The Journal of Physical Chemistry C* 111 (2): 642–651.

78 Tichit, D., Lhouty, M.H., Guida, A. et al. (1995). Textural properties and catalytic activity of hydrotalcites. *Journal of Catalysis* 151 (1): 50–59.

79 Kovanda, F., Kloušek, D., Cílová, Z., and Hulínský, V. (2005). Crystallization of synthetic hydrotalcite under hydrothermal conditions. *Applied Clay Science* 28 (1–4): 101–109.

80 Miyata, S. (1980). Physico-chemical properties of synthetic hydrotalcites in relation to composition. *Clays and Clay Minerals* 28: 50–56.

81 Kannan, S. and Jasra, R.V. (2000). Microwave assisted rapid crystallization of Mg–M (III) hydrotalcite where M (III)=Al, Fe or Cr. *Journal of Materials Chemistry* 10 (10): 2311–2314.

82 Benito, P., Labajos, F.M., Rocha, J., and Rives, V. (2006). Influence of microwave radiation on the textural properties of layered double hydroxides. *Microporous and Mesoporous Materials* 94 (1–3): 148–158.

83 Mourad, M.C., Mokhtar, M., Tucker, M.G. et al. (2011). Activation and local structural stability during the thermal decomposition of Mg/Al-hydrotalcite by total neutron scattering. *Journal of Materials Chemistry* 21 (39): 15479–15485.

84 Figueras, F., Lakshmi Kantam, M., and Manoranjan Choudary, B. (2006). Solid base catalysts in organic synthesis. *Current Organic Chemistry* 10 (13): 1627–1637.

85 Veldurthy, B., Clacens, J.M., and Figueras, F. (2005). Magnesium-lanthanum mixed metal oxide: a strong solid base for the Michael addition reaction. *Advanced Synthesis & Catalysis* 347 (6): 767–771.

86 Wang, T., Wang, J., He, X. et al. (2019). An ambient curable coating material based on the Michael addition reaction of acetoacetylated castor oil and multifunctional acrylate. *Coatings* 9 (1): 37.

87 Reddy and Verkade, J.G. (2007). Merrifield resin–$C_6H_4CH_2N_3P$ ($MeNCH_2CH_2$) 3N: an efficient reusable catalyst for room-temperature 1,4-addition reactions and a more convenient synthesis of its precursor P ($MeNCH_2CH_2$) 3N. *The Journal of Organic Chemistry* 72 (8): 3093–3096.

88 Clacens, J.M., Genuit, D., Veldurthy, B. et al. (2004). CsF supported by α-alumina: an efficient basic catalyst. *Applied Catalysis B: Environmental* 53 (2): 95–100.

89 Garcia-Raso, A., Garcia-Raso, J., Campaner, B. et al. (1982). An improved procedure for the Michael reaction of chalcones. *Synthesis* 1982 (12): 1037–1041.

90 Clark, J.H., Cork, D.G., and Robertson, M.S. (1983). Fluoride ion catalysed Michael reactions. *Chemistry Letters* 12 (8): 1145–1148.

91 Zahouily, M., Bahlaouan, B., Aadil, M. et al. (2004). Natural phosphate doped with potassium fluoride: efficient catalyst for the construction of a carbon–carbon bond. *Organic Process Research & Development* 8 (2): 275–278.

92 Zahouily, M., Abrouki, Y., Rayadh, A. et al. (2003). Fluorapatite: efficient catalyst for the Michael addition. *Tetrahedron Letters* 44 (12): 2463–2465.

93 Choudary, B.M., Kavita, B., Sreenivasa Chowdari, N. et al. (2002). Layered double hydroxides containing chiral organic guests: synthesis, characterization and application for asymmetric C–C bond-forming reactions. *Catalysis Letters* 78: 373–377.

94 Choudary, B.M., Kantam, M.L., Kavita, B. et al. (2000). Catalytic C–C bond formation promoted by Mg–Al–O–t-Bu hydrotalcite. *Tetrahedron* 56 (47): 9357–9364.

95 Corma, A., Iborra, S., Rodriguez, I. et al. (2002). MCM-41 heterogenized chiral amines as base catalysts for enantioselective Michael reaction. *Catalysis Letters* 82: 237–242.

96 Ballini, R., Marziali, P., and Mozzicafreddo, A. (1996). Amberlyst A-27, an efficient heterogeneous catalyst for the Michael reaction of nitroalkanes with β-substituted alkene acceptors. *The Journal of Organic Chemistry* 61 (9): 3209–3211.

97 Climent, M.J., Corma, A., Fornes, V. et al. (1996). Aluminophosphates oxynitrides as base catalysts: nature of the base sites and their catalytic implications. *Journal of Catalysis* 163 (2): 392–398.

98 Burange, A.S. and Gopinath, C.S. (2021). Catalytic applications of hydrotalcite and related materials in multi-component reactions: concepts, challenges and future scope. *Sustainable Chemistry and Pharmacy* 22: 100458.

3

Advanced Characterization Techniques for Solid Base Catalysts: An Overview

Neelam Sharma[1], Suman Swami[2], Sakshi Pathak[1], Aruna[1], and Rahul Shrivastava[1]

[1] *Department of Chemistry, Manipal University Jaipur, VPO-Dehmi-Kalan, Jaipur Ajmer Express Way, Jaipur, Rajasthan 303007, India*
[2] *Department of Chemistry, Chandigarh University, NH-05, Ludhiana–Chandigarh State Highway, Mohali, Punjab 140413, India*

3.1 Introduction

A chemical substance that advances the rate of a reaction while remaining unchanged after the completion of reaction is referred to as catalyst. Without affecting thermodynamic equilibrium, it can change the activation energy of a chemical reaction. The catalyst is used only for thermodynamically feasible reactions with negative Gibbs free energy change. These principles apply to all types of catalytic processes, whether they are enzymatic, homogeneous, or heterogeneous without undergoing any chemical transformation. Catalyst helps in speeding up the completion rate of a chemical equilibrium, even though it is present in minimum quantity. It stimulates several chemical processes [1]. Homogenous catalysts may not be very useful for industrial purposes because it is difficult to separate out and reuse [2, 3]. Due to extra washing steps the recovery and reuse of dissolved homogeneous catalysts within the reaction mixture make it difficult to use further in the reactions. The use of heterogeneous catalysts in the synthesis of many productions such as biodiesel has drawn a lot of attention. It can easily be separated and recycled because it is neither dissolved nor consumed during the reaction [4]. Additionally, heterogeneous catalysts perform more effectively catalytically than homogeneous catalysts. Homogenous catalysts are generally not selected owing to sluggish reaction rate and higher cost (as it employs separation issues) [5].

Solid-based catalysts can be categorized mainly into two categories: (a) solid acid catalysts and (b) solid base catalysts. Chemical reactions are started by the interactions of acids and bases among reactants and catalysts. When an acid and a base are used to catalyze a reaction, the catalyst simultaneously acts as an acid and a base for reactants [6–13]. Many materials can be used as solid base catalysts if they have been appropriately prepared. The catalyst's activity was reduced by the presence of O_2, CO_2, and H_2O, on the basic material surfaces. This is because catalysts

Solid Base Catalysts: Synthesis, Characterization, and Applications, First Edition.
Edited by Ravi Tomar, K.K. Pant, and Ramesh Chandra.

are typically prepared at moderate temperatures. Pre-treating solid base catalysts at high temperatures to remove contaminants such as CO_2, O_2, and water from the surfaces is crucial if you want to highlight their fundamental features and promote base-catalyzed reactions. When it comes to catalysis, heterogeneous catalysis is crucial for the modern chemical industry [14]. The application of homogenous acid catalysts, for example, hydrochloric acid, sulfuric acid, and phosphoric acid, is suitable raw materials with a high level of free fatty acid (FFA) such as crude vegetable oil and waste cooking oil [15]. These catalysts are less susceptible to water and environments with a high FFA level. Compared to homogeneous base catalysts, homogeneous acid catalysts have a high amount product, but the rate of reaction is substantially slower [16]. Homogeneous acid catalysts have high activity, but owing to their strong acidity, they cause serious pipeline corrosion and environmental pollution [17]. In the past few decades, homogeneous base catalysts have been the most often used in the industry. It delivers a quicker reaction time and is chemically active. By employing homogeneous base catalysts, high yields of methyl esters can be produced at benign reaction circumstances and in minimum reaction time. Although homogeneous base catalysts have these benefits of, the naturally high FFA content of feedstock such as natural oil can stimulate an unfavorable saponification side reaction, which results in the formation of a significant amount of soap byproduct and obstructs the subsequent separation and purification steps [18]. Heterogeneous solid acids or solid base catalysts have several benefits, including simple catalyst separation, eco-friendliness, recyclability, and environmental safety [19]. Strong alkalinity, large specific surface area, homogeneous pore size distribution, and ease of synthesis processing are all benefits of heterogeneous catalysts [20]. Base catalysis is an essential branch of organic synthesis and a crucial aspect of industrial relevance particularly in the petrochemical, pharmaceutical, and fine chemical industries. Commercially, homogeneous bases such as KOH, NaOH, and $Ca(OH)_2$ are frequently used in base-catalyzed processes [21, 22]. These bases are hazardous, have high operating costs, require more than stoichiometric amounts/quantities, and pose significant environmental problems related to neutralization of base, product purification and separation, waste generation, and corrosion. Significant efforts were made toward the growth of processes mediated via heterogeneous catalysts [22]. The solid base was discovered to be the best alternative for resolving the homogeneous catalyst problems as well as improving reaction selectivity [22], as the type of basic strength and basicity (Lewis or Bronsted basic sites) of the solid base could be designed [23] to meet the needs of the reactions. Due to their environmental friendliness, low-temperature operation, simplicity of product separation, and improved product selectivity, solid bases are also useful for industry. CaO, hydrotalcite, and KF/Al_2O_3 [21] are a few examples of solid base catalysts that have been extensively studied.

In 1958, Pines and Haag conducted several studies that served as the beginning of formation of solid base catalysts [24]. It has been described that alumina supported by an alkali metal exhibited catalytic activity for the double-bond isomerization of alkenes, which developed using carbanions as intermediates. In the 1970s, it was discovered that a variety of metal oxides, apart from those containing alkali metals, exhibited catalytic activity for the base-catalyzed processes. Tanabe et al., described the use of solid MgO and CaO as catalysts in the 1970s (basic strength greater than $H_2 = 26$) [25]. In contrast to solid acid catalysts, solid base catalysts took two decades to be used for industrial applications due to their better catalytic activity than liquid base catalysts. The following are some benefits of using solid base catalysts in industry applications.

- The reaction mixture does not need to be neutralized, as neutralization of base is not necessary in the solid base-catalyzed reactions. Waste produced during neutralization could be reduced by using a solid base catalyst.
- Reactions at high temperatures are possible since the temperature of reactions is constrained below boiling point under atmospheric pressure and for solid base-catalyzed reactions the temperature of reaction remains unhampered.
- It is simple to separate catalysts from products, and used catalysts could be recycled.
- The process is extremely productive when the flow reactor system is used.
- The reactor system does not need to use base-tolerant materials.
- Certain reactions are particularly completed with solid base catalysts. For example, liquid bases do not exhibit heterolytic splitting of H_2 into H^- and H^+, although it was noted for solid base catalysts and the starting step for the base-catalyzed hydrogenation reactions of conjugated dienes.
- Dual-functional (bifunctional) catalysis is possible. Acidic and basic sites can coexist on the surfaces of solid bases, although this is impossible in the liquid phase.

As illustrated in Figure 3.1, a range of solid base catalysts has been explored in the past. These catalysts are conceded as solid base catalysts owing to the following features:

- The basic site's functionality of a solid base catalyst can be easily identified via surface characterization. The presence of basic sites on the surfaces is demonstrated as acidic molecules get attached on the surface, by the change of color in the acid–base indicators, and by spectroscopy.
- The chemical processes taking place over the materials are comparable to the "base-catalyzed reactions" that are more common in homogeneous systems. Isomerization, alkylation, aldol addition, Tishchenko reaction, and Michael addition are few examples of these processes.
- The strength or quantity of basic sites could be associated with catalytic activities. Additionally, acidic chemicals such as H_2O, HCl, and CO_2 are found to be hazardous for the active sites.
- Anionic intermediates are thought to be involved in the reactions, according to mechanistic analyses of the processes and spectroscopic analysis of the surface species [26].

Accordingly, to utilize heterogenous solid base catalysts in organic transformation, number of techniques are required for their characterization. This chapter aims to summarize those characterization techniques.

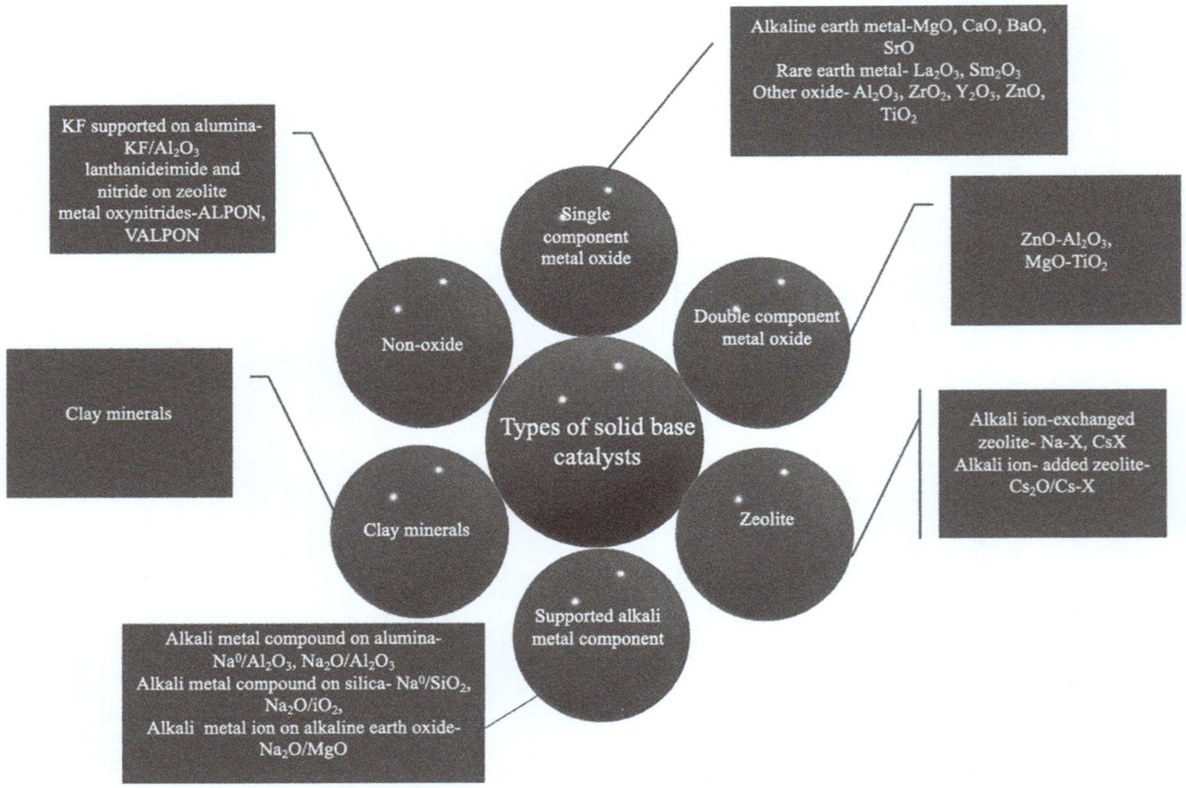

Figure 3.1 Diverse solid base catalyst. Source: Adapted from Hattori [26].

3.2 Traditional Characterization Techniques for Solid Base Catalyst

Characterization techniques for solid base catalysts are important for understanding their properties, surface chemistry, and reactivity. These techniques help researchers and scientists gain insights into the structure and performance of solid base catalysts. In the chapter, some common characterization techniques have been discussed in brief with research data.

3.2.1 Titration Method

The acid–base characteristics of the catalyst could thoroughly be investigated by titration method. As the basic sites of catalysts were determined via back titration, exhibited that the high pyridinic N ratio would result in a strong basic character, the findings of this work demonstrated that N-doping onto biochar was effective for the manufacture of a carbon-based base catalyst. In the O1s spectra, the peaks at 533 and 531.6 eV corresponded to ether-type oxygen (in anhydrides and esters) and carbonyl oxygen, respectively. The carbon surface's acidity was predominantly due to oxygenated functional groups. The C 1s spectrum showed two main peaks at 285.3 and 284.5 eV, corresponding to C—O and C—C bond, respectively. A small shoulder at 287.4 eV related to C=O bonds also been observed. The titration results verified that the biochar catalysts contained a specific quantity of basic sites is 0.247 mmol g^{-1} [27]. Similarly, the basic site content of the polyvinyl amine coated on Fe_3O_4 (n-Fe_3O_4/PVAm) nanocomposite was determined by back titration with NaOH. The nanocomposite was combined with HCl and agitated for 45 minutes. The catalysts were removed and repeatedly rinsed with DI water (deionized water). The surplus HCl was titrated using NaOH (0.1 mol l^{-1}) and phenolphthalein as an indicator. The amine concentration of the catalyst was 9.1 m equiv g^{-1} [28]. The basicity of the 50-NaHAP and hydroxyapatite (HAP) catalysts was evaluated using phenol adsorption. Estimation of basicity using phenol adsorption shows that adding $NaNO_3$ considerably improves the basic characteristics of HAP. The overall unmodified HAP basicity was 75 μmol g^{-1}; however, when the HAP was infused with $NaNO_3$(50 wt%), the total basic strength of the 50-NaHAP catalysts calcined at 800 °C gradually increased to 121 μmol g^{-1}, showing the formation of additional strong basic sites of the catalysts. These results suggest the basicity of the produced 50-NaHAP catalysts is commonly attributed to a significant basicity of alkali phosphate molecule created by the solid-state reaction in between the HAP backbone and $NaNO_3$, as well as the production of Lewis base sites on the surface of catalyst [29]. In a study, the Hammett titration method provides qualitative information on the basic characteristic properties of solid catalysts. The basic strength of Ca—M—Al (M=Ce, Mg, Y La,) layered double hydroxides (LDH) was evaluated by using Hammett indicators. To calculate total basicity, pre-treated catalyst was dispersed in cyclohexane and titrated with a benzoic acid solution using Hammett indicators. Titration studies revealed three fundamental sites: strong (18.4–26.5), medium (15.0–18.4), and weak (9.3–15.0). Adding Mg and La resulted in values in the super strong basic strength range (26.5–35.0), similar to CO_2-TPD results [30]. The basicity of pure SnO_2 and barium stannate catalyst samples (B2SO, 2BSO, and BSO) were determined using Hammett indicator in benzene carboxylic acid titration approach. The following Hammett indicator was used: phenolphthalein (H_ ¼ 9.3), neutral red (H_ ¼ 6.8), tropaeolin-O (H_ ¼ 11.1), bromothymol blue (H_ ¼ 7.2), and 2,4-dinitroaniline (H_ ¼ 15.0). The total basicity of the catalysts was computed by aggregating the individual basicities at three different pK_a ranges, which were determined using various indications. SnO_2 has no basic strength, while B2SO, 2BSO, and BSO catalysts have 0.64, 0.87, and 1.44 mmol g^{-1} in the range of 15.4–8.2 (2,4-dinitro-aniline to phenolphthalein indicator). Such finding indicates that pure Tin-oxide (SnO_2) had no basic sites to interact with substrate, whereas $BaSnO_3$ samples had sufficient basicity to operate as an effective heterogeneous catalyst in transesterification process. In contrast, BSO had the maximum basic strength, but the presence of CO_3^{-2} (carbonate species) in SnO_2 and 2BSO in B2SO reduced the basic strength more significantly. This occurred due to the acidic nature of SnO_2 and carbonate. Furthermore, it is expected the basic Ba—O link in $BaSnO_3$ will operate as an active site for potential interaction with a triglyceride molecule [31]. Further Hammett indicator was used to calculate the basic strength of KF/γ-Al_2O_3 nanoparticles-based catalyst. If the indicator changes color, the catalyst is considered stronger than the indicator; otherwise, the catalyst is considered weaker than the indicator. The catalyst changed the color of 2,4-dinitroaniline ($H\_ = 15$) from yellow to mauve and phenolphthalein ($H\_ = 8.2$) from colorless to pink, but not 4-nitro aniline ($H\_ = 18.4$). The catalyst's basic strength has been determined at $15 < H\_ < 18.4$, indicating a strong base for the transesterification process [32].

3.2.2 IR Analysis

Infrared (IR) spectroscopy is a popular analytical technique that has been commercially accessible since the 1940s. Primitive instruments were fitted with prisms functioning as dispersive elements, which were then superseded in the mid-1950s by diffraction gratings used in dispersive machines [33].

IR spectroscopy, commonly known as IR analysis, serves as an analytical technique for discerning functional groups within polymeric, organic, and certain inorganic materials. This method enables the examination of functional groups present on catalyst surfaces. In the IR and Fourier transform infrared spectroscopy (FT-IR) analysis process, infrared light is employed to scan test materials, allowing the observation of their chemical characteristics. During IR and FT-IR analysis, the instrument permits infrared radiation within the range of 10,000–100 cm^{-1} to pass through a sample. As the radiation interacts with the sample, some part of radiation is absorbed while the rest passes through the sample. The molecules in the sample convert the absorbed radiation into rotational and vibrational energy. The resulting signal manifests as a spectrum, representing the chemical fingerprint of the sample and typically spanning from 4000 to 400 cm^{-1}. This unique spectrum, indicative of the chemical structure of each molecule, distinguishes IR and FT-IR

Figure 3.2 FT-IR instrument. Source: Neelam Sharma, Central Analytical Facilities Lab at Manipal University Jaipur, India (Chapter Contributor).

analysis as a valuable tool for chemical identification, providing valuable insights into the composition of diverse materials [34]. A typical IR instrument is shown in (Figure 3.2).

Such as IR spectra of a $Sr(NO_3)_2$ loaded porous hydroxyapatite catalyst show absorption bands for PO_4^{-3} at 566, 602, 960, 1039, and 1092 cm^{-1} water linked with HAP at 1630 and 3430 cm^{-1}, CO_3^{-2} at 880 and 1463 cm^{-1}, and hydroxyl group (OH) at 3574 cm^{-1}. Additionally, a novel absorption peak at 3650 cm^{-1} that increased in intensity as $Sr(NO_3)_2$ loading increased. Sr^{2+} may have partially replaced the calcium triangles in the HAP structure. Because the bond length of Ca—O is shorter than that of Sr—O, the O—H bond was strengthened, as shown resulting in the splitting of the stretching vibration of hydroxyl group (OH) (one absorption peak remained invariable at 3574 cm^{-1}, another peak observed at 3650 cm^{-1}). Moreover, the intensity of the absorption peak at 3650 cm^{-1} steadily increased because the quantity of Sr^{2+} via which Ca^{2+} was substituted most likely grew as the loading quantity of $Sr(NO_3)_2$ increased [35]. The structural characteristics of the potassium loaded on Al_2O_3 (KNO_3/Al_2O_3) catalysts throughout a scanning range of 400–4600 cm^{-1}. An absorption band distinctive of the N—O vibration in KNO_3 was found at 1384 cm^{-1} during the IR spectroscopic analysis of KNO_3/Al_2O_3 catalysts. N—O absorption was modest in alumina without loaded KNO_3, but it became more pronounced as KNO_3 loading increased. An increase in KNO_3 breakdown and K_2O generation was shown by the weakening of the nitrate band on the 35% KNO_3/Al_2O_3 having a higher calcination temperature. The infrared spectra showed no lines related to NO_2 (1320, 1550 cm^{-1}, suggesting that KNO_3 breakdown produced K_2O instead of KNO_2. A broader and weaker IR absorption was observed at 1532 cm^{-1} in certain calcined materials, which was ascribed to carbonates. This implies that atmospheric CO_2 from the surrounding atmosphere is adsorbed on the fundamental sites of K_2O. Based on the thorough characterizations of the catalyst, it is probable that the primary active sites are the surface Al—O—K group and K_2O obtained from KNO_3 [36]. IR spectroscopy was utilized to investigate the newly discovered $Ca_{12}Al_{14}O_{33}$ and CaO-based solid base catalyst samples. Regarding O—H stretching vibration, the spectra showed a small and broad absorption peak at about 3430 cm^{-1} and a minor absorption peak at about 1630 cm^{-1} for O—H bending vibration. These absorption peaks were attributed to the absorbed H_2O molecules from the air. The absorption peak of the Al—O vibration was probably associated with two bands at 577 and 1050 cm^{-1}. The Ca—O bond stretching vibration was identified by a high absorption peak around 459 cm^{-1} and the Al—O stretching vibration bond in the tetrahedron (AlO_4) was identified by a broad peak at 839 cm^{-1}. According to the analysis, when hydrated tricalcium is calcined, Ca and Al oxides are formed [37].

3.2.3 Scanning Electron Microscopes

SEM stands for scanning electron microscopy; it is a technique used for characterization of surface morphology of the catalyst. It is a kind of electron microscope that can scan microorganism by using a beam of electrons moving toward low energy to scan and focus specimens. The inadequacy of light microscope's wavelength led to the invention of electron microscopes. Compared to light microscopes, electron microscopes have far shorter wavelengths, which allows for higher resolution power [38]. An typical SEM instrument is shown in (Figure 3.3).

Musa balbisiana Colla peels, a waste biomass material, are used to produce a highly effective and ecologically acceptable heterogeneous base catalyst for the generation of sustainable biodiesel. SEM analysis was conducted to study the morphological characteristics of the catalyst obtained from the calcinated banana peel ash (CBPA) and the burnt banana peel (BP) itself. The SEM images revealed a variety of aggregates with both microporous and mesoporous structures in both CBPA and BP. The higher surface area in the produced catalyst indicates a linear relationship between catalytic activity and surface areas. Catalysts with a higher surface area improve the interaction between reactants and catalyst-active sites

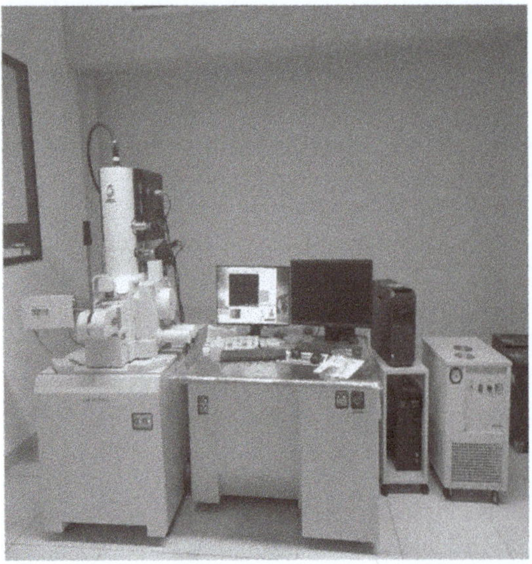

Figure 3.3 FE-SEM instrument. Source: Neelam Sharma, Central Analytical Facilities Lab at Manipal University Jaipur, India (Chapter Contributor).

[39]. KOH(32%)/ZrO_2-5 use as a solid base catalyst for transesterification reaction of *Silybum marianum* a nonedible oil with methanol to biodiesel. The traditional SEM nanographs of ZrO_2 and KOH (32%)/ZrO_2-5 particles show 100 nm crystallite particles for both samples. After adding KOH, the catalyst's surface displayed newly formed crystalline morphologies and smaller particle sizes. However, there was no noticeable difference between the ZrO_2 and KOH (32%)/ZrO_2-5 pictures, suggesting that ZrO_2 had good KOH dispersion. ZrO_2's structure held together after loading 32% KOH, which was essential to its catalytic activity. According to XRD measurements, potassium species were dispersed across the support's surface, with K_2O having the smallest particles and ZrO_2 having the biggest. Granular and porous structures, like a net, were seen in KOH (32%)/ZrO_2-5. This implies that there is more interaction in between the catalyst and the substrates, which could improve the efficiency of transesterification [40]. The surface morphology structure of NCS catalyst was observed via SEM analysis. The solid-state approach can produce consistent particle sizes of 2–3 μm with variable shapes and displays the particle as an irregular polyhedron with a distinct shape [41].

3.3 Advanced Characterization Techniques for Solid Base Catalyst

3.3.1 Fourier Transform Infrared Spectroscopy (FT-IR)

IR and FT-IR are based on the same infrared spectroscopy principles; however, FT-IR has become the favored approach due to its superior performance in terms of sensitivity, versatility, and speed. The advantages of FT-IR have led to its extensive use in a variety of industrial and scientific applications. The various kinds of basic sites present on the gasified straw slag were distinguished using IR spectroscopy of adsorbed CO_2. Four different adsorption species, bridging carbonate, unidentate carbonate hydrogen carbonate, and bi-dentate carbonate were recognized and characterized, as illustrated in Figure 3.4. The spectrum in Figure 3.4, shows a weak band at 1214 cm^{-1} and a strong one at 1448 cm^{-1}. These bands are signed to δ (C—O—H) and V_s (O—C—O) of hydrogen carbonates. It appears when CO_2 reacts with one coordinatively unsaturated cation in coordination with basic hydroxy groups, which are relatively strong basic sites. V_{as} (O—C—O) of bridged carbonate (CO_3^{-2}) and V_{as} (O—C—O) of unidentate carbonate are responsible for the bands at 1157 cm^{-1} and 1489 cm^{-1}, respectively. The bands at 1289 cm^{-1} and 1607 cm^{-1} correspond to the bidentate carbonate V_s(O—C—O) and V_{as} (O—C—O). These bands appear when carbon dioxide reacts on surfaces of low-coordination number oxygen anions (strongly basic site) whereas bridged and unidentate carbonates are generated by metal-oxygen couples, which are moderately strong basic sites [42].

The functional group and nature of potassium hydroxide-loaded neodymium oxide solid base catalyst has been analyzed by FT-IR spectra. The spectra were recorded by FT-IR spectrometer (AVATAR 360, Nicolet, Madison, USA) in between the range of 4000 and 400 cm^{-1} by using KBr pallets. In the 4000–400-cm^{-1} range,

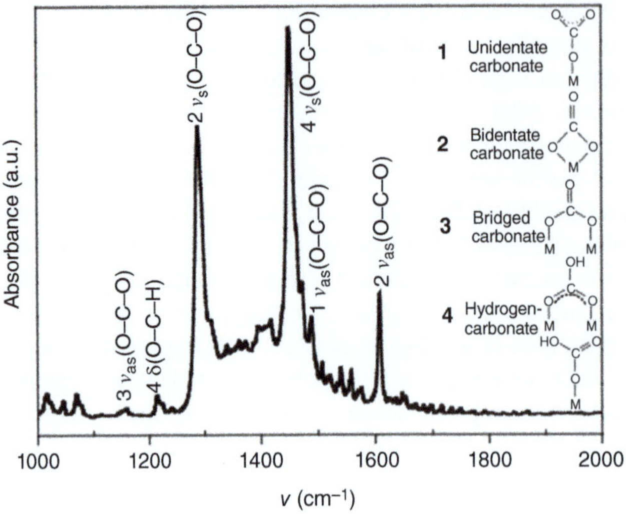

Figure 3.4 FTIR spectra of Carbon dioxide pre-adsorbed at 298 K on the gasified straw slag on desorption. Source: Reproduced from Wang et al. [42]/with permission of Elsevier.

a minimum of 32 scans were signal averaged with a resolution of 2 cm^{-1}. The spectrum of Nd_2O_3 has a high peak at 1475 cm^{-1}. From the spectra of KOH-Nd_2O_3 catalyst, the two new peaks arise at 706 and 880 cm^{-1} observed which are responsible for K—O—Nd and K—O bond (Figure 3.5). Additionally, the broader band located in the 3100–3300 cm^{-1} region might be partially assigned to stretching vibration of Nd—O—K bonds. During the activation process, K^+ ions may substitute the

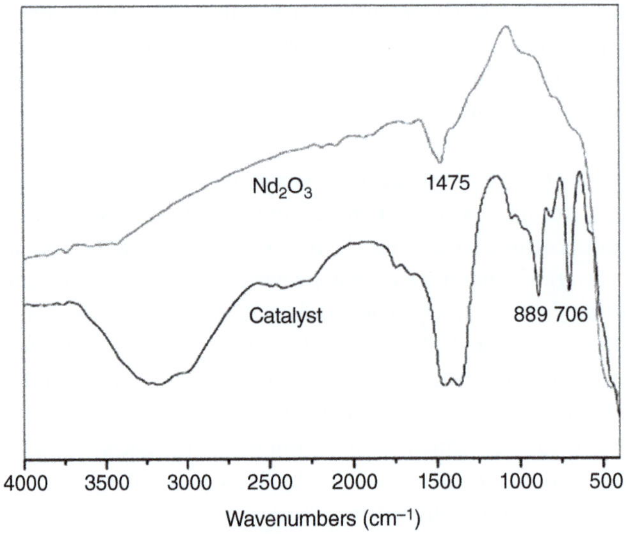

Figure 3.5 FT-IR spectra of Nd_2O_3 and catalyst. Source: Reproduced from Li et al. [43]/with permission of Elsevier.

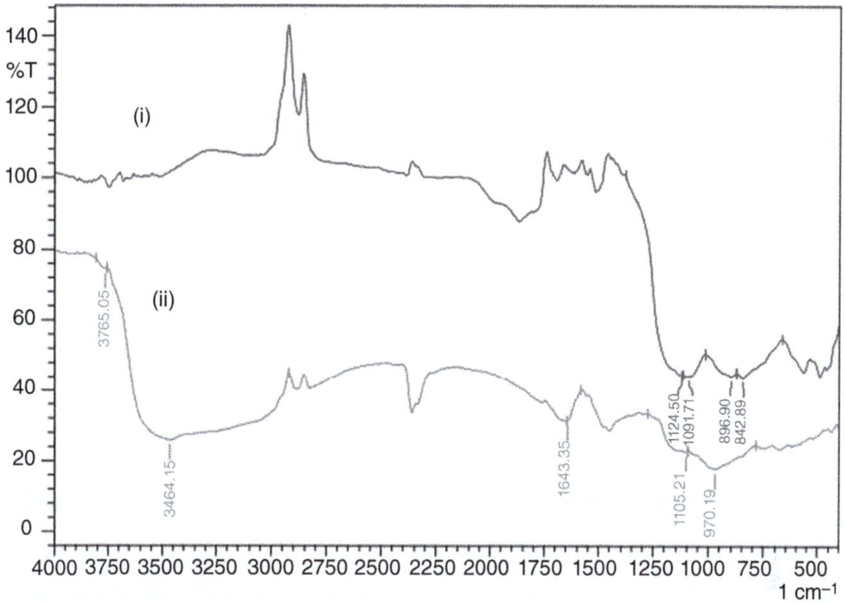

Figure 3.6 (i) FTIR of Fly ash and (ii) FTIR of solid base fly ash (SBFA). Source: Reproduced from Jain et al. [44]/with permission of Elsevier.

protons of isolated OH-groups for the construction of Nd—O—K groups which act as the active species for this catalyst [43].

The altered physiochemical properties of solid base fly ash (SBFA) have been examined via FT-IR spectroscopy. The FT-IR spectra of initial and final step of chemical activation of fly ashes exhibit a broader band in between 3400 and 3000 cm^{-1}, which are related to the surface-adsorbed water molecules and the hydroxyl groups surface of —Si—OH (Figure 3.6). The broadness of OH band can be attributed to the robust hydrogen bonding (H-bonding) present. The OH-groups are not isolated entities; rather, they exhibit a high degree of interaction because of strong H-bonding with another hydroxyl group. In the spectra of both samples, a peak at 1650 cm^{-1} is ascribed to the water molecule's bending mode (—O—H). After chemical activation, the spectrum of SBFA (Figure 3.6II) demonstrates a marked rise in the peak intensities of band for the hydroxyl group. The IR peaks' intensity, specifically those associated with Si—O—Si symmetric stretching at 798 cm^{-1}, Si—O—Si bending at 460 cm^{-1}, and T—O—Si asymmetric stretching at 1160, 1090, and 913 cm^{-1} (where T represents Si or Al), underwent noticeable changes as indicated by FT-IR investigations. In the case of SBFA, signals corresponding to the vitreous phase of unreacted mullite, quartz, and fly ash were detected at 1185, 1081, and 996 cm^{-1} [44].

Other FT-IR spectra of GO/SiO_2/PEA catalyst are demonstrated in Figure 3.7. The peak around 1578 cm^{-1} relates to carbon–carbon double bonds in the FT-IR spectra of graphene oxide. The vibration OH groups, carbonyl, and C—O, are linked to the absorption peak at 3437, 1678, and 1057 cm^{-1}, respectively. In the FT-IR spectrum of

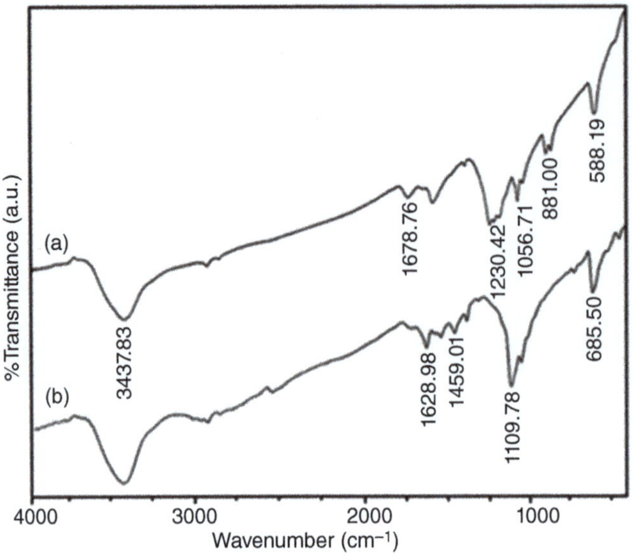

Figure 3.7 FT-IR of: (a) GO, (b) GO/SiO_2/PEA. Source: Reproduced from Toorbaf and Moradi [45]/with permission of Royal Society of Chemistry/CC BY 3.0.

GO/SiO_2/PEA catalyst vibration peaks relating to N—H, Si—O—C, C=C, and C=O at 3437, 1628, 1459, and 1109 cm^{-1} showed that the chemical attachment of PEA on the surface of GO was done via SiO_2 linker [45].

3.3.2 Field Emission Scanning Electron Microscopes (FE-SEM)

Field emission scanning electron microscope (FE-SEM) is an advanced technology for analyzing and imaging the surface morphologies of materials at high resolution. While they share similarities, there are significant variations between SEM and FE-SEM. FE-SEM has higher resolution and surface sensitivity than traditional SEM, making it especially useful for applications that need extremely detailed imaging of nanoscale structures. However, SEM is still a versatile and commonly used technology for a variety of imaging and analytical tasks. An N-doped biochar catalyst's surface had layered structure with irregular aggregates, as shown in the SEM images, which might be responsible for the catalyst's low/minimum surface area (Figure 3.8) [46].

SEM analysis has produced surface morphology images of the CaO supported on Al_2O_3, CeO_2, and ZrO_2 solid base catalysts using the FEI Quanta 200F instrument. The SEM data of CaO revealed flake-like structure, while all the developed catalysts displayed distinct structural morphologies. In the instance of CaO/CeO_2, a collection of sheet-like structures was formed via the accumulation of spherical-shaped particles. The CaO/ZrO_2 catalysts displayed an agglomerated structure of irregularly shaped particles, while the CaO/Al_2O_3 catalyst displayed a sheet-like surface that

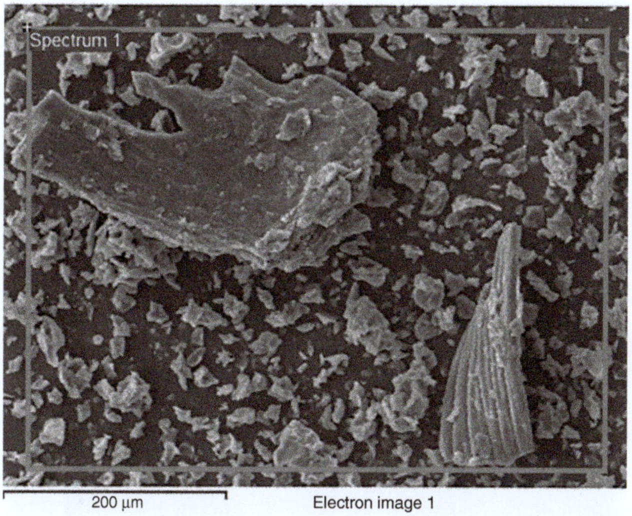

Figure 3.8 SEM images of nitrogen-doped biochar catalyst before the catalytic reaction. Source: Reproduced from Chen et al. [46]/with permission of American Chemical Society.

may have been caused by the presence of Al_2O_3. The bunch of asymmetrical particles appeared to be a section portion of the sheet [47]. A new magnetic bi-functional solid base catalyst was analyzed that was based on bamboo charcoal (K/BC–Fe_2O_3) to produce biodiesel. In Figure 3.9a the samples of bamboo charcoal (BC) shown exhibit pores of various shapes and sizes on its outer surface. Before loading, the BC was found to be smooth (Figure 3.9b), but the surface of precursor molecule was coarser, which created a protective covering on the BC surface due to the breakdown of KNO_3 molecule. The protective layer was evenly distributed on the BC and aggregation was not seen (Figure 3.9c). As seen in Figure 3.9d, the catalyst's surface is made up of many particle agglomerates. After being calcined at 500 °C, the catalyst's surface appearance changed, but its carbon skeleton and porosity structure persisted (as indicated in circle). Sintering caused some of the particles (magnetic) to become fixed in carbon skeleton. The magnetic particles of catalyst had a diameter more than 10 μm. This catalyst demonstrated higher mass and large particle size as compared to the magnetic nanocatalyst [48].

The wrinkled layered and folded structure of GO/SiO_2/PEA catalyst proved that SiO_2/PEA chemically attached to the surfaces of GO (Figure 3.10) [45].

A core–shell magnetic nitrogen-doped silica nanoparticle-based N (1.3 wt%)-MSN solid base catalyst exhibits two distinct morphologies, namely spherical and amorphous aggregates, as per the findings illustrated in SEM images (Figure 3.11). The higher hydrolysis rate of silica precursor in the reaction medium may be connected to the occurrence of aggregated formations. Certainly, the presence/appearance of DEA during the formation of silica nanoparticles influences the rate of hydrolysis of silica precursors, resulting in amorphous silica with spherical forms [49].

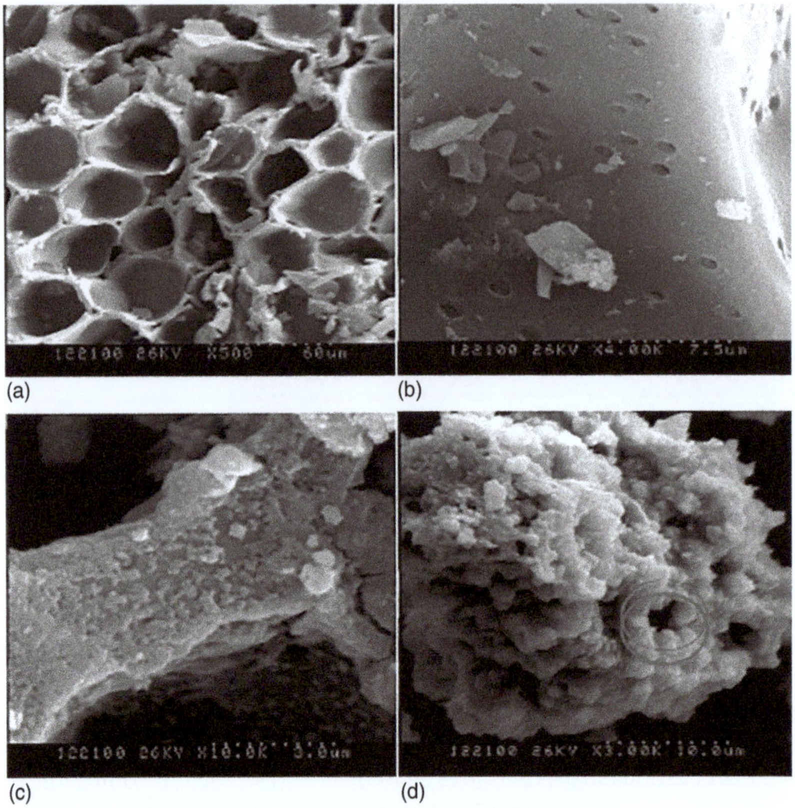

Figure 3.9 SEM images for the magnetic precursor, the BC (bamboo charcoal), and the K/BC-Fe_2O_3 magnetic catalyst. These images encompassed (a) the transverse section of the BC particles, (b) the surface structure of the BC particle, (c) the magnetic precursor, and (d) the magnetic precursor after calcination at 500 °C (d). Source: Reproduced from Liu et al. [48], with permission of Elsevier.

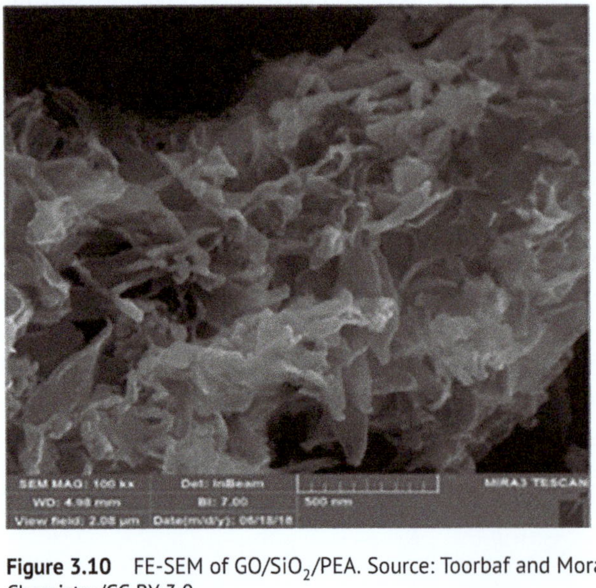

Figure 3.10 FE-SEM of GO/SiO_2/PEA. Source: Toorbaf and Moradi [45]/Royal Society of Chemistry/CC BY 3.0.

Figure 3.11 (a, b) The FE-SEM image of N (1.3 wt%)-MSN. Source: Reproduced from Neamani et al. [49], with permission of Royal Society of Chemistry.

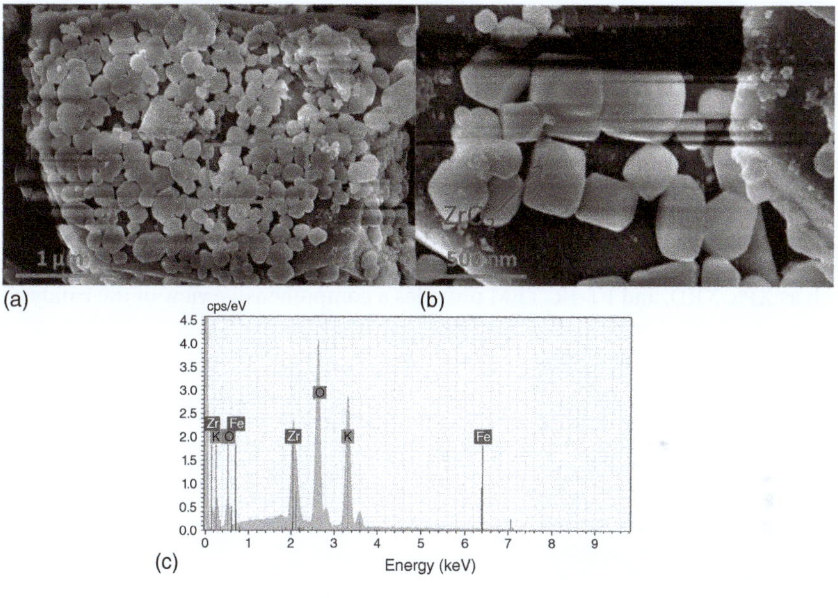

Figure 3.12 (a, b) SEM images (c) EDX of the K/ZrO$_2$/g-Fe$_2$O$_3$ nano-magnetic catalyst. Source: Reproduced from Liu et al. [50]/with permission of Royal Society of Chemistry.

SEM image of the K/ZrO$_2$/γ–Fe$_2$O$_3$ nanomagnetic solid base catalyst which was developed to be used in transesterification of soybean oil to biodiesel. Figure 3.12 illustrates the tetragonal form of the zirconia particles. These zirconia particles were equally and uniformly distributed. According to research group, the nanoparticles of iron oxide were wrapped via zirconia particles [50].

The morphology of sodium-phosphate-based solid base catalyst (NaP-400) was visualized via the SEM as shown a Figure 3.13. For NaP-400, the nanosheet and aggregated bulk structure were illustrated in Figure 3.13a; additionally, the aggregated bulk structure observed for nanoparticles as illustrated in Figure 3.13b ranged in size from 20 to 50 nm [51].

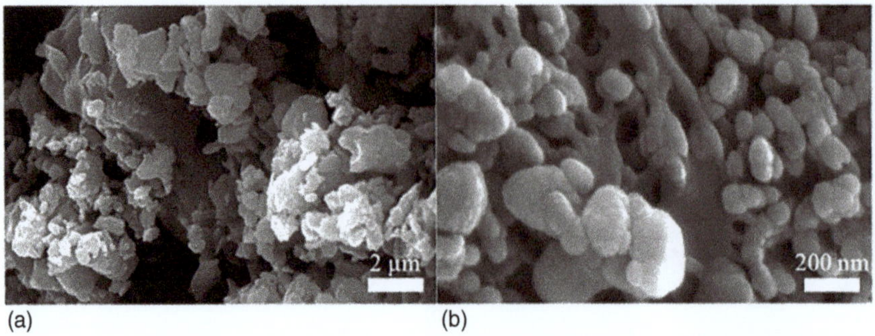

Figure 3.13 SEM data of NaP-400 under different magnification (a) at scale of 2 μm (b) at scale of 200 μm. Source: Reproduced from Zhao et al. [51], with permission of Royal Society of Chemistry.

3.3.3 Transmission Electron Microscope (TEM)

Transmission electron microscope (TEM) technique is high-resolution imaging and analytical method used for the characterization of solid-based catalysts. TEM image provides detailed information on catalyst materials including their morphology, size, structure, and composition at nanoscale. TEM is a powerful tool for characterizing solid-based catalysts, especially when a deep understanding of their nanoscale structure and composition is necessary. When combined with other analytical techniques such as XPS, XRD, and FT-IR, TEM provides a comprehensive view of the catalyst's

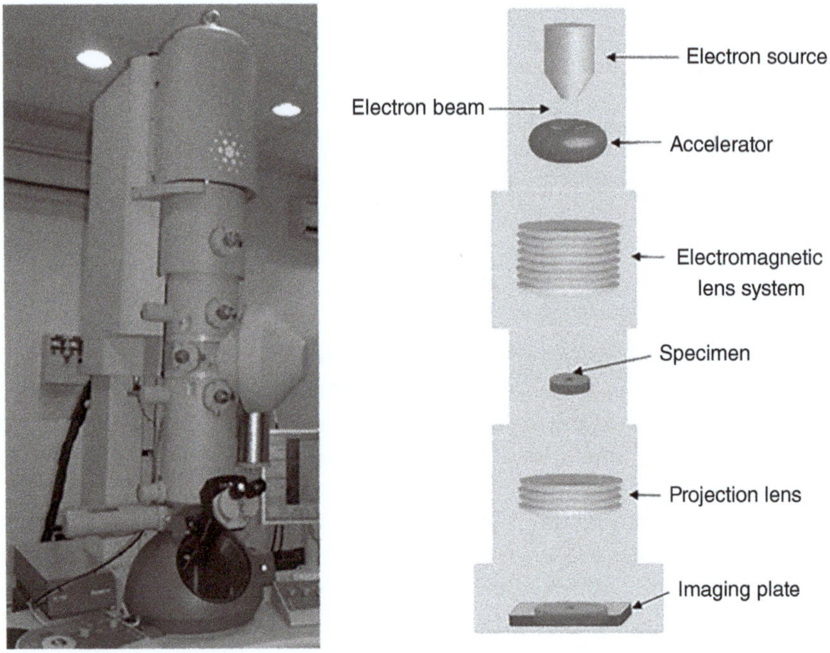

Figure 3.14 Real and schematic representation of TEM Machine. Source: Reproduced from Rostamabadi et al. [52], with permission of Elsevier.

properties, helping researchers optimize catalysts for specific applications. Such as in Figure 3.14 a real and schematic representation of a cutting-edge TEM machine is illustrated. Their instruments setups are as follow. TEM machine works on the transmission of a higher voltage beam of electrons (80–300 keV) via a nanoscale sample, which interacts with the specimen's internal structure to produce images [53]. Standard beams of electrons are mostly produced by two sources (i) field-emission sources (gun or small tungsten needles); (ii) thermionic sources (lanthanum hexaboride crystals and Tungsten filaments). Anode operating at voltage between 40 and 400 keV helps in accelerating the beam of electrons [54]. The specimen is the focus of the accelerated beam as it passes through some electromagnetic and electrostatic lenses. A portion of a specimen is transparent to the electrons, whereas the other portions may scatter or absorb the accelerated beam. Collecting the electrons and using objective lenses to magnify those results in the final image of the specimen [55]. The generated image is visible on a fluorescent monitor coated with phosphor. Thus, an image can be displayed on computer screen by attaching this system to a sensor of the charge-coupled device (CCD) camera. To get high-defined images from various nanoscale materials, several advancements have been made, such as high-technology-based CCD cameras, high-technology optical resolution imaging systems, and novel detectors that can directly gather electrons [56].

For example, in Figure 3.15, the HR-TEM image of N (1.3 wt%)-MSN is illustrated. It is evident that the acquired images support core–shell morphologies of magnetic mesoporous nitrogen-doped silica nanostructures. Furthermore, it is evident that core–shell structure is formed by magnetic nanoparticles that are securely trapped inside mesoporous silica spherical shells [49].

TEM analysis technique was used to characterize the coated structure of $K/ZrO_2/\gamma–Fe_2O_3$ nanomagnetic catalyst. Figures 3.16a,b demonstrate that the black particles of Fe_2O_3 were coated via $Zr(OH)_4$ with amorphous form in the precursor. It is evident that the sol–gel approach was successful in coating ZrO_2 on Fe_2O_3, as seen by the characteristic coated structure of $K/ZrO_2/\text{-}Fe_2O_3$ (Figure 3.16c). $K/ZrO_2/\gamma\text{-}Fe_2O_3$ has a diameter of approximately 20 nm [50].

TEM with energy-dispersive X-ray spectroscopy (TEM-EDS) was used to confirm the phases in the free gasified straw slag catalyst as well as the component of each step. TEM images were captured using an electron microscope (JEM-2100F) paired with an EDS equipment at a 200 kV accelerating voltage. Figure 3.17a displays low-magnification TEM analysis of the slag catalyst. High-resolution images of selected areas in Figure 3.17a–c, illustrate the arrangement of the particles in slag. Crystallites are represented by the larger particles (Figure 3.17c), while metal oxides are represented by the smaller particles (Figure 3.17b). The region shown by Figure 3.17c–e, when viewed at high-magnification, displays structures/patterns with inter-planar lengths of 0.198 and 0.285 nm, respectively, corresponding to (321), and (211) planes of tetragonal åkermanite. The presence of åkermanite is further confirmed by the patterns of Fast Fourier Transformation, which are displayed in the Figure 3.3d,e insets. Furthermore, minimal quantities of Al and Fe were found in the åkermanite crystallites, as demonstrated by the micro-regions' EDS analysis (Figure 3.17c). Figure 3.17g–j are displayed. The elemental mapping

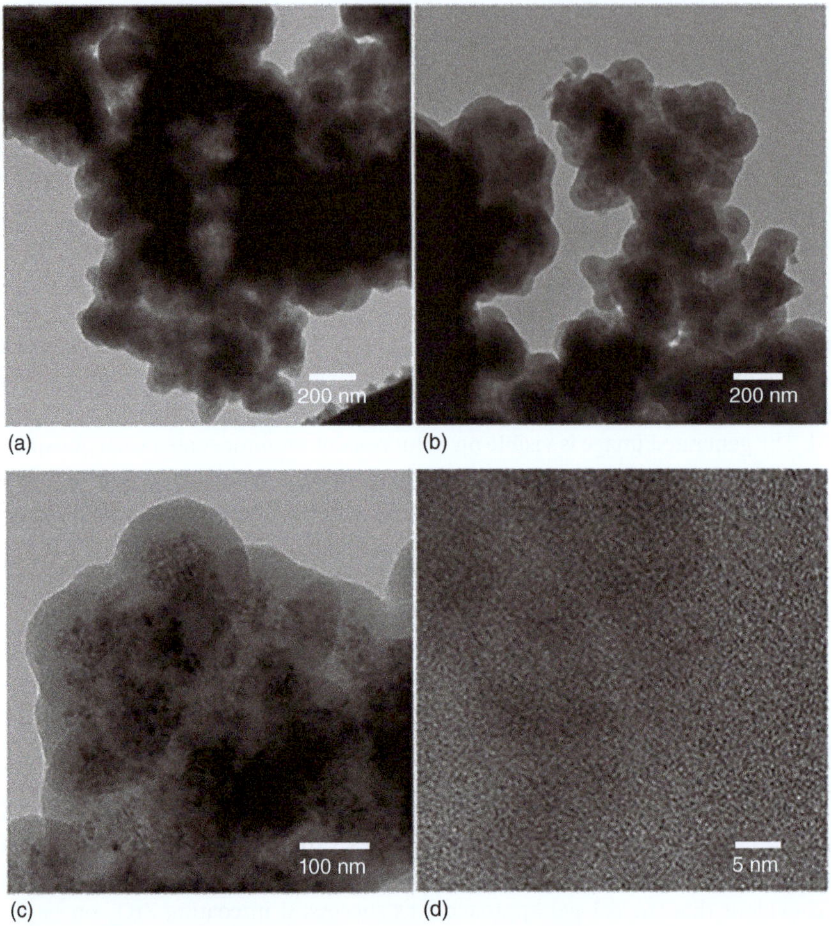

Figure 3.15 (a–c) HR-TEM image of synthesized N- (1.3 wt%) MSN. Source: Reproduced from Neamani et al. [49], with permission of Royal Society of Chemistry.

areas of Mg, Si, O, and Ca respectively, in the region depicted by Figure 3.17f, which is comparable to the region depicted by Figure 3.17a [42].

The TEM image of the MgAl–CO_3–LDH/$MgFe_2O_4$ catalyst (Figure 3.18a) illustrates the consistent hexagonal structural features characteristic of hydrotalcite-like materials. Upon reaching 773 K (during calcination), the MgAl–CO_3–LDH/$MgFe_2O_4$ underwent transformation into MgAl(O)/$MgFe_2O_4$ (Figure 3.18b), an approach involving the breakdown of interlayer CO_3^{-2} (carbonate) anions [57].

3.3.4 X-ray Diffraction (XRD) Analysis

X-ray diffraction analysis (XRD) is a nondestructive method for comprehensive details information of a material's chemical composition, crystallographic structure, and physical properties. It works by utilizing the constructive interference of a

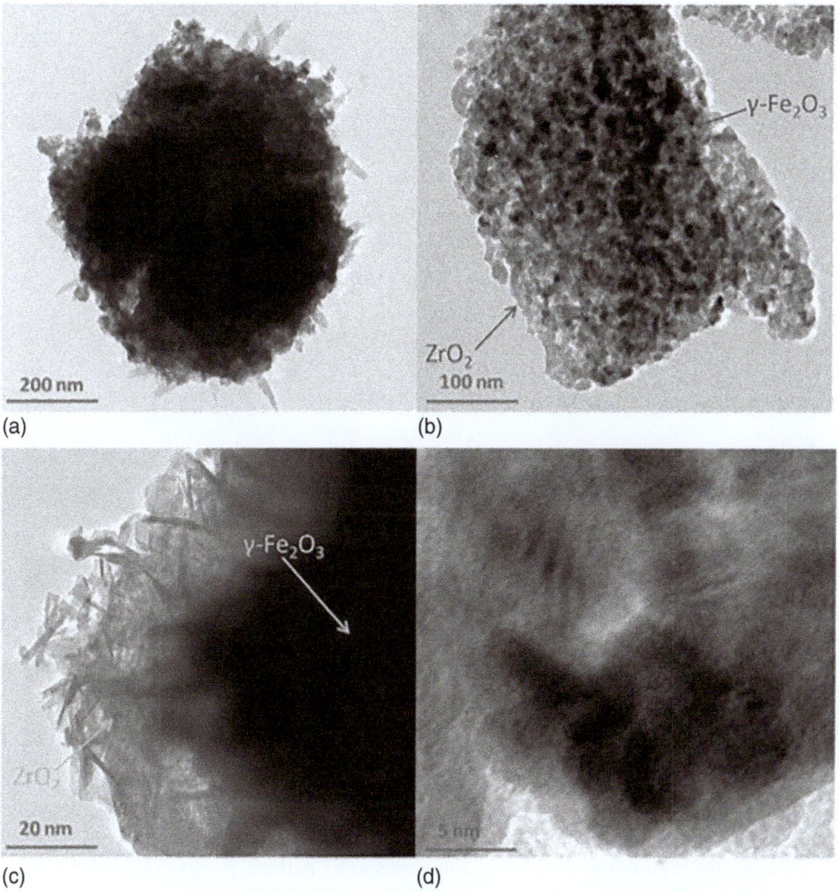

Figure 3.16 (a–d) TEM analysis of K/ZrO_2/*g*-Fe_2O_3 nano-magnetic catalyst at various scale. Source: Reproduced from Liu et al. [50], with permission of Royal Society of Chemistry.

crystalline specimen and monochromatic X-ray radiation. When charged particles with significant energy are decelerated, X-rays are formed, which are a type of electromagnetic radiation with shorter wavelength. In XRD, the produced X-rays are directed and paralleled toward a sample of nanomaterial. The sample and incoming rays combine to generate a diffracted beam, which is then detected, processed, and counted. Diffraction patterns are exhibited by plotting the diffracted ray's scattered intensity at various angles of the material. Because of the material's atomic arrangement and specific chemistry, each phase of the material creates a separate diffraction pattern [58, 59]. A typical instrument of XRD is shown in Figure 3.19.

Season Chen et al., described an *N*-doped biochar solid-based catalyst from spent coffee grounds fabricated with melamine. The biochar catalyst displayed a characteristic peak of carbon material at diffraction angle of $2\theta = 43.6°$ (100) and $2\theta = 23.5°$ (002), although they weren't as sharp as crystalline peaks. The disordering properties of non-graphitic carbons are represented by the enlarged

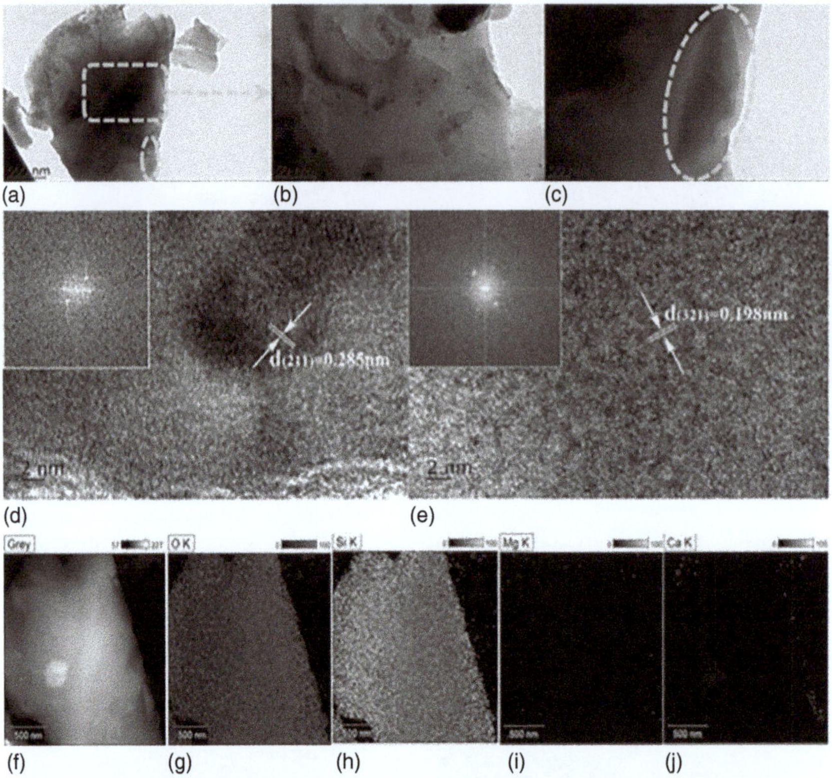

Figure 3.17 TEM-EDS image of the catalyst (gasified straw slag). (a)–(c) Low-magnification TEM data (d) and (e) high-magnification TEM analysis of the slag displaying lattice fringes and the particles crystalline nature are confirmed by FFT pattern; (f)–(j) TEM analysis of slag catalyst and the corresponding mapping of the elemental Mg, Si, O, Ca in the crystallite. Source: Reproduced from Wang et al. [42], with permission of Elsevier.

(100) peak, and the arrangement of aromatic carbon sheets within the disorderly distributed crystallites as illustrated by the (002) reflection [46]. XRD analysis of the CaO supported on Al_2O_3, CeO_2, and ZrO_2-based solid base catalysts was obtained by using Bruker D8 advance X-ray diffractometer fitted with a high-speed strip Lynx eye detector and a Cu Kα radiation source with a 0.04 step size with 4 s step time for the determining size of CaO supported on Al_2O_3, CeO_2, and ZrO_2 catalyst and structure of these oxides. A CaO supported on Al_2O_3, CeO_2, and ZrO_2 displays peaks of their various support, as well as CaO diffraction plane. In the case of CaO/Al_2O_3 catalyst detected Al_2O_3 peaks at 2θ of 43, 67, and 70°. In XRD pattern, the 2θ value for CaO/CeO_2 catalyst was found at 48, 56, 59, 69, and 77° attributed to the CeO_2 peak pattern. However, the CaO/ZrO_2 catalysts indicated the pattern of ZrO_2 peak at 56, 54, 50, 41, 32, and 28° 2θ values. The peak pattern at 24, 34, 47, 53, and 63° 2θ values indicated the existence of CaO deposited on supported metal oxides [47]. The XRD patterns of gasified straw slag are shown in Figure 3.20. The slag showed three distinct crystalline phases, which are identified as quartz, åkermanite, and leucite respectively. Quartz (SiO_2)-related diffraction

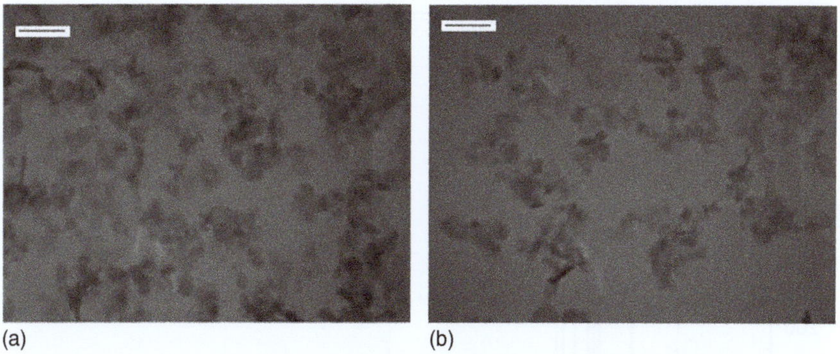

Figure 3.18 TEM images of (a) MgAl-CO_3-LDH/$MgFe_2O_4$ and (b) MgAl(O)/$MgFe_2O_4$ (in both cases Scale bar = 200 nm). Source: Reproduced from Zhang et al. [57], with permission of Elsevier.

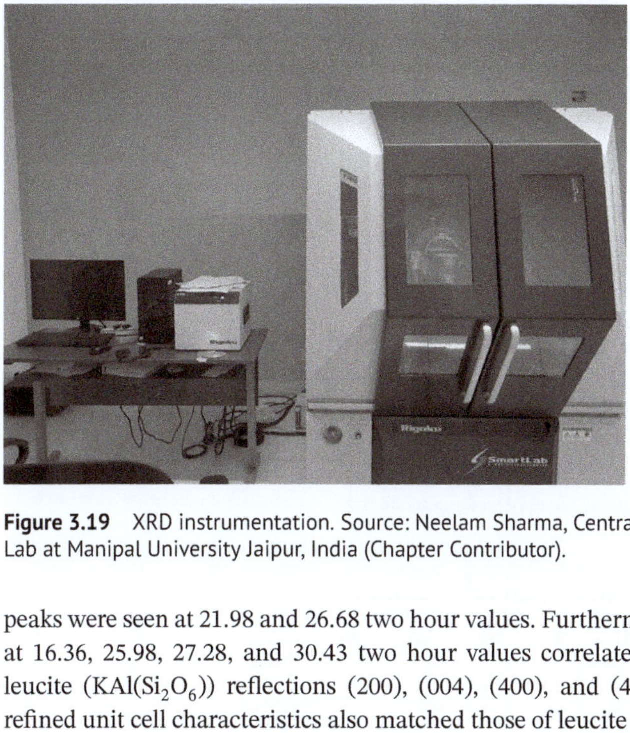

Figure 3.19 XRD instrumentation. Source: Neelam Sharma, Central Analytical Facilities Lab at Manipal University Jaipur, India (Chapter Contributor).

peaks were seen at 21.98 and 26.68 two hour values. Furthermore, diffraction peaks at 16.36, 25.98, 27.28, and 30.43 two hour values correlated with the tetragonal leucite ($KAl(Si_2O_6)$) reflections (200), (004), (400), and (420), respectively. The refined unit cell characteristics also matched those of leucite as well (a = 13.0900 Å and c = 13.7500 Å). Additional peaks detected at 23.65, 28.99, 31.17, 36.07, 44.61, and 51.95 two hour values were associated with the corresponding åkermanite reflections (111), (201), (211), (310), (212), and (312), respectively [42]. As it can be seen in Figure 3.21, fabrication with the magnetic particles causes three weak diffractions of iron oxide (γ-Fe_2O_3) to emerge at 30.1°, 35.6°, and 62.9° in the magnetic precursor's pattern. These alterations in the XRD pattern suggest that in situ precipitation occurs on the BC surface to generate the magnetic particles, which are the primary constituent of γ-Fe_2O_3. At room temperature precursor

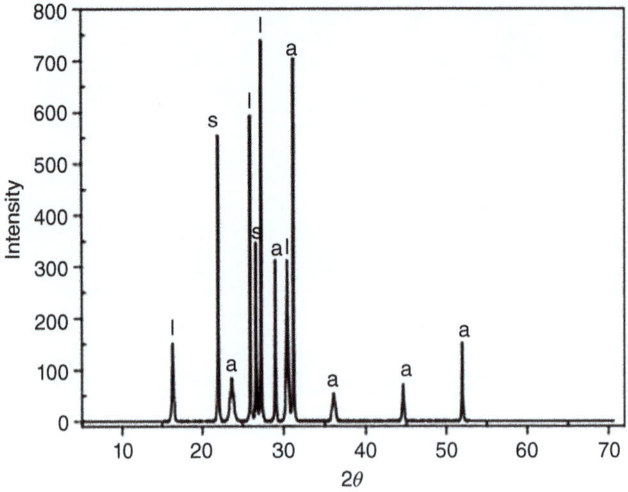

Figure 3.20 The XRD pattern of the gasified straw slag reveals distinct peaks corresponding to various phases, including Åkermanite, Leucite, and Quartz, denoted as (a), (l), and (s) respectively. Source: Reproduced from Wang et al. [42]/with permission of Elsevier.

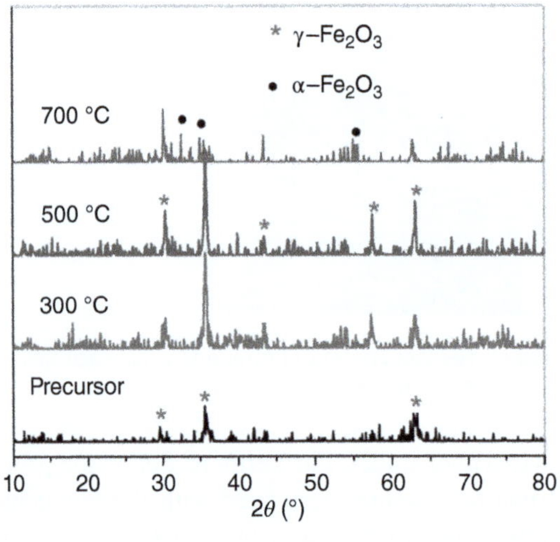

Figure 3.21 XRD patterns of magnetic precursor calcined at various temperatures. Source: Reproduced from Liu et al. [48]/with permission of Elsevier.

was black, when the temperature increased, the black precursor turned into a reddish-brown powder because iron oxide particles oxidized into γ-Fe_2O_3 during the calcinations. Simultaneously, when the temperature enhanced from 300 to 500 °C, the γ-Fe_2O_3 diffraction peaks expanded, suggesting that the γ-Fe_2O_3 content enhanced during the calcination process. The diffraction peaks of γ-Fe_2O_3 gradually vanished as calcination temperature rose to 700 °C, and a new α-Fe_2O_3 diffraction peak developed, showing that γ-Fe_2O_3 changed into α-Fe_2O_3 and Fe_3C particles. The catalysts' most of their magnetism was consequently lost (Figure 3.21) [48].

3.3.5 Thermogravimetric Analysis (TGA)

Thermogravimetric analysis (TGA) is a technique which measure thermal decomposition of sample as the rate of change in sample's weight with increase in temperature or time in nitrogen atmosphere. Loss of water molecules or moisture and volatile substances can easily be studied using TGA [60]. A commercial TGA may operate in an environment of air or another gas at temperatures above 1000 °C, with balance sensitivity (0.1 μg) and a variably regulated heat-up rate. TGA has a heat-up rate capacity ranging from 0.1 to 200 °C min^{-1}. The device is made up of an automated furnace to regulate the sample's temperature and an extremely sensitive scale to record weight variations. The balance is separated from the heat by a thermal barrier and is located above the furnace. From the balance into the furnace is suspended a highly precise hang-down wire. The sample pan is situated at the end of hang-down wire, and its location needs to be consistent. To increase the sensitivity, precision of weighing, and accuracy, the balance should be insulated from the heat influences (for example, with the help of thermostatic chamber). An easily cooled micro-furnace is a feature of the TGA device. The heating element is made up of platinum, which is dependable up to 1000 °C. The temperature might increase up to 1500 °C by using an external furnace that has a heating element composed of an alloy consisting of 30% rhodium and 70% platinum. The weight loss fraction and/or weight loss percentage is often calculated by a computer that comes with most modern machines [61]. A typical instrument of TGA is shown in Figure 3.22.

The mass degradation of *N*-doped biochar catalyst up to 1000 °C heating in argon atmosphere was approx. 23%. A temperature-based programmed desorption analysis possessing a mass spectrometer suggested that a tiny reflection point around 250 °C. The decomposition of small amounts of carboxyl-linked groups could result in the DTG curve. The high degradation rates observed between 560 and 740 °C were most likely caused by the disintegration of nitrogen-related species such as lactam-like compounds and pyrrole, as well as oxygen-containing groups such as lactones and ketones. Furthermore, it is possible that the mass degradation that is occurring at the later stage of >750 °C is mostly related to the breakdown of N-containing pyridinic, which is more stable than N in pyrrolic group (Figure 3.23) [46].

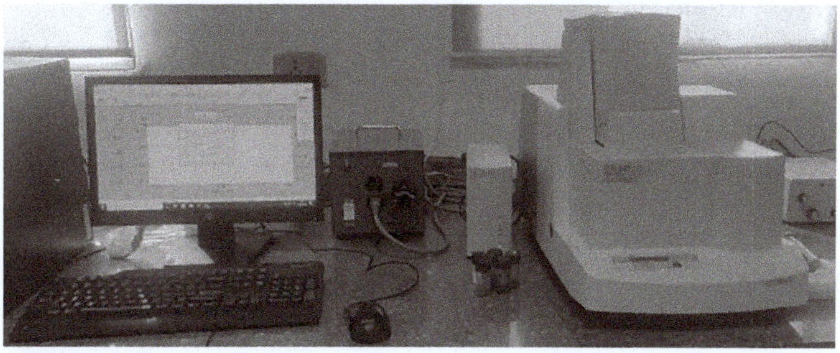

Figure 3.22 TGA instrument. Source: Neelam Sharma, Central Analytical Facilities Lab at Manipal University Jaipur, India (Chapter Contributor).

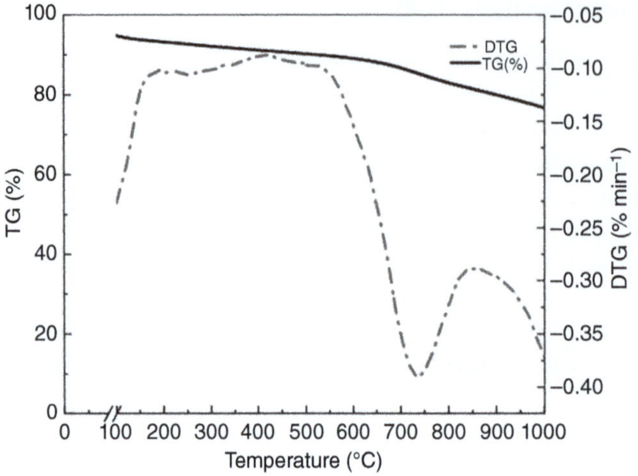

Figure 3.23 TGA analysis of nitrogen-doped biochar catalyst at 10 °C min^{-1} heating rate under Argon gas. Source: Reproduced from Chen et al. [46]/with permission of American Chemical Society.

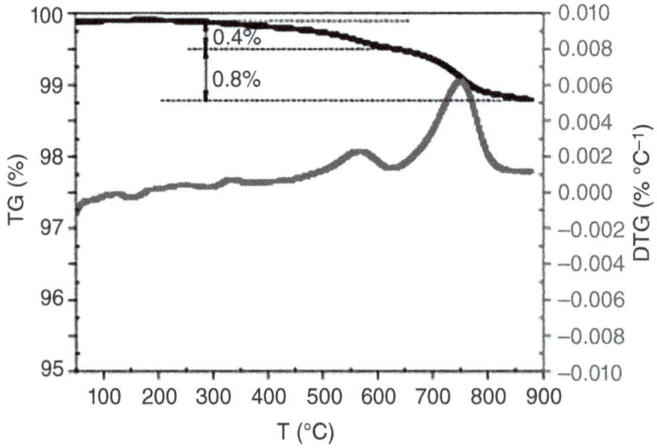

Figure 3.24 TG-DTG graph of the gasified straw slag. Source: Reproduced from Wang et al. [42]/with permission of Elsevier.

The gasified straw slag catalyst showed outstanding thermal stability. At 200 °C the weight loss of the gasified straw slag catalysts was negligible (transesterification reaction) and exhibited 1.2% mass loss at 900 °C shown in Figure 3.24. The TGA curve showed two minor mass loss stages. The associated derivative with thermogravimetry (DTG) peak centered at 567 °C, in the initial phase 0.4% mass loss was recorded between 400 and 600 °C, corresponding to hydroxide dehydration. In second step 0.8% mass loss at 600–800 °C and the corresponding DTG peak centered approximately at 752 °C corresponding to decomposition of M_2CO_3 into its oxide considered to be the heterogeneous catalyst-active component [42].

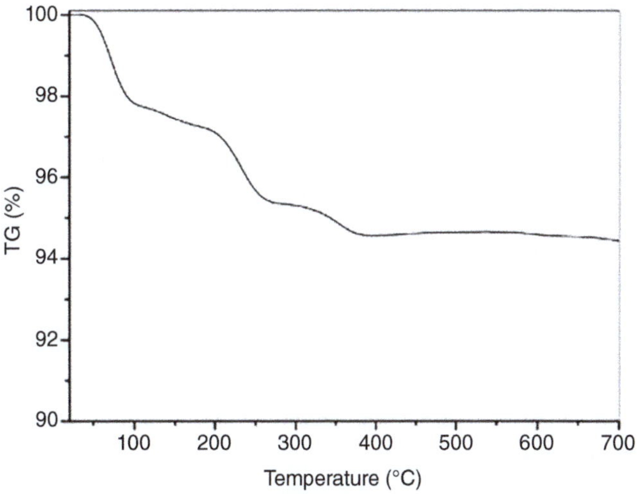

Figure 3.25 Thermogravimetric analysis of 30 wt% KOH/Nd_2O_3. Source: Reproduced from Li et al. [43]/with permission of Elsevier.

Among different potassium compounds that were loaded on Nd_2O_3, KOH seem to possess high catalytic activity and gave best yield (89.7%) of biodiesel during transesterification process. The yield of biodiesel was found to vary with the amount of KOH loaded on Nd_2O_3 and the optimal loading amount is found to be 30% at 600 °C as the calcination temperature for the catalytic activities of transesterification reaction of soybean oil. The presence of active catalytic sites for process of transesterification such as K_2O species and Nd=O=K groups was confirmed by thermal behavior, in which the first loss in weight indicates the loss of moisture while second loss at 200–400 °C corresponds to the decomposition of KOH and K_2CO_3, followed by final loss at temperature above 400 °C due to decomposition of K_2O (Figure 3.25) [43].

Employing urea–nitrate combustion, Xia and team synthesized a solid base composed of CaO–ZrO_2. Varying molar ratio of Ca/Zr ranging from 4/6 to 9/1 was also explored in preparation of CaO–ZrO_2 solid base. The preparation of CaO–ZrO_2 solid base via combustion process was confirmed by TG-DSC plot. It was observed that the 4% loss in weight ascribes the loss of water molecules, giving an endothermic peak at 30–160 °C. The simultaneous weight loss at 210 °C is attributed to the decomposition of $Zr(NO_3)_4$ which is followed by huge weight loss at around 240–550 °C temperature representing exothermic effect due to combustion between urea and nitrate, confirming the combustion reaction and formation of that Ca–Zr complex oxide in comparatively short reaction time and with no further need of calcination at high temperature. The loss is negligible above 600 °C indicating that decomposition occurs before 600 °C (Figure 3.26) [62].

Thermal stability of functionalized C60 solid base catalyst was examined using thermogravimetric analyzer and reported that weight loss of all four-amine functionalized C60 catalysts was almost similar. Although, the sudden decrease in weight at 175 °C represents the breakdown of solid base catalyst (amines functionalized C60)

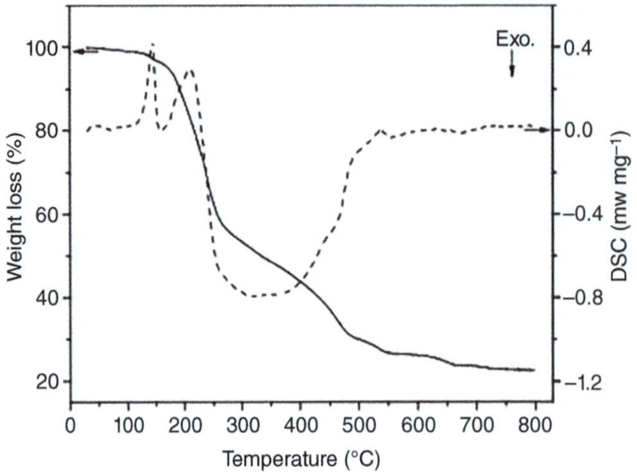

Figure 3.26 DSC and TG curve for the dried gel of C8Z2 sample. Source: Reproduced from Xia et al. [62]/with permission of Royal Society of Chemistry.

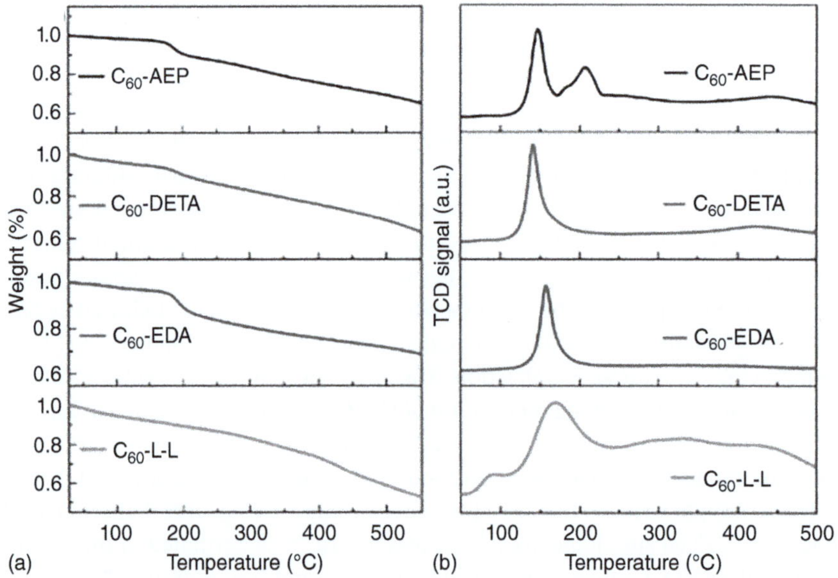

Figure 3.27 (a) TG graph and (b) CO_2-TPD spectrum of amines functionalized C60. Source: Reproduced from Sun et al. [63]/with permission of Royal Society of Chemistry.

thus indicating that catalyst can remain stable up to 170 °C in the reaction medium (Figure 3.27) [63].

Thermogravimetric analyzer to analysis thermal stability of nanomagnetic solid base catalyst ($K/ZrO_2/g\text{-}Fe_2O$) according to which it was observed that at around 100 °C the peak shows the loss of water molecule. Further peaks at around 100–500 °C and 613.23 °C signify loss of hydroxyl groups from $Zr(OH)_4$ and

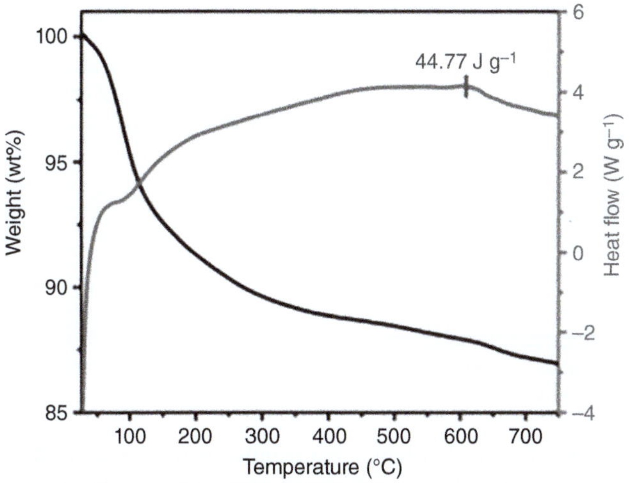

Figure 3.28 TG-DSC of the precursor. Source: Reproduced from Liu et al. [50]/with permission of Royal Society of Chemistry.

decomposition of KOH respectively. TG analysis further proved that the best calcination temperature was around 500–600 °C which led to increase in biodiesel yield (Figure 3.28) [50].

Another graphene-based solid base catalyst whose catalytic activity can be observed in synthesis of spiro-oxindole and its derivatives. The peak of thermogravimetric analysis shows that graphene oxide is stable at 220 °C. Figure 3.29a and the first weight loss occurs at about 5% due to the removal of absorbed water molecules at temperatures around 20–120 °C which is further accompanied by weight loss (about 20%) due to the attachment of SiO_2/PEA on surface of graphene oxide. However, a rapid decrease in the peak was observed at temperatures above 220 °C,

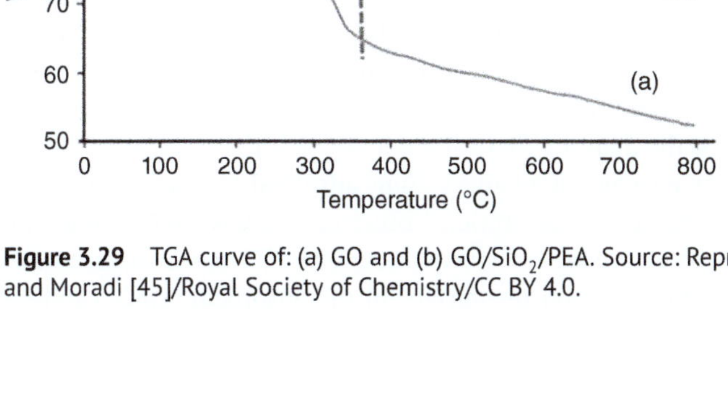

Figure 3.29 TGA curve of: (a) GO and (b) GO/SiO_2/PEA. Source: Reproduced from Toorbaf and Moradi [45]/Royal Society of Chemistry/CC BY 4.0.

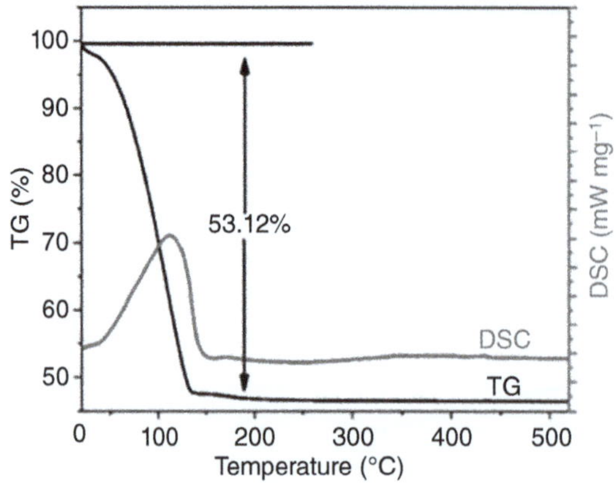

Figure 3.30 TG–DSC images of Na-T catalysts. Source: Reproduced from Zhao et al. [51]/with permission of Royal Society of Chemistry.

leading to a major weight loss of about 35% which is attributed to the decomposition of water and other oxygen-containing functional groups present on graphene oxide at temperatures around 220–300 °C (Figure 3.29b) [45].

Among various other solid base catalyst, it was noted that sodium-phosphate-based solid base catalyst found to possess high catalytic and thermal stability which was substantiate using thermogravimetric peaks plotted from 25 to 600 °C. They used sodium phosphate as a solid base catalyst to produce new ethylene glycol monomethyl ether monolaurate (EGMEML) biodiesel with the help of transesterification. Thermal stability of Na_3PO_4 solid base catalyst showed that there is significant water loss of about 53.2% due to the presence of 11.21 water molecules present in every hydrated sodium phosphate, on the other hand, there was no change observed in range of 200–500 °C pointing the fact that sodium phosphate did not decompose during the process of calcination (Figure 3.30) [51].

3.3.6 Brunauer–Emmett–Teller BET Surface Area Pore Diameter Analysis [Gas Interaction and Surface Area Measurement: (Brunauer–Emmett–Teller (BET), Barrett–Joyner–Halenda (BJH) N_2 Adsorption–Desorption Isotherms)]

Analyses of surface structure of the catalyst include study of porosity (pore size distribution and pore volume) and specific surface area by using physio-adsorption methods such as BJH and BET. The BET approach is a multilayer adsorption model that is frequently used to calculate a particular surface area by using chemically inert probing gases with material surfaces as adsorbates. Nitrogen is the most common gas used gaseous adsorbate for surface probing with BET techniques. Because of this, BET analysis is often carried out at nitrogen boiling temperature (−196 °C) based on nitrogen adsorption isotherm plot [64]. The amount of N_2 gas that adsorbs on the catalyst surface and the quantity that subsequently desorbs at a fixed temperature are measured to create a plot of relative volume adsorbed versus pressure often described as nitrogen adsorption–desorption isotherm, whereas BJH technique is used for the porosity analysis using Kelvin equation. The BET model suggests that gas is adsorbed in several layers on the surface of the adsorbent. Since capillary

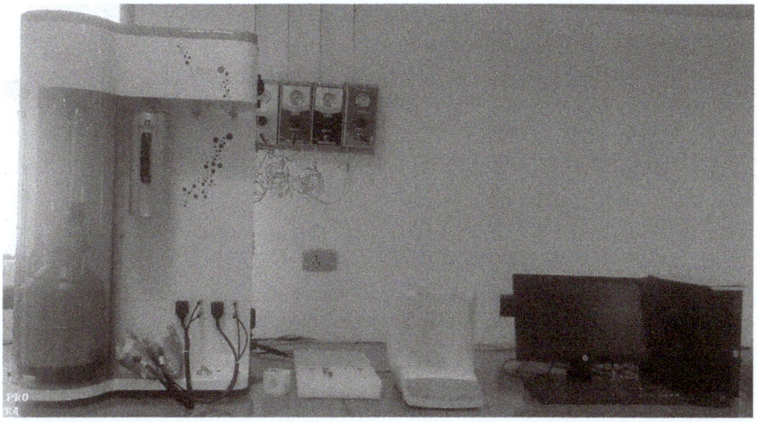

Figure 3.31 Real image of BET instrument Source: Neelam Sharma, Central Analytical Facilities Lab at Manipal University Jaipur, India (Chapter Contributor).

condensation rather than multilayer adsorption may occur in the range of 0.05–0.3 pressure zone during the surface area analysis, therefore, selection of data should be done carefully, particularly for tiny holes of 2–3 nm in diameter or poorly organized materials [65]. A typical instrument of BET is shown in Figure 3.31.

The specific surface area pore size, and volume distribution, were examined by BET and BJH method using N_2 adsorption–desorption (Type-II) isotherms. The *N*-doped biochar catalyst showed Type-II isotherm (Figure 3.32a), and the BJH curve in figure (Figure 3.32b) revealed a higher fraction of pore diameter ranging between 50–200 nm than 2–50 nm, showing the properties of macro-porous materials. In comparison to the surface areas of 100 $m^2\ g^{-1}$ in previous research, the nitrogen-enriched biochar in this examines had a lower specific surface area is

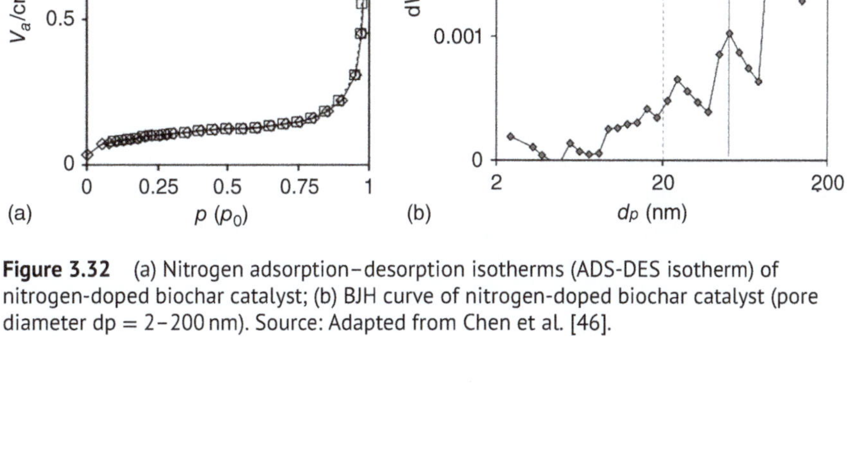

Figure 3.32 (a) Nitrogen adsorption–desorption isotherms (ADS-DES isotherm) of nitrogen-doped biochar catalyst; (b) BJH curve of nitrogen-doped biochar catalyst (pore diameter dp = 2–200 nm). Source: Adapted from Chen et al. [46].

Table 3.1 CO_2-TPD and BET characterization of the CaO supported on Al_2O_3, CeO_2, and ZrO_2 solid base catalyst.

Catalyst	Pore volume ($cm^3 g^{-1}$)	Surface area ($m^2 g^{-1}$)	Pore diameter (nm)	Basicity (μmol g^{-1})
CaO	0.04	12.2	59.6	599
CaO/Al_2O_3	0.02	17.0	44.1	21
CaO/CeO_2	0.04	16.0	44.0	58
CaO/ZrO_2	0.06	17.3	59.7	183

0.35 $m^2 g^{-1}$, most likely because the biochar aggregated after the nitrogen-doping process [46].

On the BELSORP max instrument at 77 K, the total surface area of CaO, CaO/Al_2O_3, CaO/CeO_2, and CaO/ZrO_2 was calculated using the BET method, and the average pore size and volume were calculated using BJH method. The fundamental sites were found utilizing the catalyst's temperature-programmed carbon dioxide desorption (CO_2-TPD) in conjunction with a thermal conductivity detector (TCD). Pore volume, BET surface area, and pore diameter of (CaO, CaO/Al_2O_3, CaO/CeO_2, and CaO/ZrO_2) catalysts are shown in Table 3.1. All catalysts exhibited four types of adsorption isotherm, which is consistent with their mesoporosity. There have been reports of supported metal oxide catalysts having surface areas of 18.1, 87.0, and 88.0 $m^2 g^{-1}$ for CeO_2, Al_2O_3, and ZrO_2, respectively. Whereas the CaO acquired pore diameter and surface area of 59.6 nm and 12.2 $m^2 g^{-1}$, respectively, three prepared catalysts that subdue surface area: CaO/Al_2O_3 (17 $m^2 g^{-1}$), CaO/ZrO_2 (17.3 $m^2 g^{-1}$), and CaO/CeO_2 (16.0 $m^2 g^{-1}$). CaO agglomeration on the supported oxide has been identified as the cause of the observed decrease in surface area. CaO/ZrO_2 was found to have the pore diameter (59.7 nm), and largest surface area (17.3 $m^2 g^{-1}$) of all the synthesized catalysts. This could be related to the development of new pores when CaO is added to ZrO_2. The pore volumes for CaO, CaO/Al_2O_3, CaO/CeO_2, and CaO/ZrO_2 were 0.04, 0.02, 0.04, and 0.06 $cm^3 g^{-1}$, respectively [47].

The characterization of the gasified straw slag revealed a low total pore volume of 0.01971 $cm^3 g^{-1}$ and a small surface area of 1.266 $m^2 g^{-1}$, as determined by BET surface area analysis. Figure 3.33 illustrates the pore size distribution of the gasified straw slag, demonstrating a broad range of pores with sizes ranging from 5 to 10 nm and 45 to 75 nm [42].

The magnetic solid-based catalyst can be classified as a mesoporous material due to its equal pore width exceeding 5 nm. The BET analysis of magnetic catalysts shows that it has an equal pore diameter of 14.5 nm and 28.7 $m^2 g^{-1}$ as its surface area. Because the catalyst can supply adequate basic active sites, when stirred during the reaction step, using it when reactions are in liquid phase is advantageous. This can be seen from study of pore size distribution (Figure 3.34a) that the magnetic catalyst's pore size is centered at 10–20 nm. The magnetic catalyst's type IV BET isotherm is shown in Figure 3.34b [48].

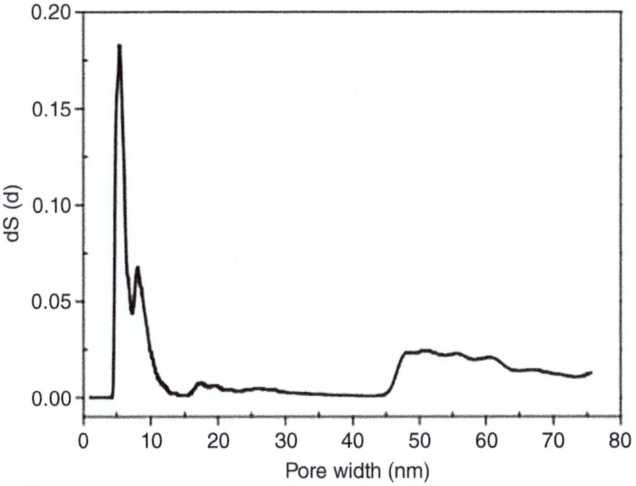

Figure 3.33 Pore size distribution of the slag. Source: Reproduced from Wang et al. [42]/with permission of Elsevier.

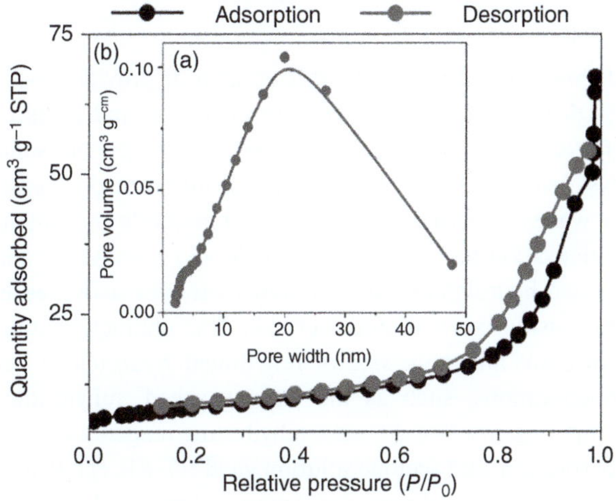

Figure 3.34 (a) Pore size distribution plot and (b) N_2 adsorption-desorption isotherm. Source: Reproduced from Liu et al. [48]/with permission of Elsevier.

Yihuai Li measured pore size for KOH/Nd_2O_3 using Type-II isotherm at −196 °C. The specific BET surface area for 30 wt% KOH/Nd_2O_3 catalyst was calculated to show that with increase in KOH loading, surface area and pore volume decrease, and potassium ions show best results as they can be easily dispersed on the surface of neodymium oxide (Nd_2O_3) thus providing active surface for base-catalyzed reactions [43]. The catalytic activities of SBFA were studied using various characterization techniques where and it showed that specific surface area was ascertained using nitrogen adsorption–desorption isotherms at 77 K while pore size and surface area were measured using BET. The high surface area of SBFA was confirmed using

Table 3.2 Nitrogen physisorption analysis of re-hydrated hydrotalcite specimen and their catalytic activities.

Entry		1	2	3	4	5
Rehydrated HT		HT-R	HT-RE	HT-RA	HT-RAA	In situ rehydrate HT-3
Rehydration condition	Rehydration reagent	H_2O	H_2O+ ethanol (1 : 1)	H_2O + acetone (1 : 1)	H_2O + acetone (1 : 1)	In situ rehydration
	Washing reagent	No	No	No	Acetone	
Pore diameter (nm)		14	9	10	12	42
Surface area ($m^2\ g^{-1}$)		23	75	80	130	82
Pore volume ($cm^3\ g^{-1}$)		0.18	0.38	0.24	0.52	0.14

BET and it was concluded that due to increase in surface area, it provides more basic or active sites for reaction to occur at its maximum capacity and leads to increase in the selectivity of product [44]. Chunli Xu prepared a heterogeneous solid base catalyst, where catalyst hydrotalcite regenerated using different processes. The rehydrated hydrotalcite was synthesized using traditional methods and in-situ reactions such as aldol condensation between benzaldehyde and acetone. It was observed that in-situ rehydrated hydrotalcite exhibits higher catalytic activities as compared to the one formed using conventional methods. The characterization results showed that the structure and pore size of rehydrated hydrotalcite get influenced by rehydration conditions, such as the stirring speed, rehydration reagent, and different washing reagents. The different rehydrating reagents such as decarbonated water, ethanol solution, and acetone solution yield HT-RE, HT-R, and HT-RA possessing different surface area and pore size. Among different rehydrating hydrotalcite HT-RAA that have been treated with additional washing steps with acetone found to have largest surface area while the pore size of the specimen is in range between 10 and 42 nm. Nitrogen physisorption data, which is calculated through the BJH method, in Table 3.2 shows that in-situ rehydrated HT-3 and HT-RA have same surface area [66].

CaO–ZrO_2 solid base catalyst was synthesized by mixing $Ca(NO_3)_2.4H_2O$ and $Zr(NO_3)_4.5H_2O$ in different molar ratio with urea thus undergoing the urea–nitrate combustion. By using Brunauer–Emmett–Teller (BET) technique it was observed that by increasing calcium amount the gradual decrease in surface area and pore volume occurs due to partial blockage of pores in CaO–ZrO_2 catalyst by CaO. Since, CaO has substantially smaller surface area than ZrO_2 therefore, as the ratio of Ca/Zr increases from 4/6 to 9/1, the CaO phase becomes more prominent while

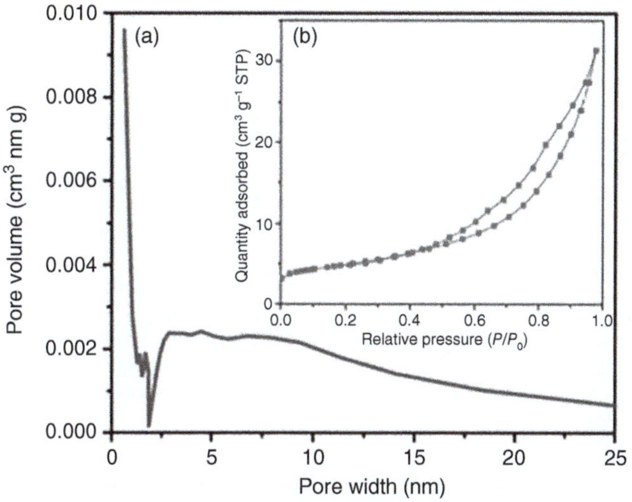

Figure 3.35 Pore size distribution graph. (b) N_2 adsorption–desorption isotherm. Source: Reproduced from Liu et al. [50]/with permission of Royal Society of Chemistry.

diminishing the ZrO_2, which is the major reason for decrease in surface area and pore volume with addition of CaO [62]. BET surface area and pores size of the $K/ZrO_2/\gamma\text{-}Fe_2O_3$ nanomagnetic solid base catalyst are shown in Figure 3.35. It is found that the pore volume of the catalyst is $0.051\,m^3\,g^{-1}$ with average pore diameters of about 10.146 and a specific surface of $20.08\,m^2\,g^{-1}$. Furthermore, mesoporous nature of the nanomagnetic solid base catalyst was confirmed by H3 hysteresis loop at 0.6–0.998 pressure calculated by using BJH method from the desorption isotherm at 196 °C after degassing all the samples at 120 °C for 12 hours under reduced pressure [50].

Similarly, a core–shell magnetic *N*-doped silica nanoparticle catalyst was doped with different weight percentages of nitrogen and the best catalytic results were obtained when 1.3% of nitrogen was doped in MSN (mesoporous silica nanostructure). It was further confirmed by BET results as the results show that surface area of independent MSN is $361\,m^2\,g^{-1}$ with $0.37\,cm^3\,g^{-1}$ pore volume and 4.1 nm a mean pore diameter whereas 1.3 wt% of nitrogen-doped MSN possess large surface area ($387\,m^2\,g^{-1}$), a smaller pore volume ($0.35\,cm^3\,g^{-1}$) and comparatively lesser mean pore diameter (3.6 nm) as compared to individual MSN when used as catalyst, thus making nitrogen-doped MSN a better and highly efficient solid base catalyst (Figure 3.36) [49].

3.3.7 X-Ray Photoelectron Spectroscopy (XPS)

X-ray photoelectron spectroscopy (XPS) is a surface-sensitive quantitative analysis technique that uses the photoelectric effect to identify the elements within a material (element composition) or on its surface and its chemical state, overall density of electronic states, and electronic structure. XPS is a significant quantitative technique

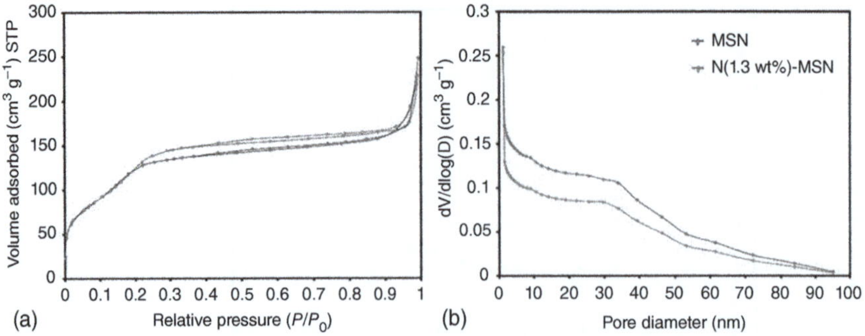

Figure 3.36 (a) N_2 adsorption–desorption isotherms (Type-II) of MSN and N(1.3 wt%)-MSN, (b) BJH graph of MSN and N(1.3 wt%)-MSN. Source: Reproduced from Neamani et al. [49]/with permission of Royal Society of Chemistry.

as it reveals not only the presence of components but also the elements to which they are linked. The XPS technique can be used to determine the elemental composition throughout the surface in a line when combined with ion-beam etching. XPS is a photoemission spectroscopy technique that uses an X-ray beam to acquire electron population spectra from a material. Chemical states are measured by the kinetic energy and number of expelled electrons. XPS needs ultra-high vacuum ($p < 10^{-7}$ Pa) or high vacuum (residual gas pressure $p \sim 10^{-6}$ Pa) conditions. Except for elements such as hydrogen and helium, all other X-ray sources can be identified using XPS technique. XPS is commonly used to analyze metal alloys, inorganic compounds, elements, polymers, catalysts, ceramics, glasses, inks, paints, woods, papers, plant parts, make-up, bones, teeth, bio-materials medical implants, coatings, ion-modified materials, viscous oils, glues, and a variety of other substances. Less frequently, XPS is also used to determine the hydrated forms of samples such as hydrogels and biological materials via freezing them in an ultrapure atmosphere and enabling multilayers of ice to sublime prior to analysis [67, 68]. The XPS spectrum of $Ca_{12}Al_{14}O_{33}$ and CaO-based solid base catalyst sample shown corroborated the presence of Si, Ca, O, and Na that was earlier identified through XRF: photoelectron spectra with binding energies of 1071.3, 531.25, 346.92, and 101.49 eV are ascribed to Na1s, O1s, Ca2p, and Si2p, respectively. HR-XPS spectra of Ca2p displayed two peaks at 350.44 and 346.88 eV corresponding to Ca2p1/2 and Ca2p3/2. The binding energy of Ca2p3/2 (346.1 eV) in CaO is lower than that of Ca2p3/2, indicating the coordination mechanism of Ca species in NCS catalyst has been altered via strong contact with the other composition. Combining the satellite spectra at 258 and 355 eV suggests that the Ca on the surface of catalyst may exist in calcium carbonate form [37]. The XPS analysis of Ca-based solid base catalyst displays a fine spectrum of O1s and C1s, confirming the presence of calcium carbonate species in the materials. The strong basic sites on NCS surface make it easier to absorb CO_2 from the air. Carbon dioxide reacts with calcium oxide to create $CaCO_3^{-2}$ species on the catalyst's surface, because of moisture in the air. The absence of a calcium carbonate diffraction peak in XRD spectrum suggests that calcium oxide carbonation occurs only at the surface layer [41]. The orange peel ash (OPA) catalyst elemental composition was determined via

Table 3.3 XPS data for elements compositions and correspondingly quantity (%) of catalyst.

Sl. no.	Elements	Atomic weight (%)
1.	C1s	33.07
2.	O1s	39.45
3.	Ca2p	7.56
4.	K2p	13.53
5.	P2p	1.23
6.	Mg1s	2.11

XPS analysis. The wide range of XPS spectra, as well as deconvoluted spectra for C1s, K2p, O1s, Mg1s, and Ca2p, are depicted through XPS data. The catalyst is made up of C, Na, K, O, Ca, Mg, and P. The deconvoluted spectra for C1s reveal two peaks at 288.53 and 283.92 eV, which correspond to CO and C—C bonds, respectively. The spectra for K2p display two peaks at 294.45 and 291.73 eV with binding energies, corresponding to K in the form of metal carbonates and metal oxides. Ca2p spectrum shows two peaks with 349.48 and 345.94 eV binding energies, indicating the existence of $CaCO_3$ and CaO. Mg1s spectrum shows one signal peak with 1303.72 eV binding energy, indicating the presence of MgO. Table 3.3 shows the atomic weight percentages of major elements contained in the OPA catalyst as determined by XPS analysis [69].

XPS technique was utilized to study the chemical composition of the magnetic nanosized solid catalyst (bio-waste *Citrus sinensis* peel ash (CSPA)@Fe_3O_4). The XPS survey data indicate the presence of C, O, Ca, K, and Mg, and the atomic weight percentage of each element is displayed. Apart from Fe, as shown, the catalyst contains very basic elements such as Ca (4.46%) and K (8.64%) in significant amounts, which is primarily responsible for the catalyst's outstanding reactivity. The deconvolution spectra of K2p show two peaks at 295.22 eV and 292.76 eV, indicating the existence of K_2CO_3 and K_2O, respectively. The deconvoluted spectra for Ca2p reveal two peaks at 350.84 eV and 346.76 eV, corresponding to $CaCO_3$ and CaO respectively. The spectra of C1s reveal two humps at 293.31 and258.18 eV, corresponding to the C—O and C—C bonds, respectively. Furthermore, the O1s spectra reveal a single hump at 531.21 eV [70].

3.3.8 X-Ray Fluorescence (XRF)

X-rays are a form of electromagnetic wave identical to visible light rays but with a much shorter wavelength ranging from 100 to 0.1 Å. X-rays, unlike other electromagnetic waves, can easily pass out through material and become stronger as the substance's atomic number through which they pass decreases. XRF analysis is a technique that employs the distinctive X-rays (fluorescent X-rays) produced

Table 3.4 NCS catalyst element composition.

Compound	Average (wt%)
O	38.77 ± 0.67
Ca	43.23 ± 0.56
Si	8.43 ± 0.11
Na	8.67 ± 0.10
Others	0.90 ± 0.11

Table 3.5 XRF analysis of OPA and OP composition.

Sl. no.	Formula of compound	OPA (wt%)
1.	CaO	25.67
2.	K_2O	51.64
3.	SiO_2	13.24
4.	P_2O_5	2.95
5.	MgO	4.76
6.	Na_2O	1.81
7.	MnO	0.033
8.	Fe_2O_3	0.070
9.	TiO_2	0.114

when X-rays irradiate a substance. XRF are electromagnetic waves produced when irradiation X-rays force constituent atoms' inner-shell electrons to an outer shell, and outer shell electrons quickly travel to inner shells to fill up the vacancies. These fluorescent X-rays (XRF) have energies specific to each type of element, allowing both quantitative studies by using the intensity (photons number) of each X-ray energy and Moseley's law analysis using qualitative study [71]. The catalyst's element content was evaluated using a wavelength-dispersive XRF spectrometer. The XRF analysis data of NCS catalyst is shown in Table 3.4. To ensure accuracy in the assay, three samples were chosen at random from the same batch of catalysts [41].

XRF analysis was used to analyze the composition of OPA catalyst. Table 3.5 demonstrates that the catalyst's composition consists primarily of alkaline CaO and K_2O are the major components in the OPA catalyst. Thus, CaO and K_2O may play a significant catalytic role in the generation of biodiesel [69].

X-ray fluorescence spectroscopy (XRF) was used to examine the presence of metallic and nonmetallic oxides in CSPA. The XRF data displayed the presence of SiO_2 (13%), CaO (24%), and K_2O (51.9%) as a major component and MgO (4.76%) and P_2O_5 (2.95%), as a minor component. Thus, these highly basic alkaline metal

Table 3.6 Comparative study of different characterization techniques.

Entry	Properties	FT-IR	SEM	TEM	XRD	TGA	BET
1.	Structural	✗x02718;	✗x02718;	✗x02718;	✗x02718;	✔	✗
2.	Thermal stability	✔	✔	✔	✔	✗x02718;	✔
3.	Surface functionalities	✗x02718;	✔	✔	✔	✔	✔
4.	Surface properties (porosity, surface area, pore volume, pore diameter)	✔	✔	✔	✔	✔	✗x02718;
5.	Morphology (size, shape, particle distribution)	✔	✗x02718;	✗x02718;	✔	✔	✔
6.	Crystallinity (crystal structure, phase composition)	✔	✔	✔	✗x02718;	✔	✔

oxides play a significant role in the synthesis of magnetic Fe_3O_4 NPs. Furthermore, coating CSPA with Fe_3O_4 NPs improves the catalyst's performance [70]. The numerous peaks in XRF spectra of *Sesamum indicum* plant-derived catalyst demonstrate the existence of K, Na, Mg, Ca, Mn, Si, Fe, Sr, and Zn as metallic elements, with the carbonates and oxides of these verified by the XRD patterns. The XRF spectra revealed a high-intensity peak for Ca and K, as well as a nearly identical peak for Mg and Si. The intensity and peak area of the XRF spectra can be used to forecast the amount of Ca and K in the calcined catalyst. The SEM-EDX analysis determines both the quantitative and elemental compositions of the components. Like XRF spectra, EDX examination reveals that Ca (33.80 wt%) and K (29.64 wt%) are the major metals, followed by Mg, Si, and Sr in comparison to other elements present in the catalyst [72]. Furthermore, an comparative study of different spectroscopic techniques is illustrated in Table 3.6.

3.4 Protocol for Characterization of Catalyst

Understanding the components, structure, and reactivity of solid base catalysts requires a variety of techniques. The specific protocol may differ based on the catalyst and the characterization goals. Here is a standard protocol for complete characterization of solid base catalysts that integrates multiple methods:

3.4.1 Sample Preparation

Prepare a representative solid base catalyst sample (pellet, powder, or supported on a substrate) for analysis.

3.4.1.1 XRD

XRD is a useful technique for identifying the crystal structure and crystalline phase of the materials. The quality of the XRD data is heavily reliant on sample preparation, thus it's critical to use the correct methodology for your material. Grind the sample until it is a fine powder (less than 75 microns). This increases the surface area exposed to X-rays, resulting in larger and more defined diffraction peaks. The grinding method will be determined by the hardness of the material and the desired particle size. Ball milling, mortar and pestle grinding, and cryogenic grinding are all common procedures. Powder-XRD is a technique used to identify the crystallinity of materials in the solid state. Amorphous (non-crystalline, disordered) materials do not diffract, while materials with any crystallinity/order, even when mixed with amorphous materials, will diffract. If your sample is amorphous, further techniques such as XRF or IR spectroscopy may be required. Remember that PXRD is a nondestructive approach, therefore the sample will remain unchanged after data collection and can be preserved and reused [73, 74].

3.4.1.2 FT-IR

ATR (attenuated total reflection)-IR spectra: In general, 20–50 mg (depends on materials properties) of sample is adequate for the observation of IR spectra in ATR mode. The sample must make good contact with the ATR crystal due to the short extension of the evanescent wave beyond the crystal. During the measurement, a mild pressure is applied to the sample. To obtain high-quality findings, the sample must have full and intimate contact with ATR crystal. The crystal's refractive index must be much larger than the sample's.

Transmittance FT-IR spectra: Sample preparation is an important step in getting high-quality transmittance FT-IR spectra. The specific method utilized is determined on the physical state of the sample: solid, liquid, or gas.

KBr pellet: For powdered particles, this is the most common method. Combine your sample with finely ground potassium bromide (KBr) and use a hydraulic press to compress the combination into a pellet.

Mull: This process is appropriate for difficult to press or grind substances into pellets and suitable for hygroscopic and moisture sensitive materials. To make a suspension, combine your sample with mineral oil or another IR-transparent liquid [75].

3.4.1.3 FE-SEM

Material must first be made conductive to be viewed using a SEM. This is performed by coating them with a very thin layer of Au or Au/Pd (1.5–3.0 nm). Furthermore, material must be able to sustain the high vacuum without affecting it, for as by losing gases or water molecules. Metals, crystals, and polymers are usually less problematic since they maintain their structure in the SEM. To prevent surface tension degradation to fine structures, chemically affixed materials must be washed and dried under the critical threshold. The coating is subsequently completed in a separate apparatus [76].

3.4.1.4 TEM

The TEM analysis protocol sample preparation can differ based on the type of sample and the information desired. Fixation process retains the sample's structure and avoids deterioration. Fixatives commonly used in TEM include formaldehyde, glutaraldehyde, and osmium tetroxide. The fixative used will be determined by the type of molecule of interest, and the fixation duration will need to be optimized for your individual sample. Before embedding the sample in resin, the fixative and any buffers must be removed [77].

3.4.1.5 BET

The BET analysis is a popular method for determining the specific surface area of porous materials. Impurities and adsorbed molecules can drastically alter the analysis, therefore good sample preparation is critical for accurate results. To avoid biased measurements, make sure your sample is homogeneous. To obtain a fine powder, gently crush or grind large materials. Avoid using too much force to avoid particle amorphization or introducing undesired contaminants. Weigh an empty, clean, and dry sample container (often a glass tube with a stopper) precisely. Transfer the homogenized sample into the container, with most BET analyzers requiring 50–100 mg. Record the weight of the container plus the sample precisely. This critical degassing process eliminates moisture and other adsorbates from the surface of the sample. Place the container in a degassing room or a vacuum chamber. Apply heat and vacuum (usually 100–300 °C) for the duration required. Handle the sample carefully once it has been degassed to avoid re-adsorption of moisture or impurities. Transfer the container directly to the BET analyzer sample port using clean tools. Reduce air exposure, especially for very hydrophilic materials.

3.4.1.6 TGA

TGA pans are commonly constructed of alumina, platinum, or ceramic. The optimum material for your experiment will be determined by the temperature range, the reactivity of your sample, and the atmosphere in which you will conduct the experiment. Sample should be in a form that allows for good heat transfer and purge gas contact. In the case of solids, this usually entails grinding the material into a fine powder. For liquids, a little droplet or film on the pan's bottom might be used. The sample weight should be in between 2 and 50 mg, but exact amount depends on the sensitivity of the TGA instrument and the projected weight loss of your sample. Place the pan containing the sample carefully on the TGA instrument's balance. Check that the pan is centered and not touching anything else. For experiment, select the temperature range, purge gas flow rate, and heating rate. These parameters will be determined by the information you hope to collect from your TGA analysis.

3.5 Characterization of Some Basic Sites of Solid Base Catalyst with Suitable Example

Understanding the basic site characterization of various solid base catalysts helps in tailoring their properties for specific catalytic applications, such as base-catalyzed reactions in organic synthesis or environmental processes such as transesterification reactions in biodiesel production.

General characterization of basic site of solid base catalyst includes acid-base titration, and other spectroscopic methods such as IR spectroscopy, XPS, SEM, and XRD.

Characterization of basic sites of solid base have been listed below with suitable examples:

1) Hydrotalcite (Mg–Al double layered hydroxide): they are the most common type of solid base catalyst in which (Mg/Al LDH) Mg/Al-layered double hydroxide with brucite-like ($Mg(OH)_2$) hydroxide layers containing octahedrally coordinated M^{3+} and M^{2+} cations which provide the base structure of hydrotalcite. XRD patterns confirm the transformation of hydrotalcites into mixture of oxides such as (MgO), as the distinctive peaks in all hydrotalcite samples can be seen due to thermal breakdown of Mg–Al double-layer hydroxides. Further IR spectroscopy confirmed the presence of IR active vibrations such as vibration due to interlayer anions, molecular vibrations of the OH-groups, and band appearance due to lattice vibrations of the octahedral layers, which confirmed the formation of Mg–Al hydrotalcite. SEM results suggest that there was no difference observed in morphology of crystals before and after calcination [78]. Other hydrotalcites with LDH of Mg^{+2}, Ni^{+2}, and Al^{+3} offer appropriate base catalysts for transesterification. The substitution of Ni^{+2} in place of Mg^{+2} affords advantages such as it increases the rate of conversion of FAME [79]. BET results show that the surface area of MgAlNi increases with decrease in Mg^{+2} ions. For efficient transesterification, it is important to have high basicity which also depends on desorption temperature, which is found to be lower for Mg_3Al than the MgAlNi catalysts. XRD patterns and similarity in FT-IR spectra of MgAlNi catalysts to that of Mg_3Al confirms hydrotalcite-like structure of MgAlNi and presence of Ni^{+2} in lattice. Hydrotalcite has been reconstructed under various factors such as stirring the calcined hydrotalcite under de-carbonated water, resulting in formation of highly active solid base catalyst with low and medium basic strength. The reconstructed hydrotalcite found to have better characteristic properties such as decreased surface area, which was confirmed using BET, whereas the enhanced intensity of the IR and XRD peaks indicated an improvement in activity of the reconstructed hydrotalcite, which was attributed to the presence of water and hydroxyl molecules [80]. Various Mg–Al double layered hydroxides were reconstructed and used for various organic reactions as a catalyst or precursor of catalyst in organic reactions such as aldol condensation [66], Michael addition [81], Friedel–Crafts reactions [82], and in isomerization reactions [83]. Due to high activity of solid base catalysts as compared to that of solid acid catalysts or other heterogenous catalysts, they can widely be used in transesterification of oils [43, 84].
2) Another example includes metal-oxide-based solid base catalyst which is formed by calcination of metal carbonate hydrates and mixing with additional metal oxides. The basic sites on the metal oxide surface are attributed to oxygen vacancies and hydroxyl groups for example magnesium oxide-based solid base catalyst which is formed by calcining of magnesium carbonate hydrates (MCH) at 500 °C

for six hours. The spectroscopic result shows that the MCH which was calcined at 500 °C, causes CO_2 and H_2O molecules to evolve and produce many Lewis and Bronsted acid and base sites on the MgO surface. FT-IR in support of XRD spectra further confirmed that carbonate species are chemisorbed on the surface of magnesium oxide, and presence of bound and free OH-groups, and different O_2-LC ions [85]. Recently, alkali-based iron oxide substituted with Mn shows higher activity as compared to unsubstituted magnesium and calcium oxides. Results of CO_2-TDP and Raman spectroscopy also confirm the effect of substitution of non-noble metal indicating the formation of weak base sites and enhancing the catalytic activity of $KFe_{11}O_{17}$ catalyst [86].

3) Other types of examples of solid base are nanoscale magnetic solid base catalyst which exhibits the properties of nanoscale material and solid base catalysts which combine basic catalytic sites with magnetic characteristics to provide effective, recyclable catalysis for a variety of chemical transformations. The catalytic activity of MgAl–OH–LDH/$MgFe_2O_4$ was analyzed on aldol condensation. XPS results of MgAl–OH–LDH/$MgFe_2O_4$ show that the magnetic $MgFe_2O_4$ spinel cores are encapsulated with MgAl–OH–LDH Bronsted base catalytic layers, bounded with Mg—O—Fe and Al—O—Fe bonds thus exposing basic sites (OH^-) on the surface making it a more favorable catalyst for organic transformations. SEM and TEM results suggest a layered structure of LDH while HR-TEM further confirms the existence of a thin coating of the LDH material encasing a ferrite core [57]. Similarly, $CoFe_2O_4$@glycine-M was prepared using three types of metal i.e. Pr, Tb, and Yb, and characterized by FT-IR, EDX, TGA, and XRD techniques. The result of FT-IR spectra indicates that $CoFe_2O_4$ phase has not been destroyed during the coating of glycine on the surface of $CoFe_2O_4$. Polycrystalline structure of $CoFe_2O_4$ and immobilization of metals such as Pr, Tb, and Yb and amount loaded of these metals were confirmed using XRD, EDX, and TGA techniques. SEM and XRD results of the reusable catalyst also confirm that there was no significant change in structure and morphology even after using catalyst multiple times [87]. Other nanomagnetic solid base catalysts such as KF/CaO–Fe_3O_4 [88], KF/γ–Al_2O_3 [32], and Ca/Al/Fe_3O_4 [89] showed catalytic activity toward transesterification reaction for synthesis of biodiesel and can be used in various other chemical transformation. Basic strength of catalyst was evaluated using characterization techniques such as FT-IR, TGA, SEM, XRD, BET, and TEM.

3.6 Conclusion

In conclusion, the characterization of solid-based catalysts is an essential aspect of catalysis research and plays a pivotal role in catalyst development and optimization. In this chapter, we have explored a range of powerful techniques that offer valuable insights into the properties, structure, and behavior of these catalysts. These techniques include FT-IR, SEM, TEM, and XRD, among others. Each of these characterization methods serves a specific purpose and provides unique information, collectively contributing to a comprehensive understanding of solid-based catalysts. FT-IR enables the analysis of surface functional groups and basicity, SEM offers detailed insights into morphology and particle size, TEM delves into the nanoscale structure, and XRD reveals the crystalline phases and lattice parameters. Through these techniques, researchers can identify catalytic active sites, assess catalyst stability, and gain knowledge about the support materials.

Acknowledgment

All the authors are highly grateful to the Chandigarh University and Manipal University Jaipur for providing the necessary e-resources and support for this work.

References

1 Jambhulkar, D.K., Ugwekar, R.P., Bhanvase, B.A., and Barai, D.P. (2020). *Chemical Engineering Communications* 209: 433.
2 Ma, C., Geng, J., Zhang, D., and Ning, X. (2020). *Journal of the Energy Institute* 93: 581.
3 Wang, W., Xu, Y., Wang, X. et al. (2018). *Bioresource Technology* 250: 474.
4 Krishnan, S.G., Pua, F., and Zhang, F. (2021). *Biomass and Bioenergy* 149: 106099.
5 Hamza, M., Ayoub, M., Shamsuddin, R.B. et al. (2021). *Environmental Technology and Innovation* 21: 101200.
6 Tomar, R., Singh, N., Kumar, N. et al. (2019). *Catalysis Letters* 149 (6): 1589–1594. https://doi.org/10.1007/s10562-019-02723-9.
7 Tomar, R., Ebitani, K., and Chandra, R. (2019). *ChemistrySelect* 4 (12): 3577.
8 Tomar, R., Kundra, P., Sharma, J. et al. (2024). *Letters in Organic Chemistry* 21 (1): 16.
9 Kumar, L., Verma, N., Tomar, R. et al. (2023). *Dalton Transactions* 52 (10): 3006.
10 Kumar, L., Verma, N., Sehrawat, H. et al. (2022). *New Journal of Chemistry* 46 (7): 3472.
11 Tomar, R., Rathee, G., Chandra, I. et al. (2018). *ChemistrySelect* 3 (6): 1645.
12 Tomar, R., Singh, N., Rathee, G. et al. (2017). *Asian Journal of Organic Chemistry* 6 (12): 1728.
13 Arya, K. and Tomar, R. (2013). *Research on Chemical Intermediates* 41 (6): 3389.
14 Hattori, H. (2004). *Journal of Japan Petroleum Institute* 47: 67.
15 Saravanan, K., Tyagi, B., and Bajaj, H.C. (2016). *Applied Catalysis B: Environmental* 192: 161.
16 Boz, N., Degirmenbasi, N., and Kalyon, D.M. (2015). *Applied Catalysis B: Environmental* 165: 723.
17 Abdullah, S.H.Y.S., Hanapi, N.H.M., Azid, A. et al. (2017). *Sustainable Energy Reviews* 70: 1040.

18 Kolvari, E., Koukabi, N., and Hosseini, M.M. (2015). *Journal of Molecular Catalysis A: Chemical* 397: 68.

19 Pasupulety, N., Gunda, K., Liu, Y. et al. (2013). *Applied Catalysis A: General* 452: 189.

20 Lee, A.F. and Wilson, K. (2015). *Catalysis Today* 242: 3.

21 Corma, A., Bee, S., Hamid, A. et al. (2005). *Journal of Catalysis* 234: 340.

22 Abello, S., Vijaya-Shankar, D., and Perez-Ramırez, J. (2008). *Applied Catalysis A: General* 342: 119.

23 Climent, M.J., Corma, A., Iborra, S., and Velty, A. (2002). *Journal of Molecular Catalysis A: Chemical* 182: 327.

24 Pines, H. and Haag, W.O. (1958). *The Journal of Organic Chemistry* 23: 328.

25 Tanabe, K., Yoshij, N., and Nattori, H. (1971). *Journal of the Chemical Society [Section] D: Chemical Communications* 464.

26 Hattori, H. (2015). *Applied Catalysis A: General* 504: 103.

27 Chen, S.S., Yu, I.K., Cho, D.W. et al. (2018). *Chemical Engineer* 6: 16113.

28 Arghan, M., Koukabi, N., and Kolvari, E. (2019). *Journal of Saudi Chemical Society* 23: 150.

29 Essamlali, Y., Amadine, O., Larzek, M. et al. (2017). *Energy Conversion and Management* 149: 355.

30 Liao, Y., Li, F., Dai, X. et al. (2017). *Chinese Journal of Catalysis* 38: 1860.

31 Roy, T., Sahani, S., Madhu, D., and Sharma, Y.C. (2020). *Journal of Cleaner Production* 265: 121440.

32 Shahraki, H., Entezari, M.H., and Goharshadi, E.K. (2015). *Ultrasonics Sonochemistry* 23: 266.

33 Yurdakal, S., Garlisi, C., Özcan, L. et al. (2019). *Heterogeneous PC*, 87. Elsevier.

34 FTIR Analysis | RTI Laboratories. (2023). https://rtilab.com Techniques.

35 Chen, W., Huang, Z., Liu, Y., and He, Q. (2008). *Catalysis Communications* 9: 516.

36 Xie, W., Peng, H., and Chen, L. (2006). *Applied Catalysis A: General* 300: 67.

37 Wang, B., Li, S., Tian, S. et al. (2013). *Fuel* 104: 698.

38 Microbe Notes. (2023) *Oncogenic Viruses-An Overview*. https://microbenotes.com/scanning-electron-microscope-sem/on (accessed 12 October 2023).

39 Gohain, M., Devi, A., and Deka, D. (2017). *Industrial Crops and Products* 109: 8.

40 Takase, M., Zhang, M., Feng, W. et al. (2014). *Energy Conversion and Management* 80: 117.

41 Liu, X., Xing, S., Yang, L. et al. (2021). *Fuel* 302: 121094.

42 Wang, J., Xing, S., Huang, Y. et al. (2017). *Applied Energy* 190: 703.

43 Li, Y., Qiu, F., Yang, D. et al. (2011). *Biomass and Bioenergy* 35: 2787.

44 Jain, D., Khatri, C., and Rani, A. (2011). *Fuel* 90: 2083.

45 Toorbaf, M. and Moradi, L. (2021). *RSC Advances* 11: 21840.

46 Chen, S.S., Yu, I.K., Cho, D.W. et al. (2018). *ACS Sustainable Chemistry & Engineering* 6: 16113.

47 Biswas, B., Kumar, A., Fernandes, A.C. et al. (2020). *Bioresource Technology* 307: 123232.

48 Liu, K., Wang, R., and Yu, M. (2018). *Renewable Energy* 127: 531.

49 Neamani, S., Moradi, L., and Sun, M. (2020). *RSC Advances* 10: 35397.
50 Liu, K., Wang, R., and Yu, M. (2017). *RSC Advances* 7: 51814.
51 Zhao, Z., Wu, W., Jia, L., and Guo, X. (2023). *RSC Advances* 13: 26700.
52 Rostamabadi, H., Falsafi, S.R., and Jafari, S.M. (2020). Introduction to characterization of nanoencapsulated food ingredients. In: *Characterization Nanoencapsulated Food Ingredients*. p. 4, 53.
53 Tiede, K., Tear, S.P., David, H., and Boxall, A.B. (2009). *Water Research* 43: 3335.
54 Jackson, C.L., Chanzy, H.D., Booy, F.P. et al. (1998). *Macromolecules* 31: 6259.
55 Kuntsche, J., Horst, J.C., and Bunjes, H. (2011). *International Journal of Pharmaceutics* 417: 120.
56 Koning, R.I. and Koster, A.J. (2009). *Annals of Anatomy-Anatomischer Anzeiger* 191: 427.
57 Zhang, H., Qi, R., Evans, D.G., and Duan, X. (2004). *Journal of Solid State Chemistry* 177: 772.
58 Raja, P.B., Munusamy, K.R., Perumal, V., and Ibrahim, M.N.M. (2022). *Nano-bioremediation: Fundamentals and Applications*, 57. Elsevier.
59 Ali, A., Chiang, Y.W., and Santos, R.M. (2022). *Minerals* 12 (2): 205.
60 Nasrollahzadeh, M., Atarod, M., Sajjadi, M. et al. (2019). *An Introduction to Green Nanotechnology* 199.
61 Ebnesajjad, S. (2006). *Surface Treatment of Materials for Adhesion Bonding* 43.
62 Xia, S., Guo, X., Mao, D. et al. (2014). *RSC Advances* 4: 51688.
63 Sun, Y., Cao, C., Huang, P. et al. (2015). *RSC Advances* 5: 86082.
64 Jaroniec, M., Kruk, M., and Sayari, A. (1998). Mesoporous Molecular Sieves Proceedings of the 1st International Symposium, 325.
65 Chen, Z., Qing, H., Zhou, K. et al. (2019). *Progress in Materials Science* 108: 100618.
66 Xu, C., Gao, Y., Liu, X. et al. (2013). *RSC Advances* 3: 793.
67 Andrade, J.D. (1985). *Surface and Interfacial Aspects of Biomedical Polymers: Surface Chemistry and Physics* 1: 105–195.
68 Ray, S. and Shard, A.G. (2011). *Analytical Chemistry* 83: 8659.
69 Changmai, B., Sudarsanam, P., and Rokhum, L. (2020). *Industrial Crops and Products* 145: 111911.
70 Changmai, B., Rano, R., Vanlalveni, C., and Rokhum, L. (2021). *Fuel* 286: 119447.
71 Pinto, A.H. (2018). *Aspects in Mining and Mineral Science* 1 (2).
72 Nath, B., Kalita, P., Das, B., and Basumatary, S. (2020). *Renewable Energy* 151: 295.
73 Harris, K.D., Tremayne, M., and Kariuki, B.M. (2001). *Angewandte Chemie International Edition* 40 (9): 1626–1651.
74 Cook, H.E., Johnson, P.D., Matti, J.C., and Zemmels, I. (1975). *Methods of Sample Preparation and X-Ray Diffraction Data Analysis, X-Ray Mineralogy Laboratory, Deep Sea Drilling Project*. University of California, Riverside.
75 Berna, F. (2017). FTIR microscopy. *Archaeological soil and sediment micromorphology* 411–415.
76 Ruska, E. (1987). *Reviews of Modern Physics* 59 (3): 627.

77 Peng, W., Lu, K., Lai, S. et al. (2013). *Bio-Protocol* 3: 816.
78 Brito, A., Borges, M.E., Garín, M., and Hernández, A. (2009). *Energy & Fuels* 23: 2952.
79 Wang, Y.B. and Jehng, J.M. (2011). *Journal of Chemical Engineering* 175: 548.
80 Sharma, S.K., Parikh, P.A., and Jasra, R.V. (2010). *Applied Catalysis A: General* 386: 34.
81 Prescott, H.A., Li, Z.J., Kemnitz, E. et al. (2005). *Journal of Catalysis* 234: 119.
82 Choudhary, V.R., Jha, R., and Narkhede, V.S. (2005). *Journal of Molecular Catalysis A: Chemical* 239: 76.
83 Kishore, D. and Kannan, S. (2002). *Green Chemistry* 4: 607.
84 Liu, Q., Wang, B., Wang, C. et al. (2014). *Green Chemistry* 16: 2604.
85 Sutradhar, N., Sinhamahapatra, A., Pahari, S. et al. (2011). *Journal of Physical Chemistry C* 115: 12308.
86 Ikeue, K., Miyamoto, Y., and Ando, E. (2023). *Chemical Physics Letters* 830: 140816.
87 Tamoradi, T., Arash, G.C., and Ghadermazi, M. (2018). *Solid State Sciences* S1293255818307830.
88 Hu, S., Guan, Y., Wang, Y., and Han, H. (2011). *Applied Energy* 88: 2685.
89 Tang, S., Wang, L., Zhang, Y. et al. (2012). *FPT* 95: 84.

4

Advanced Solid Catalysis for Biomass Conversion into High Value-Added Chemicals

Urja, Amanpreet Kaur Jassal

Guru Nanak Dev University, Department of Chemistry, UGC Sponsored-Centre of Advance Studies-II, Amritsar, Punjab 143005, India

4.1 Introduction

The term solid catalyst is the combination of two words "solid" and "catalyst," which means that the catalysts are in a solid state. Being catalysts, they have the ability to increase the rate of reaction without making any alterations to the overall Gibbs free energy (ΔG) of the reaction [1]. Solid catalysts are also known as heterogenous catalysts because the catalysts, reactants, and products are present in different phases. Heterogenous catalysis was discovered by *Johann Wolfgang Dobereiner* in 1812 while working on American platinum ore [2]. It can be well said that almost all solid catalysts are heterogenous, but not all heterogenous catalysts are solids. For example, in the solution of two immiscible liquids, one can be the catalyst and the other can be the reactant; thus, heterogeneous catalysts can also be in a liquid state [3]. Being a heterogenous catalyst, these bear great advantages over many liquid and gaseous catalysts, which are most likely homogenous in nature (i.e. catalysts and reactants or products are in the same phase) [4]. Although many homogenous catalysts are being used in various fields due to their high selectivity along with high catalytic performance, they easily get dissolved in reaction media and make all their catalytic sites available. Regardless of these advantages, homogenous catalysts are being used in only 20% of industrial processes because they are tough to separate from reaction mixtures, resulting in product contamination [5]. Moreover, it is difficult to reuse and recycle them; therefore, they are not cost-efficient. On the contrary, solid catalysts, being heterogenous, can be easily separated from reactants and products due to their different phases; thus, they are reusable and cost-efficient [6]. Also, these are bifunctional, thus providing both acidic and basic sites in a single catalyst, thus making it possible to carry out various types of reactions in a one-pot process [7].

The total surface area of the catalyst is a significant factor for efficient catalytic performance. More is the specific surface area of a catalyst, more is approachability of active sites, and more will be its catalytic activity [8]. Nanoporous materials (e.g. mesoporous and microporous), due to their large surface area, can act as good

Solid Base Catalysts: Synthesis, Characterization, and Applications, First Edition.
Edited by Ravi Tomar, K.K. Pant, and Ramesh Chandra.

solid catalysts as they have high selectivity. Zeolites (microporous solids) are being widely used in industries. Solid catalysts can be classified into four classes based on their structures and properties of substrate activation. These four classes of solid catalysts are named metal oxides, micro- and mesoporous materials, sulfonated polymers, and metal catalysts [9]. Solid catalysts have applications in various sectors, such as environmental protection, energy production, chemical synthesis [10–19], oil refining [20], and water treatment [21]. For many years, solid catalysts have found their remarkable applications in petrochemical industries, in which they have been involved in the manufacture of many high value-added products (including biofuels, chemicals, etc.) [21]. Solid catalysts can be easily recycled, and they avoid the generation of toxic substances [5]. They are considered more environmentally friendly materials to produce value-added products.

For decades, fossils have been used for chemical and energy production. These are nonrenewable energy resources, which are expensive and cause environmental hazards. It is recommended to find an alternative to these materials that is environmentally friendly, renewable, and cost-efficient. Biomass fits into the aforementioned category very well [22]. Biomass is the term used to describe the organic matter that originates from plants. Both terrestrial and aquatic vegetation can be considered as biomass. Energy is stored in the biomass, which is 1% of sun's energy and can easily be transformed into chemical energy via photosynthesis [23]. There are various resources from which biomass can be obtained, such as energy crops, manure, municipal waste, wastewater, forest residues, algae, straw, paper waste, and seed waste [23, 24]. Biomass is composed of five important components: hemicellulose, lignin, ash, cellulose, extractives, and volatiles. Cellulose, hemicellulose, and lignin are major components of biomass, named lignocellulose, and they constitute 90% of the total biomass. In herbaceous biomass, 80% of the biomass is composed of lignocellulose, whereas other components are present in minor quantities and have a major impact on conversion processes. The composition of biomass may vary from one type to another and within the same feedstock [25].

There are top 12 building block chemicals (Figure 4.1) that can be further transformed into high value-added products or chemicals with excellent industrial applications. For example, 2,5-FDCA can be converted into polyethylene 2,5-furandicarboxylate, which is a biodegradable plastic. This material can replace polyethylene terephthalate (PET), which is a nonbiodegradable plastic [26]. Before converting this biomass into chemicals, it must be pretreated to improve its chemical composition and produce a consistent product. Some pretreatment methods are hydrothermal carbonization, leaching (acidic and alkaline), pelletization, softening of wood using steam, etc. There are two major conversion methods for biomass that have been reported so far, i.e. thermochemical and biochemical methods [25]. The former is a more efficient method since it takes fewer reaction times and is better at destroying organic constituents [22]. Thermochemical methods are further divided into two types: pyrolysis and gasification, which give bio-oil and syngas, respectively.

Various sugars and aromatics in bio-oils can be further transformed into top value-added chemicals (for instance, levulinic acid (LA) can be prepared from

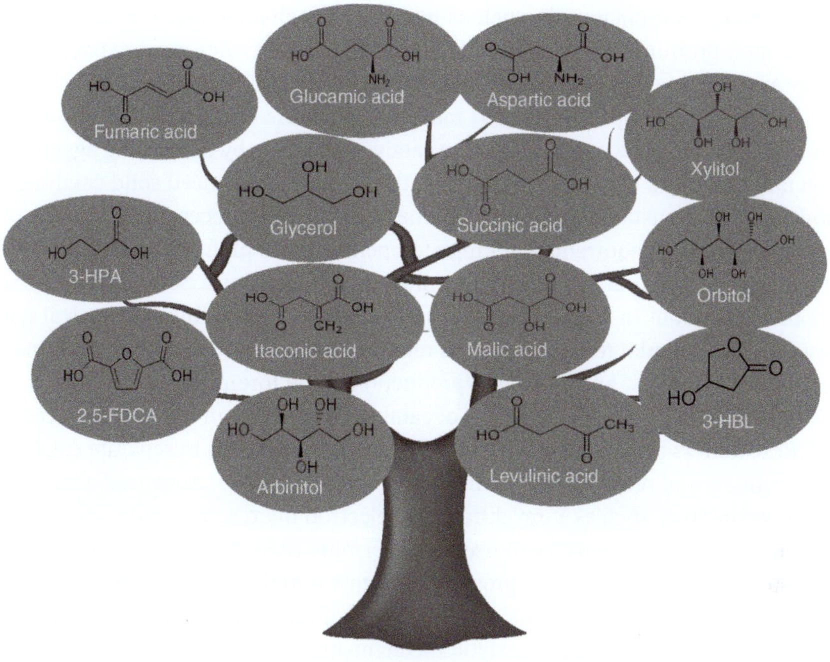

Figure 4.1 Top 12 building block chemicals.

bio-oils by acid catalysis) [27]. Syngas can also be converted into top value-added chemicals by one of the classical routes, i.e. Fischer–Tropsch synthesis (FTS) [28]. The biochemical method involves the cleavage of cellulose into simpler sugars, which can be further converted into value-added chemicals [22]. After the pretreatment, biomass feedstock-derived substances can be transformed into high value-added products or chemicals with the help of solid catalysts, as these catalysts have advantages over homogeneous catalysts. The core aim of this chapter is to discuss the role of suitable solid catalysts and the best synthetic approaches for biomass translation into various value-added chemicals. Additionally, the discussion related to the structural properties, activity, and selectivity of solid catalysts toward various chemical reactions has been done. Recent developments and prospects in the process of biomass conversion into value-added chemicals are also discussed.

4.2 Advanced Solid Catalysis

In the last few decades, solid catalysis has become a subject of extensive research and development. Scientists have focused on the design of solid catalysts with specific properties to tailor the chemicals. The solid catalyst provides an effective surface area where reactants can adsorb, react, and then desorb as products [5, 29, 30]. Along with this, information related to the pore structure, chemical composition, and active sites of the catalyst has been collected to enhance its catalytic activity and selectivity. Nanomaterials, such as nanoparticles, nanowires, and nanosheets, have gained prominence in advanced solid catalysis that offer great catalytic efficiency and improved activity due to high surface area and unique properties. Solid catalysts are often supported by inert materials, such as alumina, silica, or zeolites, because these supports can enhance the catalyst's stability by preventing agglomeration and providing better dispersion of active sites [9]. Advanced solid catalysis is known as a more sustainable and environmentally friendly process because this process can reduce the requirement for high temperature and high pressure conditions, thus saving energy and decreasing greenhouse gas emissions. Solid catalysts find various applications in industries (e.g. in petrochemicals, oil refining, chemical synthesis), environmental remediation, and renewable energy production (e.g. biomass conversion, fuel cells, and hydrogen production). The foremost challenge in solid catalysis is the recovery and recycling of catalysts, especially in industrial processes [20]. Researchers are working on developing various methods to separate catalysts from products and reuse them to reduce their costs and waste. Advanced characterization techniques, such as X-ray diffraction, electron microscopy, and surface analysis for solid catalysts, have been used to determine their structure and properties at atomic- and nanoscale levels, providing insights into their catalytic mechanisms. High-performance computational modeling and simulations play a substantial role in understanding of complex catalytic mechanisms. It helps to design novel solid catalysts with enhanced performance [5]. Poisoning by impurities or coke formation can reduce the activity of catalysts over time. Understanding the reasons behind the catalyst deactivation mechanisms and finding ways to mitigate them are essential in advanced solid catalysis. Thus, advanced solid catalysis is an interdisciplinary field that combines chemistry, material science, and engineering to develop solid catalysts that improve the efficiency and sustainability of chemical processes. Ongoing study in this area aims to discover new catalyst materials, optimize existing ones, and develop innovative solutions for complex reactions and environmental challenges. Information related to the different categories of solid catalysts and methods for their preparation is given in the next section:

4.2.1 Types of Solid Catalysts

As mentioned earlier, solid catalysts are grouped into four categories based on their structure and properties of activated substrate (Figure 4.2a–d): (1) metal oxides, (2) sulfonated polymer, (3) supported metal catalysts, and (4) meso- and microporous materials [5]. Metal oxides (e.g. transition metal oxides) are chemical compounds containing metal-to-oxygen bonds that are present in a solid state and have a large surface area; thus, they catalyze heterogenous reactions.

Transition metal oxides are highly stable due to their stable metal oxidation states. The order of the stability of various categories of metal oxide as catalysts is given in Figure 4.3 [29]. Metal oxide catalysts became eminent in the middle of the 20th century due to their ability to catalyze different categories of reactions and act as bifunctional catalysts [6, 30]. SNb_2O_5 is a metal oxide that is utilized as a heterogeneous catalyst for glucose transformation into 2-hydroxymethylfurfural (2-HMF) [30].

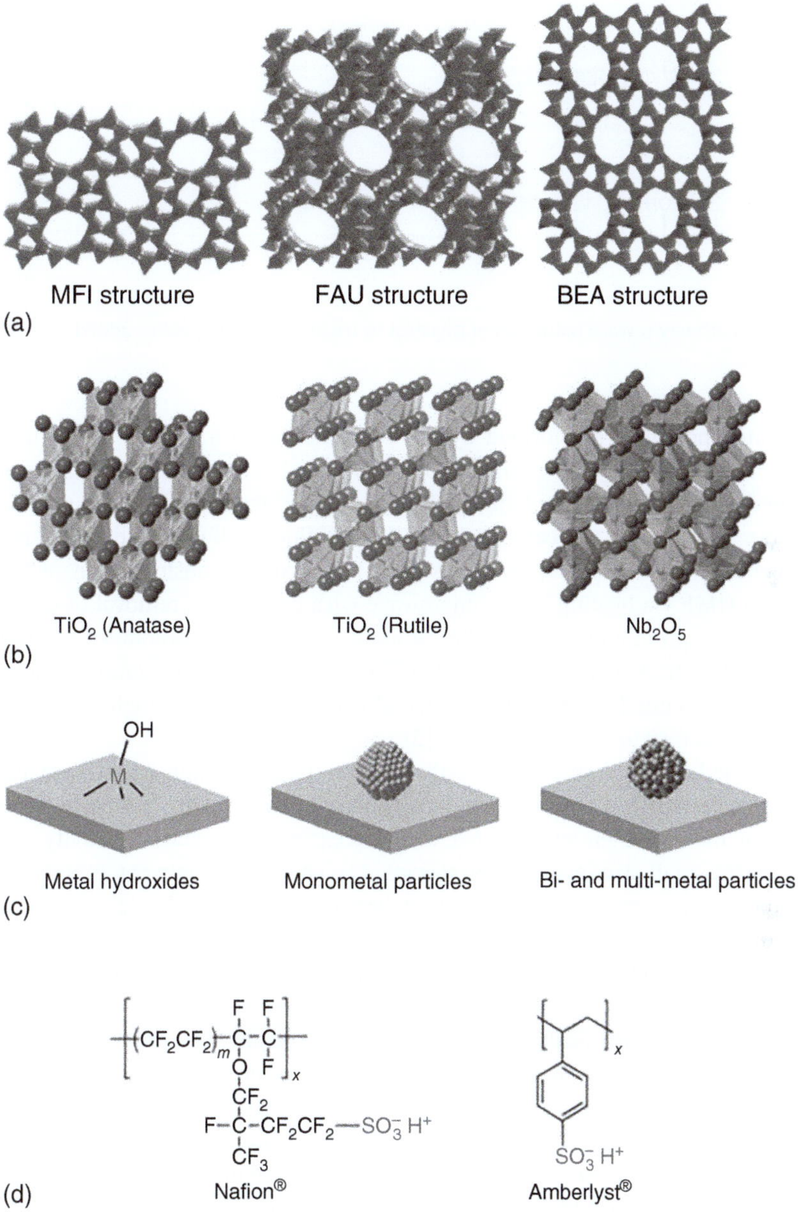

Figure 4.2 Examples of the materials used in biomass conversion. (a) Micro- and mesoporous materials. (b) Metal oxides. (c) Metal supported catalysts. (d) Sulfonated polymers. Source: Reproduced with permission from Ref. [5].

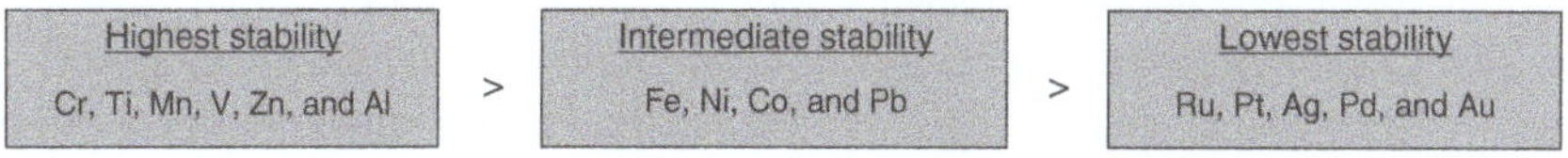

Figure 4.3 Order of stability in various transition metal oxides.

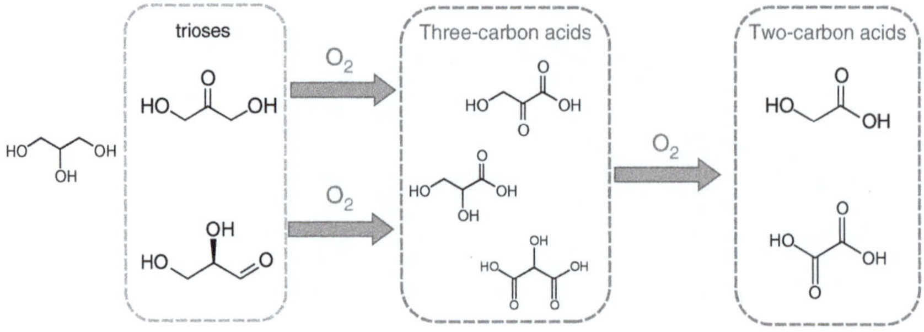

Figure 4.4 Pathway toward oxidation of glycerol to trioses (including value-added chemicals).

Sulfonated polymers are formed by the introduction of a sulfone group into the polymer, which can change their properties [31]. There are some examples of sulfonated polymers, such as Amberlyst-15, Dow-type resins (sulfonated styrene polymers), and Nafion (a catalyst based on perfluorosulfonic acid). Styrene-divinylbenzene sulfonated copolymer is used for catalytic transformation of the fructose to HMF via heating in the microwave with continuous removal of water. There was no sign of a decrease in the catalytic activity of the sulfonated catalyst after multiple reuses, which shows that sulfonic acid present in catalyst remains stable in experimental conditions. Sulfonated catalysts also show high selectivity for HMF in fructose-to-HMF conversion [5].

In supported catalysts, small and highly active materials (usually metals) are dispersed over the surface of a less active substance called support. Supporting the metals may increase their surface area, thus increasing their catalytic activity and stabilizing the small-sized active metal species [32]. This also helps in obtaining polyfunctional catalysts, as supporting metal catalysts to introduce new acidic and basic sites. Nanoparticles of gold (Au) supported on titanium dioxide (TiO_2) show exceptional performance in the oxidation of CO [5]. Metal catalysts can also be utilized in the oxidation of glycerol to various trioses that have been included in the category of high value-added chemicals [33] (Figure 4.4). The selectivity of product depends on kind of support provided to the metal catalyst. For instance, glycerol oxidation on gold, palladium, and platinum nanoparticles supported on carbon and graphitic surfaces gives glyceric acid as a major and glyceraldehyde as a minor product. Among the three, Au supported on either carbon or activated graphite has the highest selectivity (~100%) for glyceric acid. Contrarily, Pd/C has 4% and Pt/C has 13% selectivity toward glyceric acid [33].

Meso- and microporous materials are solid with different pore sizes and specific surface areas that help to increase their catalytic performance. According to International Union of Pure and Applied Chemistry (IUPAC), these can be classified into two categories based on pore sizes. According to the IUPAC notation, materials with pore diameters less than 2 nm are classified as microporous, while those with pore diameters greater than 50 nm are classified as microporous, and the mesoporous category (diameter sizes between 2 and 50 nm) thus lies in the center of these

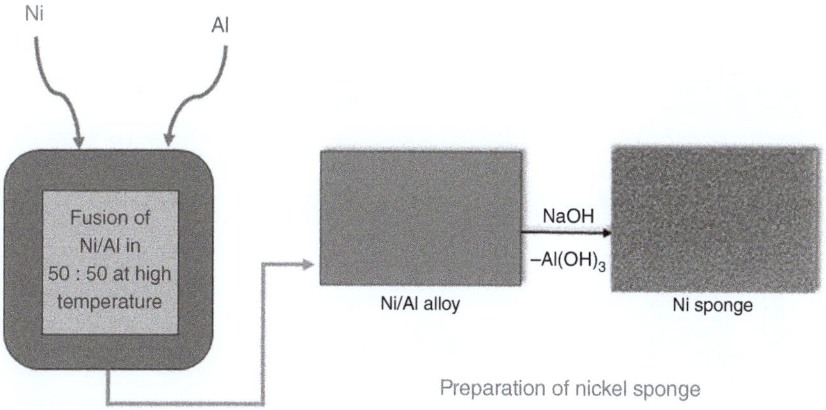

Figure 4.5 Diagrammatic representation of the preparation of nickel sponge. Source: Reproduced from Cesário et al. [39], with permission of Elsevier.

materials [34]. Zeolites are well-established microporous materials [35], which are being used in chemical industries, petrochemical refineries as separators for drying and purification, waste treatment, radioactive waste storage, etc. There are 235 types of structures of zeolites, of which 228 are ordered and the rest are partially ordered. These structures have three letter codes (e.g. MFI, BEA). From these structures, only MFI, FER, FAU, MOR, and BEA are part of the so-called "big five" [36]. Shape selectivity is considered very vital for chemical reactions and can be easily achieved by using highly shaped and size-selective zeolites [37].

4.2.2 Methods for the Synthesis of Solid Catalysts

As the technique of synthesis and nature of precursors may directly affect the properties of solid catalysts, it is important to choose right method for their synthesis [38]. One of the most important catalysts prepared by the *leaching of alloys technique* is Raney nickel, which can be used in hydrogenation reactions [38]. In 1924, Murray Raney fused nickel (Ni) and aluminum (Al) in a 50 : 50 ratio (Figure 4.5) and then leached Al out of it with the help of NaOH [39]. The nickel catalyst obtained is known as "nickel sponge" or "skeletal catalyst." These catalysts have more importance compared to other catalysts available at that time because of the large surface area and the fact that they do not require any pretreatment.

The precipitation method (Figure 4.6) involves the preparation of an aqueous solution of metal salts (e.g. $Al(NO_3)_3$, $Mg(NO_3)_2$), and in these solutions, a suitable precipitating agent (e.g. Na_2CO_3, $NaHCO_3$, NaOH, and $(NH_4)_2CO_3$) [40] is added dropwise. Maintaining appropriate conditions such as temperature, pH, and concentration of salt during precipitation can help to induce the precipitation process. The precipitates so obtained are filtered, washed, and dried to obtain a solid mass. Types of precipitating agents can directly affect the properties of solid catalysts obtained by the precipitation method. The last step is calcination, in which salt and hydroxide forms are converted into oxides by reaction under

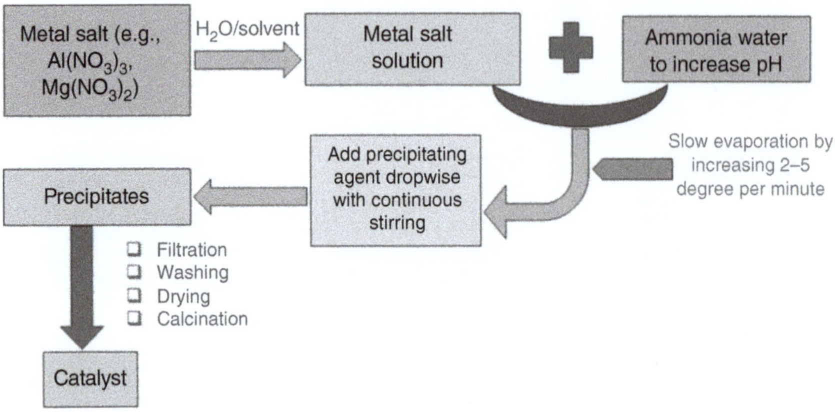

Figure 4.6 Steps involved in the precipitation methods.

oxygen at an appropriate temperature [41]. There are many advantages to the precipitation method, such as the fact that the synthesized solid catalysts have high catalytic activity and a uniform pore size [42]. This method also has some disadvantages, such as the loss of a large amount of catalyst while using the solvent and precipitating agent. The extra cost is required for the separation of catalysts from solvent.

The impregnation method is used mostly for the preparation of supported catalysts [38, 40, 41] in which the solution of catalytic precursors has been encountered with solid support. This is followed by drying of solution mixture to remove the solvent, which gets imbibed in the catalyst. Impregnation methods can be classified as dry or wet, depending on the amount of solvent used for the impregnation. The concentration of precursor in the solution directly controls the amount of precursor that can be loaded on the support. It depends on the temperature conditions; with increasing temperature, solubility of precursor will be high, and wetting time will be higher [41]. The foremost advantages of this approach are controlled properties of solid catalysts at a low price compared to other methods. The dry impregnation method is better as compared to the wet impregnation method because the latter causes a considerable loss of solvent. In addition to these two impregnation methods, a vacuum impregnation is also available, which can be performed in a vacuum [28, 43]. In this method, medium is dried first and then placed under vacuum to create pores, and then support is added according to the volume of pores.

Zeolization is the process of synthesizing zeolites by converting compounds of aluminum and silica into aluminum silicates (zeolites). This process involves different stages, such as nucleation and crystal growth. Nucleation is first step in zeolite formation [38], in which atoms are suspended in the zeolite precursor solution and accumulate to form quasi-solid structures called nuclei. Supersaturation acts as a driving force for nucleation and helps to bring the suspended particles of the solution together, resulting in nucleation. With increasing temperature, the entropy of the solution will increase, which helps in supersaturation. Nucleation occurs only at a certain degree of supersaturation, which is called the critical supersaturation zone [44, 45]. After nucleation, followed by aging or heating under pressure, crystalline structure is obtained, which is the purest form of the solid catalysts.

4.3 Biomass, Its Composition, and Properties

Biomass is a plant-based material obtained from the reaction between carbon dioxide in air, water, and sunlight via photosynthesis. Biomass contains wheat straw, wood, barley straw, lignite, and bituminous coal, in which the maximum percentage of moisture content and volatile matter are present in the wood [46]. Photosynthesis produces carbohydrates, which are the building blocks of biomass, and converts less than 1% of existing sunlight to stock chemical energy [47]. After processing biomass, either chemically or biologically, the obtained carbon content is oxidized to produce carbon dioxide and water. So, this is a cyclic process in which existing carbon dioxide is used to harvest new biomass. Biomass is usually a mixture of hydrocarbon materials comprising carbon, hydrogen, and oxygen, with traces of nitrogen, sulfur, and other minerals. In broad terms, biomass has included crops (agricultural and forestry), wood or wood waste, animal residues, aquatic plants, algae, and seaweeds [48]. Scientists have tried to characterize biomass in diverse ways, but one simple method is to define four main types: woody plants, herbaceous plants and grasses, aquatic plants, and manures. Biomass is the only renewable carbon source capable of being converted into convenient solid, liquid, and gaseous fuels, as well as value-added chemicals. The biomass resource can be processed to generate energy through two major transformation technologies, such as thermochemical and biochemical methods [49–51]. A greater understanding of the physicochemical properties of biomass is useful for its sustainable utilization in energy and chemical production. Various parameters, e.g. chemical composition, contents of moisture, ashes, and inorganic substances, account for disparities among biomass species.

In biomass feedstock, lignocellulose (carbohydrates), triglycerides, terpenes, and proteins have been included as main components [52]. Lignocellulose is composed of 20–34% of lignin, 10–35% of hemicellulose, 35–55% of cellulose, 0–2% of ash, and 1–6% of extractives by weight of biomass (Figures 4.7 and 4.8). The three key components of lignocellulose, namely cellulose, lignin, and hemicellulose, are totally intertwined together. These three components can be separated from each other by some pretreatments to make chemically reactive polymers that can be further utilized to produce high value-added chemicals and fuels [51, 53]. The comparative proportions of cellulose and lignin in the biomass are one of the influential factors in identifying the suitability of plant species for subsequent processing as energy crops [51–56]. Lignin is a 3D polyphenolic biopolymer that has a nonuniform structure and an amorphous nature. It has a high molecular weight that conveys rigidity to the plant cell walls. However, it is a challenging task to selectively convert into aromatic components. Hemicellulose is a mixture of pentose and hexose, methyl glucuronic, and galacturonic acids that bind together via non-covalent bonding interactions. Cellulose is a polymer of glucose units joined by β-glycosidic bonds, with an average

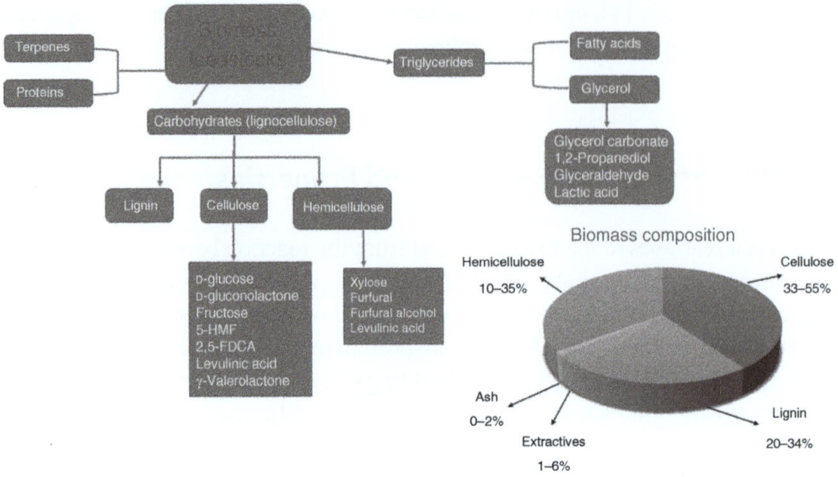

Figure 4.7 Biomass composition and various components present in the biomass feedstock.

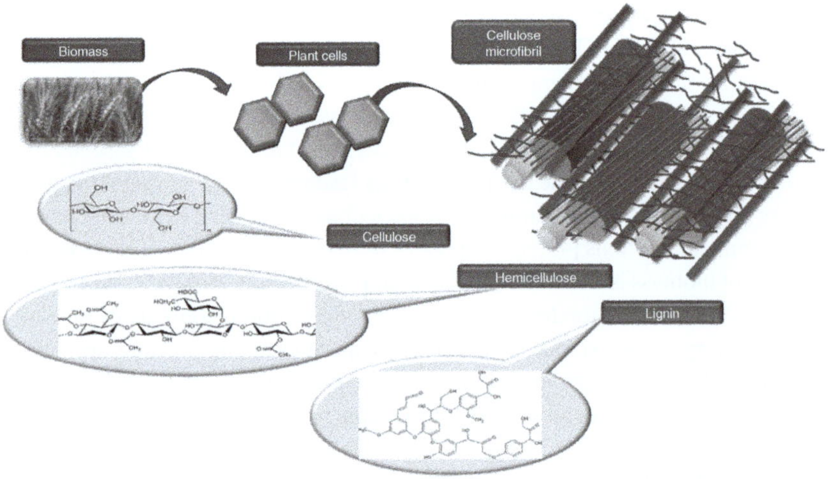

Figure 4.8 Components (cellulose, hemicellulose, and lignin) and their structures in biomass. Source: AngieC/Adobe Stock.

molecular weight of around one lakh [56]. Although lignocellulose can easily be obtained from plant material, it is difficult to exploit it because of its rigid structure and composite nature. Of the three major components of lignocellulose, cellulose is crystalline, whereas lignin and hemicellulose are amorphous in nature. Therefore, pretreatment is useful to remove amorphous polymers, especially lignin, from the crystalline component. Hemicellulose is strongly bonded to lignin by strong covalent bonding interactions, so it also gets removed in some cases. Thus, pretreatment increases the crystallinity of biomass and exposes the cellulose for further processing with enhanced reactivity. Table 4.1 shows the chemical composition of

Table 4.1 Chemical composition of typical biomass feedstock.

Source	Composition (%)			
	Cellulose	Hemicellulose	Lignin	Extractives
Hardwood	43–47	25–35	16–24	2–8
Softwood	40–44	25–29	25–31	1–5
Bagasse	40	30	20	10
Coir	32–43	10–20	43–49	4
Corn cobs	45	35	15	5
Corn stalks	35	25	35	5
Cotton	95	2	1	0.4
Flax (retted)	71	21	2	6
Flax (unretted)	63	12	3	13
Hemp	70	22	6	2
Henequen	78	4–8	13	4
Istle	73	4–8	17	2
Jute	71	14	13	2
Kenaf	36	21	18	2
Ramie	76	17	1	6
Sisal	73	14	11	2
Sunn	80	10	6	3
Wheat straw	30	50	15	5

Source: Adapted from Zugenmaier [56].

biomass feedstock from various sources (e.g. hardwood, softwood, bagasse, coir) and the percentage composition of cellulose, hemicellulose, lignin, and extractives in this feedstock. There are various techniques, such as energy dispersive X-ray fluorescence spectroscopy, near-infrared spectroscopy, solid-state NMR (nuclear magnetic resonance), field emission scanning electron microscopy, transmission electron microscopy, wide-angle and small-angle X-ray scattering, and atomic force microscopy, which can be utilized for the characterization of biomass feedstock.

4.4 Biomass Conversion into High Value-Added Chemicals

The conversion of biomass (which includes plant materials, agricultural residues, forestry by-products, and organic waste) into high value-added chemicals is an imperative area of research and development in the fields of renewable energy, sustainable chemistry, and bioeconomy. There are various types of value-added chemicals that can be synthesized from sustainable biomass feedstock [5]. The key approaches toward biomass conversion into high value-added chemicals include

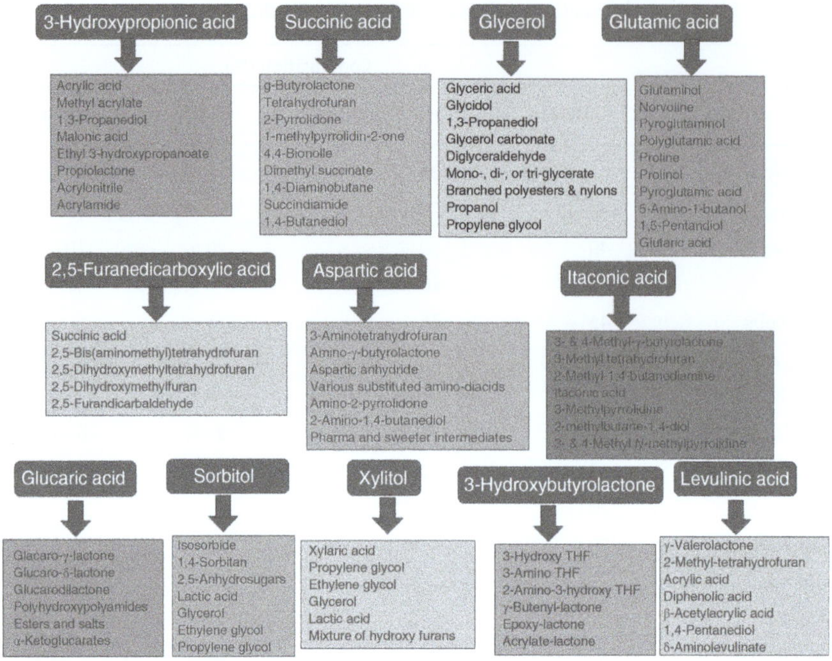

Figure 4.9 List of 12 high value-added chemicals and products obtained from chemicals.

biological conversion (e.g. fermentation), microbial production, thermochemical conversion (pyrolysis), gasification, hydrothermal liquefaction (HTL), and catalytic conversion (hydrogenation) [57–59]. The list of 12 high value-added chemicals and products obtained from them by different chemical reactions is given in Figure 4.9. From them, there are five industrially useful chemicals (5-HMF, LA (linoleic acid), GLA (gamma-linolenic acid), DHA (docosahexaenoic acid), and 2,5-FDCA) as bio-based monomers and their intermediate species are well known [60, 61]. In the fermentation process, chemicals such as ethanol, butanol, and organic acids are produced. In chemical processing, such as biorefinery, integrated biorefinery aims to convert various components of biomass (e.g. cellulose, hemicellulose, and lignin) into a range of valuable chemicals, including bio-based polymers, organic acids, and platform chemicals [62]. Enzymes and other biocatalysts can be used to selectively transform biomass components into specific high-value chemicals (e.g. cellulase convert into cellulose and then into glucose), which can further be used to prepare biofuels or other chemicals [63, 64]. Electrochemical processes, such as electrolysis and electrosynthesis, can be used to convert biomass-derived feedstock into high-value chemicals (e.g. electrochemical transformation of lignin to valuable aromatic compounds) [65–67]. Hybrid approaches, such as biological and chemical processes, are also effective for making value-added chemicals. Both biological and chemical processes are combined to enhance the efficiency of biomass transformation and yield of high-value chemicals [68].

From aforementioned methods, three key catalytic paths have been utilized to convert biomass into valuable chemicals. These routes are pyrolysis, gasification,

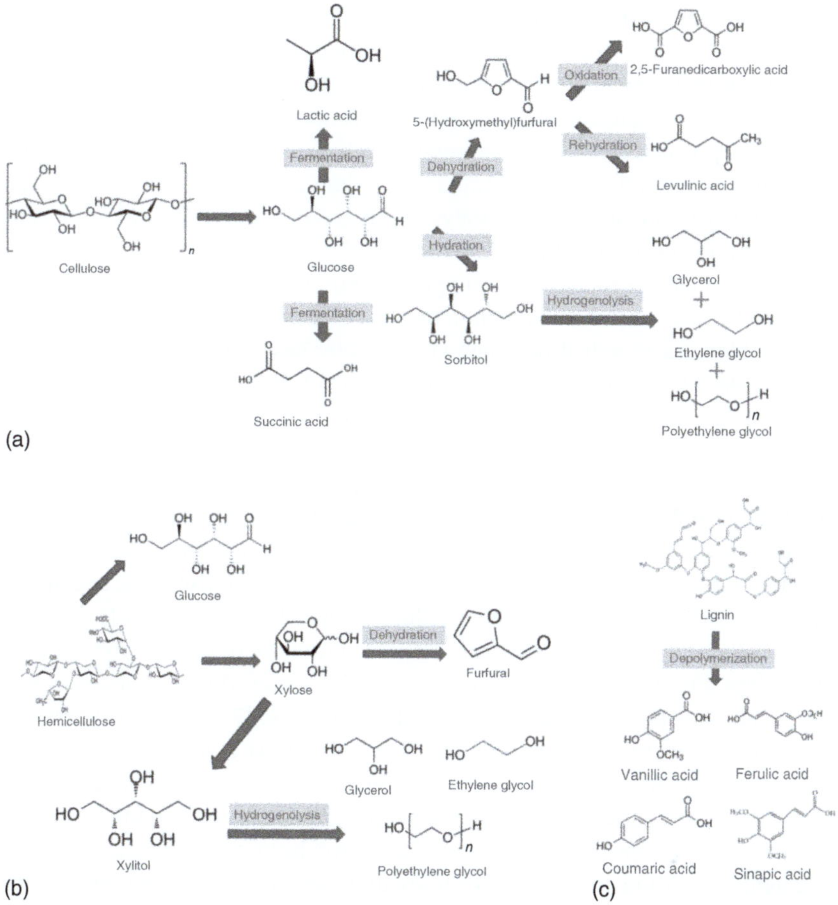

Figure 4.10 (a) Conversion of cellulose, (b) conversion of hemicellulose, (c) conversion of lignin into high value-added chemicals.

and hydrolysis [58]. The focus is on the hydrolysis process that has been utilized to cleave the lignocellulose into essential parts (cellulose, hemicellulose, and lignin) (Figure 4.10) and their hydrolyzed products [69]. Lignocellulose biomass has been converted into cellulose, hemicellulose, and lignin by pretreatment methods such as physical, chemical, physicochemical, and biological pretreatments. Mechanical and irradiation are two physical pretreatment methods that generally decrease the particle size and overall crystallinity of biomass [70]. These methods do not help in delignification, and so additional methods are required for removing lignin or hydrolyzing cellulose or hemicellulose [71]. Microwave pretreatment is a commonly used method because of its easy operation, low price, low generation of inhibitors, and high efficiency. This pretreatment combines physical and chemical methods; as in this case, lignocellulosic biomass is soaked in acidic or alkali solvents before pretreatment. Higher temperatures used for microwave pretreatment increase cellulose concentration in biomass and decrease lignin concentration. If high temperature is used, it can destroy the sugar content. While using the sonication process, small cavitation bubbles are formed, which can break hemicellulose and cellulose parts, thus exposing cellulose to breaking into simple sugars [72]. Apart from that stream explosion, physicochemical pretreatment is also used to process lignocellulosic biomass. The powdered biomass is exposed to high pressure steam at a high temperature, which has expanded the cell walls of the biomass. This will ultimately result in hydrolysis with the help of acid catalysts such as sulfuric acid and sulfur dioxide, thus increasing the cellulose's accessibility to the cellulase enzyme. In another method, steam can be replaced by compressed hot water for the hydrolyzation of hemicellulose and the delignification of biomass feedstock. If there is a high lignin content in biomass, then hydrogen peroxide or water has been used along with oxygen to process the biomass via wet oxidation [73]. In chemical treatment, acidic (e.g. oxalic, maleic, or sulfuric acid) or alkaline (ammonium, sodium, potassium, and calcium salts) reagents can be used for the acidic or basic hydrolysis of hemicellulose and lignin. After the acidic and basic chemical pretreatment, biomass needs to be neutralized to reduce the corrosive effect on equipment [71]. In another chemical pretreatment, organic solvents (e.g. methanol, ethylene, and acetone) can be used at high temperature to break strong covalent bonds between hemicellulose and lignin, which further depolymerize the hemicellulose into oligosaccharides [74]. Apart from that, there is another cost-effective and eco-friendly biological pretreatment that involves the utilization of microbes (fungi and bacteria) for delignification of the biomass. This process is known as bio-delignification [75]. Biological pretreatment also involves the utilization of lignolytic enzymes (e.g. laccase) without the generation of any inhibitors. It is better to use enzymes than microorganisms, as they require a wide range of optimum temperature and pressure conditions.

Furthermore, cellulose and hemicellulose have been transformed into their fundamental building units (e.g. pentoses and hexoses). These building block monosaccharides and oligosaccharides can be further utilized as precursors to synthesize biofuels and platform commodity chemicals (e.g. furans, polyols, and acids) via chemocatalytic conversion processes. For example, cellulose can be easily hydrolyzed to form glucose monomers at high temperatures, along with using H_2SO_4 as a catalyst [61]. The depolymerization of cellulose and hemicellulose into fundamental building units without the involvement of any kind of catalyst or acid is called autohydrolysis. This process has some problems as there is the generation of inhibitors at high temperatures, and the process will be slow at low temperatures. Autohydrolysis is a cost-effective and eco-friendly process, and it does not release any hazardous waste. This is slower as compared to typical catalytic processes, and it takes a long time for a reaction to complete. Sodium chloride helps to increase the rate of autohydrolysis by absorption of H^+ ions on the surface of cellulose and further released by the degradation of cellulose itself. This H^+ ion acts as a catalyst and helps to increase the rate of autohydrolysis [76].

Besides, glucose can be transformed into valuable chemicals via oxidation to form gluconic and glucaric acid (GA); hydrogenation to synthesize sorbitol, which is trailed by acid-catalyzed dehydration to form sorbitan and isosorbide; and dehydration to form HMF. Glucose also works as a monomer for the preparation of biodegradable polyamides with exceptional structural and chemical properties. The acid-catalyzed dehydration of fructose prepared by glucose isomerization gives HMF. This can be used as a pioneer to produce numerous chemicals, polymers, and biofuels [77]. The oxidation of HMF has given 2,5-FDCA, which is a structural analog of terephthalic acid monomer. Furthermore, HMF has reacted with water in acidic media to give LA, along with formic acid. LA and its hydrogenated product, γ-valerolactone (GVL), are important platform chemicals and probable building units for polyesters [78]. Additionally, the transformation of triglyceride feedstock into biodiesel and platform products has fascinated the unified triglyceride biorefinery. A mixture of fatty acid methyl esters and glycerol has been produced by transesterifying triglycerides with methanol. These chemicals can be exploited commercially as biodiesel, bio-oils, biolubricants, and biosurfactants. Glycerol can be transformed into valuable chemicals such as propylene glycol and trioses (e.g. glyceraldehyde and 1,3-dihydroxyacetone). Trioses can be converted into lactic acid in acidic and basic media, which is a bio-based platform chemical with useful applications (e.g. bioplastic) [5]. The third component of lignocellulosic biomass is lignin, which can be depolymerized to obtain oligosaccharides and monosaccharides, which can be further converted by various reactions into high value-added chemicals (Figure 4.10c). As previously mentioned, lignin is one of the most abundant components of biomass [79]. Therefore, the method of conversion of lignin into value-added chemicals is of great concern. For transforming lignin, the main step is the breakage of the C—O or the C—C bond present in lignin and then the insertion of a heteroatom leading to the construction of cyclic ketones, aldehydes, phenols, arenes, etc. The depolymerization of lignin can be performed by various thermal, chemical, and biological methods. Chemical and biological methods are the same as those discussed in pretreatment methods. The thermal method involves various methods such as combustion, gasification, pyrolysis, and HTL [80]. Combustion and gasification occur at relatively high temperatures, and so pyrolysis and HTL are more widely used. In pyrolysis, lignin is treated at high temperatures, leading to its depolymerization. Between the temperature range 200–400 °C, cleavage of the C—O bond occurs at higher temperature. Pyrolysis can be slow pyrolysis or fast (flash) pyrolysis. HTL is a good option for biomass with high moisture content as there is no requirement for pre-drying, which saves a lot of energy and time. Moreover, in this case, water is used as a solvent under subcritical conditions of high temperature and pressure [80].

4.5 Utilization of Solid Catalysts for Biomass Conversion into High Value-Added Chemicals

Solid catalysts are an ideal choice for the conversion of biomass feedstock into value-added products, as they do not involve the use of any metals [81]. Many scientists have worked in this field with a variety of choices for these heterogeneous catalysts with advanced properties. In literature, various noble and non-noble metals supported on carbon substrate, Ni–Cu bimetallic, Cu–Zn–Al trimetallic,

Cu–Ni–Al hydrotalcite, gold nanoparticles (Au NPs), Cu–Ni bimetallic NPs, have been reported as catalysts to synthesize hydrocarbon fuels, renewable solvents, and agrochemicals [82]. Solid base catalysts are good candidates for biomass conversion into valuable products. Generally, solid base catalysts are favorable for interacting with acidic functional groups containing reactants. The basic sites of this catalyst can be activated by removal of water or CO_2 molecules from the surface. Apart from that, basicity can be exhibited by using various materials, such as metal oxides (e.g. CaO, MgO, ZnO, and TiO_2), mixed metal oxides (e.g. MgO-$A1_2O_3$ and ZnO–SiO_2), zeolites, and hydrotalcite clay. Additionally, strong basicity can be attained by using lanthanides (Ce, Nd), and alkaline earth metals (Ca, Mg) as promoters on the hydrotalcite surface for the desorption of carbon dioxide at high temperatures. Apart from solid base catalysts, some solid acid catalysts have reported in literature for biomass conversion. For example, Cao et al. reported transformation of glucose into HMF by using ammonium resin [83]. Nevertheless, this process required a high temperature (~140 °C) along with 70% DMSO (dimethyl sulfoxide) in water, and the maximum yield will be 56%. The acidic sites on ammonium resin influence the conversion of glucose into 5-HMF with a series of dehydration and rearrangement reactions. 5-HMF is known as a valuable chemical that can be further converted into bio-based value-added chemicals and fuels. Bhaumik et al. utilized sulfonated triphenyl aniline polymer as an acidic organo-catalyst [84], wherein the preparation of catalyst required toxic triphenyl aniline polymerization under anhydrous conditions, followed by sulfonation [85]. Afterward, Verma et al. described the sustainable pathway for transformation of carbohydrates into furanics from biomass by using a heterogeneous sulfonated graphitic carbon nitride (Sg–CN) [86]. This catalyst shows its capability as a vastly acidic solid catalyst that can be used to break carbohydrates into biologically significant scaffolds and chemicals at elevated temperature conditions. Zhang et al. [87] have reported carbon-SO_3H as a catalyst for the conversion of xylose and corn stalk into furfural and GVL with ~78.5% and ~60.6% yields, respectively [88]. Sulfonated clay and zeolite can also be used to produce furfural with 87% yield from xylose at 180 °C that exhibits good catalysis activity and recyclability [89]. The corncob feedstock can be successfully converted into furfural with ~72% yield by using H-ZSM-5 zeolite as a catalyst in the presence of GVL as a solvent. Additionally, HSO_3-ZSM-5 (functionalized zeolite) showed a better yield of furfural ~89% in the mixture of THF/H_2O [90]. Gupta et al. [91] developed an amorphous Nb_2O_5 catalyst with hydrophilic Bronsted and Lewis acid sites for the conversion of xylose into furfural with ~66% yield in the mixture of water/toluene at ~100 °C. Lately, bi-metallic mesoporous heteropoly acid-based salt has been used as a heterogeneous solid catalyst for breakage of xylose into furfural with 63% yield at 200 °C for three hours in a DMSO and H_2O mixture due to its strong acidity and low cost [92]. This heteropoly acid-based solid catalyst has a broad surface area, excellent porosity, diverse structures and can be easily utilized for biomass valorization. Apart from that, MIL-101(Cr)-SO_3H with Lewis and Brønsted acid sites has been established to transform xylose into furfural.

4.6 Electrocatalytic Conversion of Biomass into High Value-Added Chemicals

Electrocatalytic conversion of various components into valuable chemicals represents an operative kit for many significant industrial applications. For example, organic pollutants and antibiotic waste materials are present in wastewater systems, and after giving a potential difference between two electrodes (oxidation at anode and reduction at cathode), chemical reactions can be driven to convert them into useful products. This process offers several advantages, including reduced greenhouse gas emissions, reduced carbon footprints, improved energy efficiency, cost-effective process, and the potential to produce valuable chemicals from renewable resources.

Biomass electroreforming embodies an auspicious green and supportable path for synthesis of value-added chemicals with negligible harm to atmosphere [93–95]. By using electrocatalytic processes, electrocatalysts (typically metals, metal alloys, metal oxides, or other conductive materials, etc.) may facilitate the conversion of biomass components into desired products such as alcohols, organic acids, or other valuable chemicals [95]. By controlling the reaction conditions and selecting appropriate electrocatalysts, researchers can steer the conversion of biomass components toward specific target chemicals. The experimental conditions of electroreforming are habitually milder as compared to the thermochemical method. In this method, the control of product formation is better achieved by tuning electrochemical cell parameters (e.g. pH and applied potential). The electroreforming of cellulose derivatives, such as LA, HMF, and glucose, has been done by using versatile metal-based electrocatalysts that provide improved products with better yields [96]. In cellulose electroreforming, two different types of desired products can be produced, such as valorized chemicals at anode and H_2 gas at cathode. Through electrochemical oxidation at different pH conditions, these valorized chemicals can further be transformed into other beneficial compounds by oxidation and hydrogenation reactions. In the past, a study related to the passing of electricity through the cellulose was described in 1947 by O'Sullivan. They have inspected the electricity that is passing through recycled cellulose with changing salt and humidity contents, thus understanding the conductivity of cellulose. Murphy et al., in 1963, achieved first electrolysis in cellulose and examined that H_2 has been generated at anodic surface [96]. Although reports of cellulose electrolysis remained scarce for a longer period until the twenty-first century, the ever-growing claims for sustainable and green chemistry appeared.

Sugano et al. accomplished the cellulose electro-oxidation at a polycrystalline gold electrode with a highly basic electrolyte (pH ~14) in 2010 [97]. In the same year, Bin et al. performed the glucose conversion to GA and gluconic acid in a flow-through electrochemical cell by using nano-MnO_2-loaded tubular and porous titanium electrodes [98]. Stimulatingly, pH has a less significant role in glucose conversion as it increases from 2 to 10, resulting in a slight increase in yield from

90% to 93%. As the pH increases from 2 to 7, there is an increase in the yield of gluconic acid and GA from 87% to 94%. Afterward, yield decreases to 78% at pH 10. With 5% MnO_2 loading, glucose conversion of 98% has been attained, with the selectivity toward gluconic acid and GA being 43% and 55%, respectively. The increase in current density ~6 mA cm^{-2} led to the 99% conversion of glucose into gluconic acid and GA, with increasing selectivities of 15% and 84%, respectively. Liu et al. utilized bimetallic Ni–Fe oxide or nitride-based electrodes to produce gluconic acid (~4.7% yield) and GA (~11.6% yield) from glucose in 1 M KOH at an applied voltage of 1.0 V [99]. According to the technoeconomic analysis, from this approach, GA has been produced at 54% less price compared to conventional approaches. Neha et al. designed a Pt–Bi alloy-based electrocatalyst on a glassy carbon electrode for the conversion of glucose into gluconic acid, followed by the conversion of methyl-glucoside into methyl-glucuronate [100]. From chronoamperometric measurements, at a fixed potential (~0.3 V) after 90 min, the current is halved because some poisoning has occurred. Electrode poisoning is a substantial task in the scale-up electroreforming of glucose. This is alleged to be prompted by the act of reaction intermediate species [101]. Liu et al. designed the Ni/Fe-layered double hydroxide nanosheets as electrocatalysts for the electroreforming of 5-HMF into 2,5-FDCA [102]. Weidner et al. explored the 5-HMF electrooxidation into 2,5-FDCA by using bimetallic cobalt-derived alloys to exchange the oxidation reduction reactions in overall water splitting [103]. Nam et al. employed an alkaline electrolyte (pH ~13) for electroreforming 5-HMF into FDCA [104]. In 2020, Cai et al. inspected the oxidation of 5-HMF into 2,5-FDCA by using Ni(II)-altered covalent-organic framework based film, i.e. TpBpy-Ni@FTO [105]. In another report, 1,3-butanediol and 1-butanol have been produced electrocatalytically from 4-OH-2-butanone and octane has been obtained from valeric acid [106]. Shi et al. reported an electrocatalytic reforming strategy to upcycle waste plastics (e.g. PET) into high value-added products (e.g. terephthalate and carbonate) at the anode and hydrogen fuel at the cathode [65] with a Pd-modified Ni foam as an electrocatalyst (Figure 4.11). Nilges et al. reported the dual-step electrochemical transformation of LA in water at room temperature to produce octane via valeric acid [67]. LA is considered an important intermediate to produce renewable chemicals and fuels for transportation. Based on the conversion of LA, derivates of valeric acids, such as valeric esters (also known as valeric biofuels), have been projected as biofuels for future transportation.

The transformation of LA into hydrocarbons (n-octane) or valeric esters is frequently attained by a multistep method, with a major reaction being the hydrogenation of LA (Figure 4.12). This process comprises two kinds of reactions that are electrochemical reduction and oxidative decarboxylation. Zhou et al. developed a Ni-modified cobalt phosphide electrocatalyst in KOH electrolyte for catalytic upcycling of PET to valuable commodity chemicals, such as potassium diformate and terephthalic acid, and hydrogen fuel (Figure 4.13). Preliminary analysis advocates the profitability of this process when ethylene glycol (EG) module of PET is electrooxidized to formate [107]. To conclude, the electrocatalytic transformation of biomass into value-added products represents an innovative

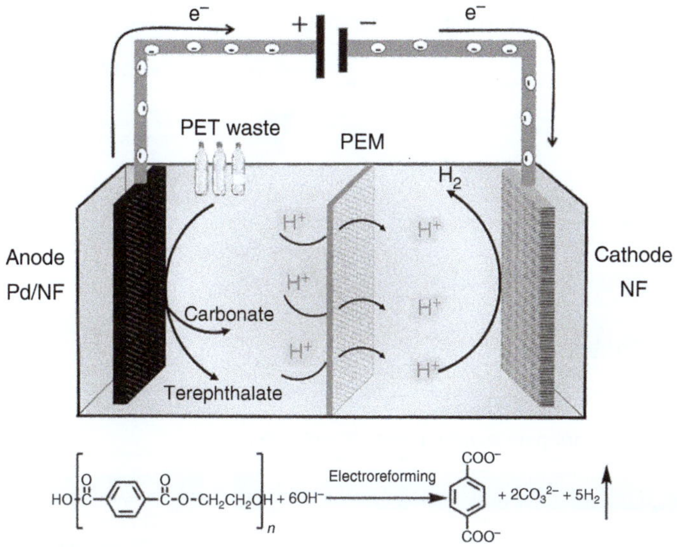

Figure 4.11 Schematic electroreforming of PET into high value-added products and hydrogen fuel. Source: Reproduced from Shi et al. [65]/with permission of Royal Society of Chemistry.

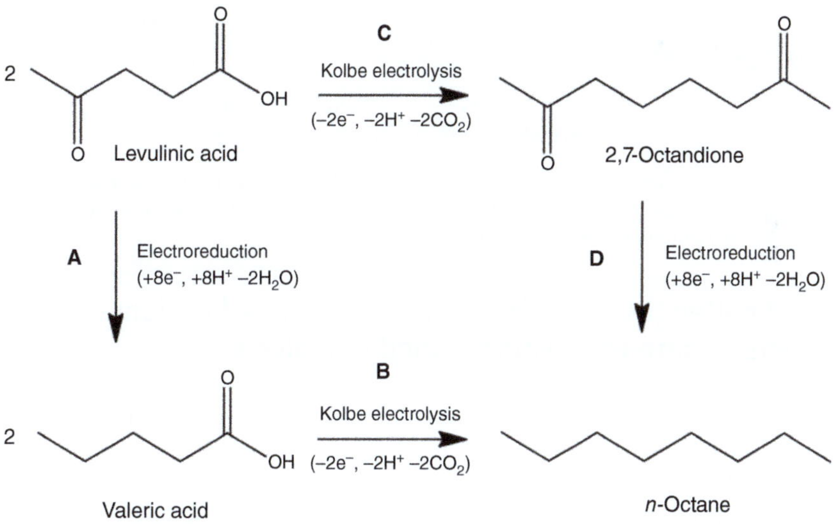

Figure 4.12 This figure shows two electrochemical routes for transformation of levulinic acid to *n*-octane.

approach toward sustainable chemistry and green energy. It offers the potential to transform renewable biomass resources into valuable products while reducing the environmental impact of chemical production. Ongoing research in this field aims to advance the efficiency, selectivity, and scalability of electrocatalytic methods for biomass conversion.

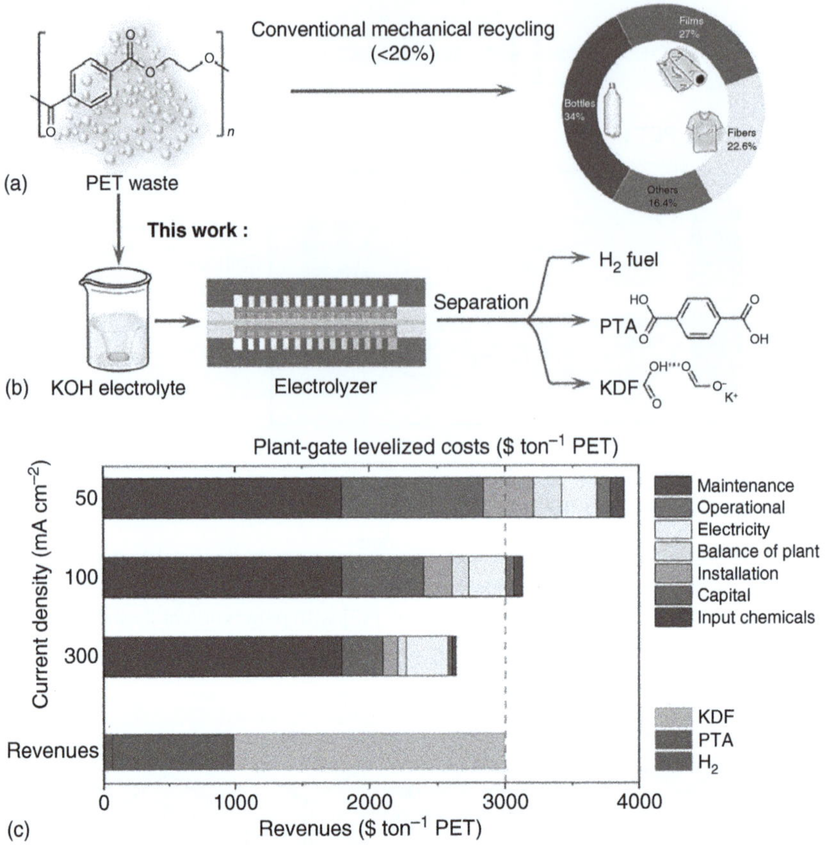

Figure 4.13 (a) Conventional mechanical route for PET recycling. (b) Electrocatalytic PET upcycling. (c) Technoeconomic analysis at diverse current density. Source: Reproduced from Zhou et al. [107]/Springer Nature/CC by 4.0.

4.7 Challenges in Design of Solid Catalysts for Biomass Conversion into High Value-Added Chemicals

Designing solid base catalysts for biomass transformation into value-added products is a complex and multidisciplinary task that involves several challenges (Figure 4.14) [108–111]. Biomass is an auspicious feedstock to yield renewable chemicals and fuels, but its conversion often requires the growth of efficient and selective catalysts. The key challenges in the design of solid catalysts for biomass transformation involve diverse biomass feedstock, complex reaction pathways, catalytic stability, product selectivity, catalyst design and synthesis, catalyst characterization, scale-up commercialization, environmental and economic sustainability, catalyst poisoning, catalyst recycling, and regulatory and safety considerations.

Although biomass combustion produces pollutants such as CO_2, SO_2 and NO_x, and dust, the quantity of these pollutants obtained from biomass combustion is far less than that secreted from combustion of fossil fuels [112–115]. Biomass utilization

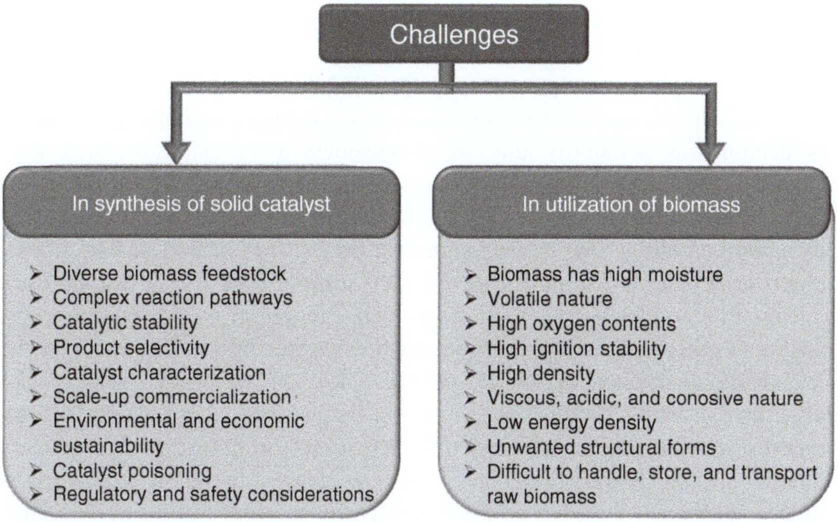

Figure 4.14 List of challenges in the designing of solid catalysts and utilization of biomass conversion into high value-added chemicals.

is mainly confronted by its low energy density and unwanted structural forms. It is difficult to handle, store, and transport raw biomass as compared to fossil fuels. The technologies for biomass conversion seek to elevate their properties for utilization in various fields [116]. Biomass-derived fuels have low heating points as compared to fossil fuels due to their high moisture, volatile, and oxygen contents. Along with this, they have high ignition stability, high density, and a viscous, acidic, and corrosive nature [117]. Biomass feedstock has been chosen as a source of energy based on its calorific value, the content of moisture, carbon, volatile matter, alkali metal, residual matter, and cellulose–lignin ratio.

Diverse biomass feedstock comprises a wide range of feedstock, including lignocellulosic materials, algae, and waste biomass. Each feedstock has a unique composition, making it challenging to design catalysts that can efficiently convert different types of biomass. Biomass conversion involves multiple chemical reactions, including depolymerization, dehydration, hydrogenation, and isomerization [118, 119]. Designing a catalyst that can selectively encourage desired reactions while minimizing undesirable side reactions is a significant challenge. Biomass conversion reactions often occur at high temperatures and under harsh reaction conditions. Catalysts must be stable over extended periods to maintain their activity and selectivity. Deactivation mechanisms, such as carbon deposition and sintering, need to be addressed. Achieving high selectivity for the desired value-added chemicals is crucial for the economic viability of biomass conversion processes. Catalyst design should minimize the construction of unwanted by-products and maximize the yield of target products. Developing catalysts with the desired properties requires a deep understanding of catalytic mechanisms and materials science [120, 121]. Catalysts must be tailored at the atomic or molecular level to optimize their performance. Accurate characterization of catalyst structures and active sites is essential for understanding their catalytic behavior. Techniques such as spectroscopy, microscopy, and surface analysis are critical but can be challenging to apply to complex heterogeneous catalysts [122]. Moving from laboratory to commercial-scale biomass conversion processes is a significant challenge. Catalysts must demonstrate scalability and cost-effectiveness to be viable for large-scale production. The design of solid base catalysts for biomass transformation should consider environmental sustainability, including minimizing waste and energy consumption. Additionally, the economics of catalyst production and its impact on the overall process economics are essential considerations. Biomass feedstock may contain impurities or contaminants that can poison catalysts over time. Strategies to mitigate catalyst poisoning, such as pretreatment of feedstock or the development of robust catalysts, are needed [123–127]. Developing methods for catalyst recovery and recycling is important for reducing cost and waste in biomass conversion processes. Catalysts used in biomass conversion must meet regulatory standards, and safety considerations related to catalyst handling and disposal should be addressed. Addressing these challenges requires collaboration among chemists, materials scientists, chemical engineers, and environmental scientists. In conclude, advances in catalyst design and development for biomass conversion will play a crucial role in the transition to more sustainable and eco-friendly chemical production processes [128].

4.8 Advantages of High Value-Added Chemicals

Biomass derived value-added products have advantages, such as the fact that they often demand higher prices in the market compared to their basic counterparts, which leads to increased revenue and profit for the manufacturers. Producing value-added chemicals allows manufacturers to diversify their product offerings, reducing dependence on a single product or market [129–131]. This can enhance business stability and resilience. Value-added chemicals typically have higher profit margins due to their enhanced functionality or unique properties. This can lead to improved profitability for chemical producers. These chemicals often offer superior performance compared to their basic counterparts. This can make them more attractive to customers, seeking improved product quality or efficiency. Manufacturers can tailor value-added chemicals to meet customers' specific requirements and enhance their satisfaction and loyalty. Companies that innovate and produce value-added chemicals gain a competitive edge in the marketplace. They can differentiate themselves from competitors who only offer basic chemicals. These chemicals can be developed using sustainable practices, such as utilizing renewable feedstock or reducing waste. Innovations in value-added chemical processes and formulations can be protected by patents, providing legal protection against imitation and encouraging continued investment in research and development. These chemicals with unique properties or benefits can have greater export potential, allowing companies to tap into international markets and expand their global reach. The development of value-added chemicals often involves collaboration with research institutions, suppliers, and customers. This can foster innovation and create valuable partnerships within the industry. Some value-added chemicals may be formulated to meet specific regulatory requirements, ensuring compliance with environmental, safety, or health standards. In summary, value-added chemicals offer numerous advantages, including increased market value, improved margins, enhanced performance, and a competitive edge. They enable chemical manufacturers to meet customer demands, explore new markets, and contribute to sustainable practices, while fostering innovation and collaboration within the industry.

4.9 Summary and Future Prospectus

Although most of the solid catalysts can be effortlessly recycled via filtration and centrifugation after the reaction, the effectual illustrations of fully recovered and recycled solid catalysts without losing any catalytic performance are still limited. Additionally, high temperature and pressure conditions for a chemical reaction, requirement of a high quantity of additives, and microwave irradiation provide less stability and ion leaching in solid catalysts. Subsequently, imminent goals in this field will entail innovative approaches for the design of solid catalysts to overcome these issues. Such perceptions will start a new path for the growth of heterogeneous catalysts that can comfortably work in mild situations for the conversion of biomass. These will include the challenging reactions for direct synthesis of value-added products with high yield from waste biomass and selectively transforming the lignocellulose into beneficial aromatic components. The properties of biomass feedstock can vary depending on their sources, including agriculture waste, energy crops, and organic residues. So, it is difficult to replace the biorefinery process with the petroleum refinery process in the present state. It is aimed at establishing biorefinery processes to mark a significant stride toward sustainable and environmentally friendly practices in the development of fuels, chemicals, and other valuable materials.

Acknowledgments

Dr. Amanpreet Kaur Jassal, assistant professor at Guru Nanak Dev University located in Amritsar, Punjab, India, is whole heartedly thankful to the university and especially the Department of Chemistry for providing top-notch infrastructure and cutting-edge facilities.

References

1 Panah, H.S. (2021). Gibbs free energy change using Ru/Al_2O_3 catalyst: an application in supercritical water gasification process. *International Journal of Hydrogen Energy* 46 (54): 27444–27458.

2 Wisniak, J. (2010). The history of catalysis—from the beginning to Nobel prizes. *Educación Química* 21 (1): 60–69.

3 Grubera, M., Wagnera, M., Heidenreicha, R. et al. (2004). Controlled distribution of solid catalysts into two immiscible liquid phases—synthesis and catalytic applications. *Catalysis Letters* 94 (3–4): 177–180.

4 Silva, B.A., Ribeiro, L.F.B., González, S.Y.G. et al. (2021). SiOC and SiCN-based ceramic supports for catalysts and photocatalysts. *Microporous and Mesoporous Materials* 327: 111435.

5 Hara, M., Nakajima, K., and Kamata, K. (2015). Recent progress in the development of solid catalysts for biomass conversion into high value-added chemicals. *Science and Technology of Advanced Materials* 16 (3): 034903.

6 Miceli, M., Frontera, P., Macario, A. et al. (2021). Recovery/reuse of heterogeneous supported spent catalysts. *Catalysts* 11 (5): 591.

7 Rinaldi, R. and Schüth, F. (2009). Design of solid catalysts for the conversion of biomass. *Energy & Environmental Science* 2: 610–626.

8 Storck, S., Bretinger, H., and Maier, W.F. (1998). Characterization of micro- and mesoporous solids by physisorption methods and pore-size analysis. *Applied Catalysis A: General* 174 (2): 137–146.

9 Singh, B.K., Kim, Y., Kwon, S. et al. (2021). Synthesis of mesoporous zeolites and their opportunities in heterogeneous catalysis. *Catalysts* 11: 1541.

10 Tomar, R., Kundra, P., Sharma, J. et al. (2024). An overview: synthesis of menthol using heterogeneous catalysis. *Letters in Organic Chemistry* 21 (1): 16–28. https://doi.org/10.2174/1570178620666230623114308.

11 Kumar, L., Verma, N., Tomar, R. et al. (2023). Development of bioactive 2-substituted benzimidazole derivatives using and MnO_x/HT nanocomposite catalyst. *Dalton Transactions* 52 (10): 3006–3015. https://doi.org/10.1039/d2dt02923e.

12 Tomar, R., Jain, S., Yadav, P. et al. (2022). Conversion of limonene over heterogeneous catalysis: an overview. *Current Organic Synthesis* 19 (3): 414–425.

13 Kumar, L., Verma, N., Sehrawat, H. et al. (2022). Successive oxidation–condensation reactions using a multifunctional gold-supported nanocomposite (Au/MgCe–HDO). *New Journal of Chemistry* 46 (7): 3472–3481.

14 Tomar, R., Singh, N., Kumar, N. et al. (2019). Base-free Suzuki–Miyaura coupling reaction using Palladium (II) supported catalyst in water. *Catalysis Letters* 149 (6): 1589–1594.

15 Tomar, R., Singh, N., Rathee, G. et al. (2017). Synthesis and characterization of hybrid $Mg(OH)_2$ and $CeCO_3OH$ composite with improved activity towards henry reaction. *Asian Journal of Organic Chemistry* 6 (12): 1728–1732.

16 Tomar, R., Sharma, J., Nishimura, S., and Ebitani, K. (2016). *Chemistry Letters* 45 (7): 843–845.

17 Arya, K. and Tomar, R. (2013). Microporous zeolite catalyst system: an eco-approach for regioselective synthesis of pyrimidobenzimidazoles. *Research on Chemical Intermediates* 41 (6): 3389–3400.

18 Tomar, R., Ebitani, K., and Chandra, R. (2019). Hydrotalcite-supported ceria nanoparticles as a heterogeneous catalyst for one–pot synthesis of imines under atmospheric air. *ChemistrySelect* 4 (12): 3577–3581.

19 Tomar, R., Rathee, G., Chandra, I. et al. (2018). Synthesis and characterization of magnesium hydroxide & cerium oxide composite: application in organic transformation. *ChemistrySelect* 3 (6): 1645–1649.

20 Ralphs, K., Hardacre, C., and James, S.L. (2013). Application of heterogeneous catalysts prepared by mechanochemical synthesis. *Chemical Society Reviews* 42: 7701–7718.

21 Centi, G. and Perathoner, S. (2005). Use of solid catalysts in promoting water treatment and remediation technologies. *ChemInform* 18 (18): 46–71.

22 Muh, E., Tabet, F., and Amara, S. (2020). Biomass conversion to fuels and value-added chemicals: a comprehensive review of the thermochemical processes. *Current Alternative Energy* 4: 1–23.

23 McKendry, P. (2002). Energy production from biomass (part 1): overview of biomass. *Bioresource Technology* 83: 37–46.

24 Soffian, M.S., Abdul Halim, F.Z., Aziz, F. et al. (2022). Carbon-based material derived from biomass waste for wastewater treatment. *Environmental Advances* 9: 100259.

25 Williams, C.L., Emerson, R.M., and Tumuluru, J.S. (2017). Biomass compositional analysis for conversion to renewable fuels and chemicals. In: *Biomass Volume Estimation and Valorization for Energy* (ed. J.S. Tumuluru). InTech.

26 Meraldo, A. (2016). Introduction to bio-based polymers. In: *Multilayer Flexible Packaging, Science Direct*, 2e, 47–52. Academic Press.

27 Hu, X., Mourant, D., Gunawan, R. et al. (2012). Production of value-added chemicals from bio-oil via acid catalysis coupled with liquid-liquid extraction. *RSC Advances* 2: 9366–9370.

28 Bentaharl, N., Khelassi, S., and Abdelrazek, F.M. (2017). Preparation of solid catalysts and its application for conversion of heavy petroleum fractions. *Archives of Petroleum & Environmental Biotechnology* 106: 1–17.

29 Mabate, T.P., Maqunga, N.P., Ntshibongo, S. et al. (2023). Metal oxides and their roles in heterogeneous catalysis: special emphasis on synthesis protocols, intrinsic properties, and their influence in transfer hydrogenation reactions. *SN Applied Sciences* 5: 196.

30 Védrine, J.C. (2017). Heterogeneous catalysis on metal oxides. *Catalysts* 7: 341.

31 Schrickel, J. (2013). *Sulfonated Polymers: Simple Chemistry for High Tech Materials & Applications*. Speciality Chemicals Magazine.

32 Hagemeyer, A. and Volpe, A. (2016). Catalysts: combinatorial catalysis. In: *Reference Module in Materials Science and Materials Engineering*. Elsevier.

33 Carrettin, S., McMorn, P., Johnston, P. et al. (2003). Oxidation of glycerol using supported Pt, Pd and Au catalysts. *Physical Chemistry Chemical Physics* 5: 1329–1336.

34 Schwanke, A.J., Balzer, R., and Pergher, S. (2017). Microporous and mesoporous materials from natural and inexpensive sources. In: *Handbook of Ecomaterials*. Cham: Springer.

35 Ekpe, I. (2017). Zeolite synthesis, characterisation and application areas: a review. *International Research Journal of Environmental Sciences* 6 (10): 45–59.

36 Bingre, R., Louis, B., and Nguyen, P. (2018). An overview on zeolite shaping technology and solutions to overcome diffusion limitations. *Catalysts* 8 (4): 163.

37 Yilmaz, B. and Müller, U. (2009). Catalytic applications of zeolites in chemical industry. *Topics in Catalysis, Springer Nature* 52: 888–895.

38 González-Cortés, S.L. and Imbert, F.E. (2013). Fundamentals, properties and applications of solid catalysts prepared by solution combustion synthesis (SCS). *Applied Catalysis A: General* 452: 117–131.

39 Cesário, M.R., Barros, B., Aouad, S. et al. (2022). Chapter 1—Understanding heterogeneous catalysis: a brief study on performance parameters. In: *Heterogeneous Catalysis*, 1–18. Elsevier.

40 Zhang, Y. (2020). Preparation of heterogeneous catalysts based on CWAO technology. *Journal of Physics: Conference Series, IOP Publishing* 1549: 032052.

41 Deraz, N.M. (2018). The comparative jurisprudence of catalysts preparation methods: I. Precipitation and impregnation methods. *Journal of Industrial Environmental Chemistry* 2 (1): 19–21.

42 Shimura, K., Yoshida, S., Oikawa, H. et al. (2022). Preparation of NiO_x/SiO_2–Al_2O_3 catalysts by a homogenous precipitation method and their catalytic activity for ethylene oligomerization. *Microporous and Mesoporous Materials* 338: 111955.

43 Wu, D., Shen, X., Zhou, L.Q. et al. (2020). A vacuum impregnation method for synthesizing octahedral Pt_2CuNi nanoparticles on mesoporous carbon support and the oxygen reduction reaction electrocatalytic properties. *Journal of Colloid and Interface Science* 564: 245–253.

44 Chen, H., Wydra, J., Zhang, X. et al. (2011). Hydrothermal synthesis of zeolites with three-dimensionally ordered mesoporous-imprinted structure. *Journal of American Chemical Society* 133: 12390–12393.

45 Cubillas, P. and Anderson, M.W. (2010). Synthesis mechanism: crystal growth and nucleation. *Zeolites and Catalysis: Synthesis, Reactions and Applications* 1: 1–55.

46 Corma, A., Iborra, S., and Velty, A. (2007). Chemical routes for the transformation of biomass into chemicals. *Chemical Reviews* 107 (6): 2411–2502.

47 Kamm, B., Gruber, P.R., and Kamm, M. (2006). Biorefineries—industrial process and products: status quo and future directions. *Chemistry International-Newsmagazine for IUPAC* 29 (3): 25.

48 Boerjan, W., Ralph, J., and Baucher, M. (2003). Lignin biosynthesis. *Annual Review of Plant Biology* 54: 519–546.

49 Chen, H. and Wang, L. (2016). *Technologies for Biochemical Conversion of Biomass*. Elsevier Inc.

50 García-Velásquez, C.A. and Cardona, C.A. (2019). Comparison of the biochemical and thermochemical routes for bioenergy production: a techno-economic (TEA), energetic and environmental assessment. *Energy* 172: 232–242.

51 Tekin, K., Karagöz, S., and Bekta, S. (2014). A review of hydrothermal biomass processing. *Renewable & Sustainable Energy Reviews* 40: 673–687.

52 Hayes, D.J.M. (2013). Biomass composition and its relevance to biorefining. In: *The Role of Catalysis for the Sustainable Production of Biofuels and*

Bio-chemicals (ed. K.S. Triantafyllidis, A.A. Lappas, and S. Michael), 27–65. Elsevier.

53 Demirbas, A. (2010). Biorefineries, for biomass upgrading facilities. In: *Green Energy and Technology*, 1–240. Springer.

54 Ragauskas, A.J., Williams, C.K., Davison, B.H. et al. (2006). The path forward for biofuels and biomaterials. *Science* 311 (5760): 484–489.

55 Vassilev, S.V., Baxter, D., Andersen, L.K. et al. (2010). An overview of the chemical composition of biomass. *Fuel* 89 (5): 913–933.

56 Zugenmaier, P. (2008). *Crystalline Cellulose and Derivatives: Characterization and Structures*. Berlin, Germany: Springer-Verlag.

57 Spivey, J. and Egbebi, A. (2007). Heterogeneous catalytic synthesis of ethanol from biomass-derived syngas. *Chemical Society Reviews* 36: 1514–1528.

58 Leung, D.Y.C., Caramanna, G., and Maroto-Valer, M.M. (2014). An overview of current status of carbon dioxide capture, and storage technologies. *Renewable and Sustainable Energy Reviews* 39: 426–443.

59 Yue, H., Ma, H., and Gong, J. (2014). An alternative synthetic approach for efficient catalytic conversion of syngas to ethanol. *Accounts Chemistry Resources* 47 (5): 1483–1492.

60 Sheldon, R.A. (2014). Green and sustainable manufacture of chemicals from biomass: state of the art. *Green Chemistry* 16: 950–963.

61 van Putten, R.-J., van der Waal, J.C., de Jong, E. et al. (2013). Hydroxymethylfurfural, a versatile platform chemical made from renewable resources. *Chemical Reviews* 113 (3): 1499–1597.

62 Geboers, J.A., Van de Vyver, S., Ooms, R. et al. (2011). Chemocatalytic conversion of cellulose: opportunities, advances and pitfalls. *Catalysis Science & Technology* 1: 714–726.

63 Bhosale, S.H., Rao, M.B., and Deshpande, V.V. (1996). Molecular and industrial aspects of glucose isomerase. *Microbiological Reviews* 60 (2): 280–300.

64 Tewari, Y. (1990). Thermodynamics of industrially-important, enzyme-catalyzed reactions. *Applied Biochemistry and Biotechnology* 23 (3): 187–203.

65 Shi, R., Liu, K.-S., Liu, F. et al. (2021). Electrocatalytic reforming of waste plastics into high value-added chemicals and hydrogen fuel. *Chemical Communication* 57 (94): 12595–12598.

66 Gouda, L., Sévery, L., Moehl, T. et al. (2021). Tuning the selectivity of biomass oxidation over oxygen evolution on NiO–OH electrodes. *Green Chemistry* 23: 8061–8068.

67 Movil-Cabrera, O., Rodriguez-Silva, A., Arroyo-Torres, C. et al. (2016). Electrochemical conversion of lignin to useful chemicals. *Biomass and Bioenergy* 88: 89–96.

68 Velvizhi, G., Balakumar, K., Shetti, N.P. et al. (2022). Integrated biorefinery processes for conversion of lignocellulosic biomass to value added materials: paving a path towards circular economy. *Bioresource Technology* 343: 126151.

69 Deng, W., Feng, Y., Fu, J. et al. (2023). Catalytic conversion of lignocellulosic biomass into chemicals and fuels. *Green Energy & Environment* 8: 10–114.

70 Xu, H., Che, X., Ding, Y. et al. (2019). Effect of crystallinity on pretreatment and enzymatic hydrolysis of lignocellulosic biomass based on multivariate analysis. *Bioresource Technology* 279: 271–280.

71 Putro, J.N., Soetaredjo, F.E., Lin, S.-Y. et al. (2016). Pretreatment and conversion of lignocellulose biomass into valuable chemicals. *RSC Advances* 6: 46834–46852.

72 Flores, E.M.M., Cravotto, G., Bizzi, C.A. et al. (2021). Ultrasound-assisted biomass valorization to industrial interesting products: state-of-the-art, perspectives and challenges. *Ultrasonics Sonochemistry* 72: 105455.

73 Watkins, D., Nuruddin, M., Hosur, M. et al. (2015). Extraction and characterization of lignin from different biomass resources. *Journal of Materials Research and Technology* 4: 26–32.

74 Bharadwaj, A.V.S.L.S., Dev, S., Zhuang, J. et al. (2023). Review of chemical pretreatment of lignocellulosic biomass using low-liquid and low-chemical catalysts for effective bioconversion. *Bioresource Technology* 368: 128339.

75 Kumar, A.K. and Sharma, S. (2017). Recent updates on different methods of pretreatment of lignocellulosic feedstocks: a review. *Bioresources and Bioprocessing* 4: 7.

76 Huang, Y.-B. and Fu, Y. (2013). Hydrolysis of cellulose to glucose by solid acid catalysts. *Green Chemistry* 15: 1095–1111.

77 Rosatella, A., Simeonov, S.P., Frade, R.F.M. et al. (2011). 5-Hydroxymethylfurfural (HMF) as a building block platform: biological properties, synthesis, and synthetic applications. *Green Chemistry* 13: 754–793.

78 Kim, J., Byun, J., Ahn, Y. et al. (2018). Catalytic production of gamma-valerolactone from two different feedstocks. *Computer Aided Chemical Engineering* 44: 295–300.

79 Roy, R., Rahman, M.S., Amit, T.A. et al. (2022). Recent advances in lignin depolymerization techniques: a comparative overview of traditional and greener approaches. *Biomass* 2: 130–154.

80 Ciuffi, B., Loppi, M., Rizzo, A.M. et al. (2021). Towards a better understanding of the HTL process of lignin-rich feedstock. *Scientific Reports* 11: 15504.

81 Kundu, S.K. and Bhaumik, A. (2015). Novel nitrogen and sulfur rich hyper-cross-linked microporous poly-triazine-thiophene copolymer for superior CO_2 capture. *ACS Sustainable Chemistry & Engineering* 3: 1715–1723.

82 Nishimura, S., Takagaki, A., and Ebitani, K. (2013). Characterization, synthesis, and catalysis of hydrotalcite-related materials for highly efficient materials transformations. *Green Chemistry* 15: 2026–2042.

83 Cao, X., Teong, S.P., Wu, D. et al. (2015). An enzyme mimic ammonium polymer as a single catalyst for glucose dehydration to 5-hydroxymethylfurfural. *Green Chemistry* 17: 2348–2352.

84 Mondal, S., Mondal, J., and Bhaumik, A. (2015). Sulfonated porous polymeric nanofibers as an efficient solid acid catalyst for the production of 5-hydroxymethylfurfural from biomass. *ChemCatChem* 7: 3570–3578.

85 Varma, R.S. (2016). Greener and sustainable trends in synthesis of organics and nanomaterials. *ACS Sustainable Chemistry & Engineering* 4: 5866–5878.

86 Verma, S., Baig, R.B.N., Nadagouda, M.N. et al. (2017). Sustainable pathway to furanics from biomass *via* heterogeneous organo-catalysis. *Green Chemistry* 19: 164–168.

87 Zhang, T., Li, W., Xu, Z. et al. (2016). Catalytic conversion of xylose and corn stalk into furfural over carbon solid acid catalyst in-valerolactone. *Bioresource Technology* 209: 108–114.

88 Qi, Z., Wang, Q., Liang, C. et al. (2020). Highly efficient conversion of xylose to furfural in a water–MIBK system catalyzed by magnetic carbon-based solid acid. *Industrial & Engineering Chemistry Research* 59: 17046–17056.

89 Wang, R., Liang, X., Shen, F. et al. (2020). Mechanochemical synthesis of sulfonated palygorskite solid acid catalysts for selective catalytic conversion of xylose to furfural. *ACS Sustainable Chemistry & Engineering* 8: 1163–1170.

90 Hoang, P.H. and Cuong, T.D. (2021). Simultaneous direct production of 5-hydroxymethylfurfural (HMF) and furfural from corncob biomass using porous HSO_3-ZSM-5 zeolite catalyst. *Energy Fuels* 35: 546–551.

91 Gupta, N.K., Fukuoka, A., and Nakajima, K. (2017). Amorphous Nb_2O_5 as a selective and reusable catalyst for furfural production from xylose in biphasic water and toluene. *ACS Catalysis* 7: 2430–2436.

92 Guo, X., Guo, F., Li, Y. et al. (2018). Dehydration of D-xylose into furfural over bimetallic salts of heteropolyacid in DMSO/H_2O mixture. *Applied Catalysis A: General* 558: 18–25.

93 Theerthagiri, J., Karuppasamy, K., Park, J. et al. (2023). Electrochemical conversion of biomass-derived aldehydes into fine chemicals and hydrogen: a review. *Environmental Chemistry Letters* 21: 1555–1583.

94 Biddinger, E.J., Gutierrez, O.Y., and Holladay, J. (2021). Electrochemical routes for biomass conversion. *Journal of Applied Electrochemistry* 51: 1–3.

95 Dolle, C., Neha, N., and Coutanceau, C. (2022). Electrochemical hydrogen production from biomass. *Current Opinion in Electrochemistry* 31: 100841.

96 Lai, Z.I., Lee, L.Q., and Li, H. (2021). Electroreforming of biomass for value-added products. *Micromachines* 12: 1405.

97 Sugano, Y., Vestergaard, M., Yoshikawa, H. et al. (2010). Direct electrochemical oxidation of cellulose: a cellulose-based fuel cell system. *Electroanalysis* 22: 1688–1694.

98 Bin, D., Wang, H., Li, J. et al. (2014). Controllable oxidation of glucose to gluconic acid and glucaric acid using an electrocatalytic reactor. *Electrochimica Acta* 130: 170–178.

99 Liu, W.J., Xu, Z., Zhao, D. et al. (2020). Efficient electrochemical production of glucaric acid and H_2 via glucose electrolysis. *Nature Communications* 11: 1–11.

100 Neha, N., Kouamé, B.S.R., Rafaïdeen, T. et al. (2021). Remarkably efficient carbon-supported nanostructured platinum-bismuth catalysts for the selective electrooxidation of glucose and methyl-glucoside. *Electrocatalysis* 12: 1–14.

101 Rao, J.R., Richter, G.J., von Sturm, F. et al. (1976). The performance of glucose electrodes and the characteristics of different biofuel cell constructions. *Bioelectrochemistry and Bioenergetics* 3: 139–150.

102 Liu, W.J., Dang, L., Xu, Z. et al. (2018). Electrochemical oxidation of 5-hydroxymethylfurfural with NiFe layered double hydroxide (LDH) nanosheet catalysts. *ACS Catalysis* 8: 5533–5541.

103 Weidner, J., Barwe, S., Sliozberg, K. et al. (2018). Cobalt-metalloid alloys for electrochemical oxidation of 5-hydroxymethylfurfural as an alternative anode reaction in lieu of oxygen evolution during water splitting. *Beilstein Journal of Organic Chemistry* 14: 1436–1445.

104 Nam, D.H., Taitt, B.J., and Choi, K.S. (2018). Copper-based catalytic anodes to produce 2,5-furandicarboxylic acid, a biomass-derived alternative to terephthalic acid. *ACS Catalysis* 8: 1197–1206.

105 Cai, M., Ding, S., Gibbons, B. et al. (2020). Nickel(II)-modified covalent-organic framework film for electrocatalytic oxidation of 5-hydroxymethylfurfural (HMF). *Chemical Communications* 56: 14361–14364.

106 Nilges, P., Santos, T.R.D., Harnisch, F. et al. (2012). Electrochemistry for biofuel generation: electrochemical conversion of levulinic acid to octane. *Energy & Environmental Science* 5: 5231–5235.

107 Zhou, H., Ren, Y., Li, Z. et al. (2021). Electrocatalytic upcycling of polyethylene terephthalate to commodity chemicals and H_2 fuel. *Nature Communications* 12: 4679.

108 Mukhtar, A., Saqib, S., Lin, H. et al. (2022). Current status and challenges in the heterogeneous catalysis for biodiesel production. *Renewable and Sustainable Energy Reviews* 157: 112012.

109 Zaera, F. (2012). New challenges in heterogeneous catalysis for the 21st century. *Catalysis Letters* 142: 501–516.

110 Liu, L. and Corma, A. (2021). Structural transformations of solid electrocatalysts and photocatalysts. *Nature Reviews Chemistry* 5: 256–276.

111 Kodama, K., Nagai, T., Kuwaki, A. et al. (2021). Challenges in applying highly active Pt-based nanostructured catalysts for oxygen reduction reactions to fuel cell vehicles. *Nature Nanotechnology* 16: 140–147.

112 Demirbas, A. (2005). Potential applications of renewable energy sources, biomass combustion problems in boiler power systems and combustion related environmental issues. *Progress in Energy and Combustion Science* 31 (2): 171–192.

113 Khan, A.A., Jong, W.D., Jansens, P.J. et al. (2009). Biomass combustion in fluidized bed boilers: potential problems and remedies. *Fuel Processing Technology* 90 (1): 21–50.

114 Fouilland, T., Grace, J.R., and Ellis, N. (2010). Recent advances in fluidized bed technology in biomass processes. *Biofuels* 1 (3): 409–433.

115 Jia, Y., Li, Z., Wang, Y. et al. (2021). Visualization of combustion phases of biomass particles: effects of fuel properties. *ACS Omega* 6 (42): 27702–27710.

116 Baruah, D. and Baruah, D.C. (2014). Modeling of biomass gasification: a review. *Renewable and Sustainable Energy Reviews* 39: 806–815.

117 Zhang, L., Xu, C., and Champagne, P. (2010). Overview of recent advances in thermo-chemical conversion of biomass. *Energy Conversion and Management* 51 (5): 969–982.

118 Kumar, M., Oyedun, A.O., and Kumar, A. (2018). A review on the current status of various hydrothermal technologies on biomass feedstock. *Renewable and Sustainable Energy Review* 81: 1742–1770.

119 Gollakota, A.R.K., Kishore, N., and Gu, S. (2018). A review on hydrothermal liquefaction of biomass. *Renewable and Sustainable Energy Review* 81: 1378–1392.

120 Chen, Y.-Y., Kunz, M.R., He, X. et al. (2022). Recent progress toward catalyst properties, performance, and prediction with data-driven methods. *Current Opinion in Chemical Engineering* 37: 100843.

121 Xia, Y., Campbell, C.T., Cuenya, B.R. et al. (2021). Introduction: advanced materials and methods for catalysis and electrocatalysis by transition metals. *Chemical Reviews* 121 (2): 563–566.

122 Pujals, S., Feiner-Gracia, N., Delcanale, P. et al. (2019). Super-resolution microscopy as a powerful tool to study complex synthetic materials. *Nature Reviews. Chemistry* 3: 68–84.

123 Cimino, S. and Lisi, L. (2019). Catalyst deactivation poisoning and regeneration. *Catalysts* 9 (8): 668.

124 Lange, J.P. (2015). Renewable feedstocks: the problem of catalyst deactivation and its mitigation. *Angewandte Chemie International Edition* 54: 13187–13197.

125 Kim, S., Tsang, Y.F., Kwon, E.E. et al. (2019). Recently developed methods to enhance stability of heterogeneous catalysts for conversion of biomass-derived feedstocks. *Korean Journal of Chemical Engineering* 36: 1–11.

126 Hu, J., Galvita, V.V., Poelman, H. et al. (2018). Advanced chemical looping materials for CO_2 utilization: a review. *Materials* 11 (7): 1187.

127 Dou, B., Zhang, H., Song, Y. et al. (2019). Hydrogen production from the thermochemical conversion of biomass: issues and challenges. *Sustainable Energy & Fuels* 3: 314–342.

128 Lin, L., Han, X., Han, B. et al. (2021). Emerging heterogeneous catalysts for biomass conversion: studies of the reaction mechanism. *Chemical Society Reviews* 50: 11270–11292.

129 Sodhi, A.S., Sharma, N., Bhatia, S. et al. (2022). Insights on sustainable approaches for production and applications of value added products. *Chemosphere* 286 (1): 131623.

130 Kover, A., Kraljić, D., Marinaro, R. et al. (2022). Processes for the valorization of food and agricultural wastes to value-added products: recent practices and perspectives. *Systems Microbiology and Biomanufacturing* 2: 50–66.

131 Jõgi, K. and Bhat, R. (2020). Valorization of food processing wastes and by-products for bioplastic production. *Sustainable Chemistry and Pharmacy* 18: 100326.

5

Applications of Solid Basic Catalysts for Organic Synthesis

Aditi Tiwari[1], *Anirudh Singh Bhathiwal*[2], *and Anjaneyulu Bendi*[3,*]

[1] *Eurofins Analytical Services, Seetharampalya, Hoodi, Bengaluru – 560048, Karnataka, India*
[2] *Department of Chemistry, MMV, Banaras Hindu University, Varanasi, UP, India*
[3] *Presidency University, Department of Chemistry, Rajanukunte, Itgalpura, Bangalore 560064, India*

5.1 Introduction

A catalyst is defined as a chemical compound that is used to expedite or delay the reaction rate without being utilized in the chemical reaction. A catalyst functions by changing the activation energy within a reaction without influencing the thermodynamic equilibrium of the chemical reaction. A catalyst is used for only those reactions that have negative Gibbs free energy change. A catalyst is capable of increasing the completion rate of a chemical reaction even being available in small quantities. Catalysts used in different organic reactions can be categorized as either solid-acid catalysts (SAC) or solid-base catalysts (SBCs). SBCs are first pretreated at elevated temperatures, which results in the elimination of CO_2, O_2, H_2O, etc., from the superficies to furnish basic traits and restorative base-catalyzed chemical reactions [1–11]. Homogenous solid catalysts possess high activity along with high selectivity but with the least stability over heterogeneous solid catalysts. Homogenous solid catalysts are difficult to reuse because they cannot be easily separated, while heterogenous solid catalysts are much easier to recycle [12]. Heterogenous solid-based catalysts are vital to modern chemical industries. The traditional methods for liquid-based catalysts are normally associated with complications such as being highly toxic, corrosion, catalytic wastage, hassle in separation of products, and producing other desired products. SBCs can address these issues by providing various benefits, as outlined in the following:

(a) Ability to reutilize the catalyst with good ease of product separation.
(b) Environmentally friendly as being noncorrosive, safe, and producing less wastage.
(c) Due to its bifunctional nature, the SBC provides coexistence of basic and acidic positions on the exterior, which is not possible with liquid-based catalyst.
(d) Highly advantageous and effective when used in flow reactor systems.

*Corresponding Author.

Solid Base Catalysts: Synthesis, Characterization, and Applications, First Edition.
Edited by Ravi Tomar, K.K. Pant, and Ramesh Chandra.

(e) Removal or reduction of waste produced by nullification of reaction mixtures.
(f) The possibility of proceeding with reactions at high temperatures is generally not possible with liquid base catalysts as they may evaporate if the system is not performed below their boiling points.
(g) Uniqueness for carrying out novel reactions via heterolytic splitting of H_2, which is not possible with liquid base catalysts.

The exploration of solid-based catalysts commenced in 1958 through a sequence of analyses carried out by Pines and Haag [13]. Their investigations outlined the use of alkali metal-supported alumina as a catalyst isomerization of double bonds within an alkene which is progressed through the formation of carbanions as intermediates.

The following classification summarizes the different types of SBCs used for organic synthesis [14]:

(1) Metal oxides.
    (a) Metal oxides containing single component
    (b) Metal oxides containing double component
(2) Non-oxides
(3) Zeolites
(4) Metal compounds with supported alkali compounds
(5) Mineral clay

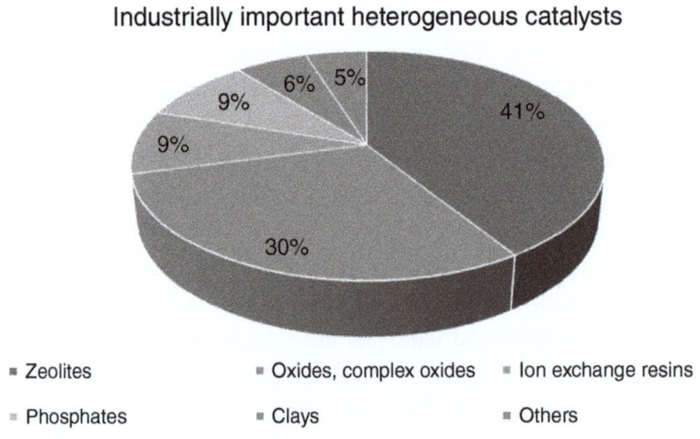

Characteristics of a catalyst for being an SBC are mentioned as follows [14]:

(1) **SBCs have basic sites on their surfaces**: The alteration in the color of acid or base indicators, acidic molecules being adsorbed, and various spectroscopic analyses suggest that there is the existence of basic positions on the exterior of a catalyst.
(2) **SBCs have catalytic activities such as the amount or strength of basic sites**: The active sites of SBCs are incorporated by acidic molecules such as carbon dioxide, water, and hydrochloric acid.

(3) Base-catalyzed reactions are widely known in homogeneous systems and involved in organic reactions such as isomerization, Tishchenko, aldol addition, Michael addition, and alkylation reaction.
(4) Spectroscopic observations for the exterior sites of base catalysts indicate the formation of anionic intermediates proceeded by SBCs.

5.2 Solid-Based Catalyst for Organic Synthesis

5.2.1 Metal Oxides

Among all other different types of solid-based catalysts, metal oxides perform one of the most prominent heterogeneous catalysis as it consists of the majority of industrial processes. The use of metal oxides as catalysts began in the mid-twentieth century when they were reported to effectively catalyze oxidation, acid–base reactions, and various other reactions [15]. Metal oxides as catalysts are widely used in industrial processes such as petrochemicals, intermediate synthesis, fine and pharmaceutical chemicals, biomass transformation using C–O hydrogenolysis and hydrodeoxygenation, and various synthetic conversions [16–24]. Metal oxides containing a single component can crystallize at room temperature with different frameworks such as being isotropic, anisotropic, or amorphous in nature. Examples of single-atom catalysts are Fe_2O_3, ZnO, WO_3, CuO, Co_3O_4, Al_2O_3, CeO_2, MgO, etc. [25]. Among metal oxide catalysts, catalysts associated with transition metals occupy a significant role in industrial applications due to their cheap production, being easily renewable, and selective activity and hence are widely used in a variety of organic reactions such as isomerization, dehydration, dehydrogenation, and oxidation. Different types of metal oxide catalysts include:

Metal oxides with transition metals are astronomically used as heterogenous catalysts due to their high stability administered by the coherence of the metal oxidation states [26]. The commonly used transition metals as catalysts in heterogenous catalysis include lead (Pb), manganese (Mn), vanadium (V), titanium (Ti), chromium (Cr), nickel (Ni), zinc (Zn), iron (Fe), cobalt (Co), and aluminum (Al). The unstable oxides cannot form firm bulk oxides under normal temperatures and thus are used as nanoparticles, which include metals such as ruthenium (Ru), gold (Au), platinum (Pt), silver (Ag), and palladium (Pd) oxides [27]. Single-component metal oxides constitute only one cation and are hence also referred to as binary oxides. The simplest example of binary oxides is a transition metal bonded with oxygen atoms to constitute an oxide compound and are used as catalysts for reduction, oxidation, and ammoxidation reactions [28]. Binary oxides with different proportions of cations belonging to the same atom and bearing dissimilar oxidation states in their structures are known as spinel oxides [29]. Single metal oxides have positive aspects such as varied valence electronic states, flexibility in redox reactions, being inexpensive, readily available, and economically friendly [30]. Despite various applications and admirable features of the oxide catalysts, single metal oxides have drawbacks such as low catalytic activity. Another drawback is that metal oxides in their nonporous states only react with atoms or molecules at their outer surface using their interior atoms [30].

Metal oxides when mixed with two or more varieties of cationic metals form mixed metal oxides (MMO). They are of greater notable use in scholastic and industrial use due to their acidic–basic and redox properties. Various utilization of MMO as catalysts in organic synthesis is shown in the following [31] (Schemes 5.1–5.4).

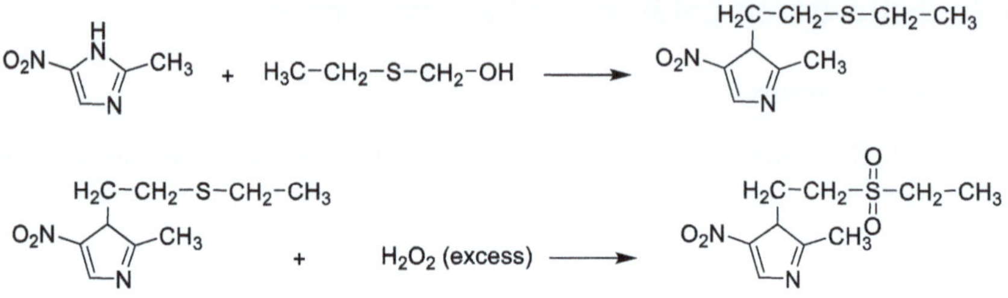

Scheme 5.1 Production of pharmaceutically active tinidazole using MoO_3/SiO_2. Source: Adapted from Hong et al. [14].

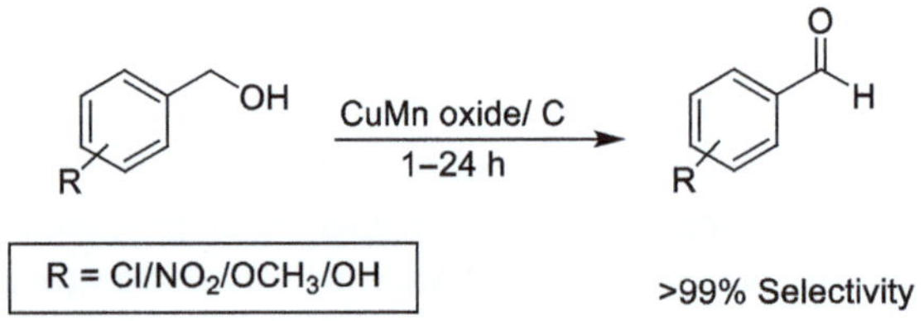

Scheme 5.2 Selective oxidation of alcohols using MMO.

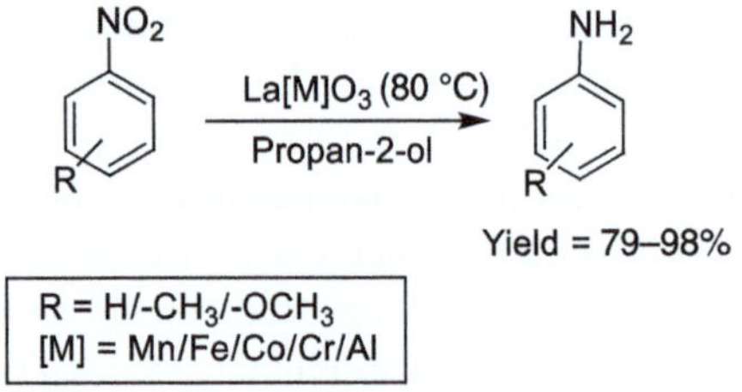

Scheme 5.3 $La[M]O_3$ perovskites for reducing reactions.

MMO also find their use in green chemistry as well for generating complex targets in an exceptionally short lifespan. The use of MMO in green chemistry includes conversions such as the following [31] (Schemes 5.5–5.8):

Yinyun Lu et al. used a porous nanocage catalyst ($Co_{50}Zn_{50}$-350) built from the mixture of Co_3O_4 and ZnO for the production of glycerol carbonate (GLC). The use of a rhombic dodecahedral Co_3O_4/ZnO nanocrystal catalyst for the synthesis has

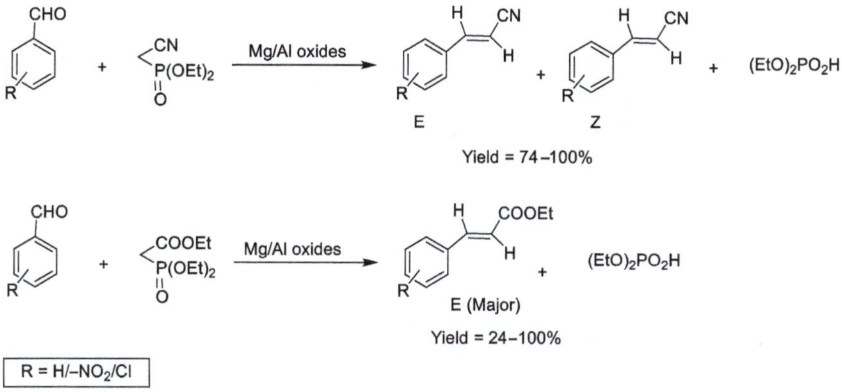

Scheme 5.4 Reaction of aldehydes with phosphonates (Wadsworth–Emmons).

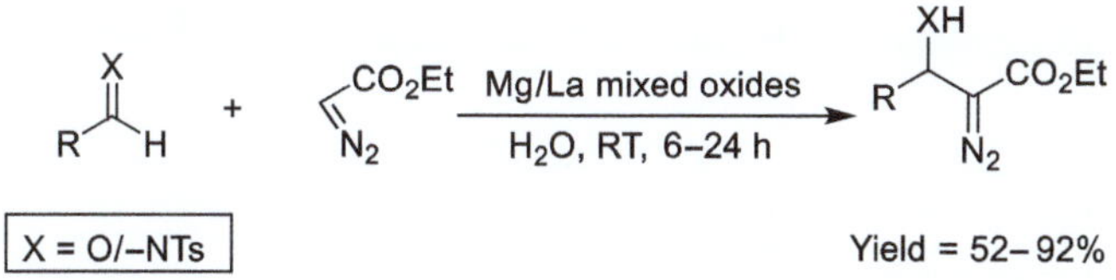

Scheme 5.5 (a) Schematic approach for β-α-diazo carbonyl compounds. (b) Condensation of aldehydes following Knoevenagel condensation

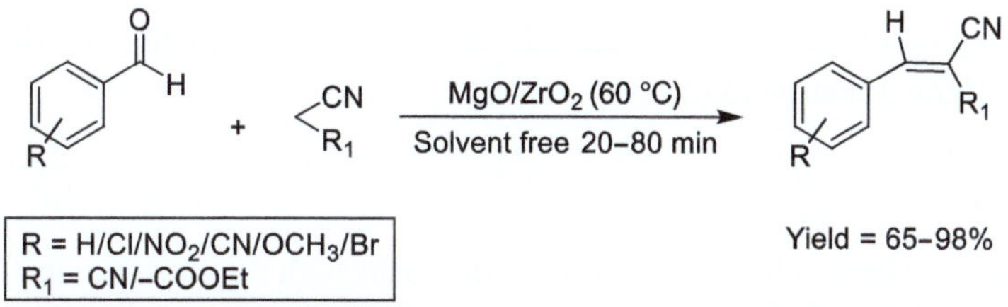

Scheme 5.6 MgO/ZrO_2 for Knoevenagel condensation of aldehydes.

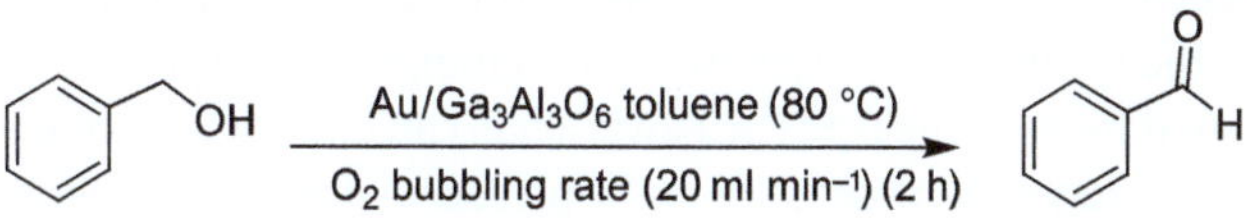

Scheme 5.7 Use of MMO for selectively oxidizing benzyl alcohol to benzaldehyde.

improved the yield to 85.2%, and conversion to 93.3% and approximately 91% GLC selectivity [32] (Scheme 5.9).

Tongdong Shen et al. synthesized magnesium-doped ZnO for the catalytic ozonation of isoniazid. The use of MgZnO catalyst when used in organic synthesis has improved the efficiencies of isoniazid which was not achieved by pure ZnO. The MgZnO also possessed stronger basicity which further showed enhanced isoniazid and total organic carbon removal efficiencies with greater stability.

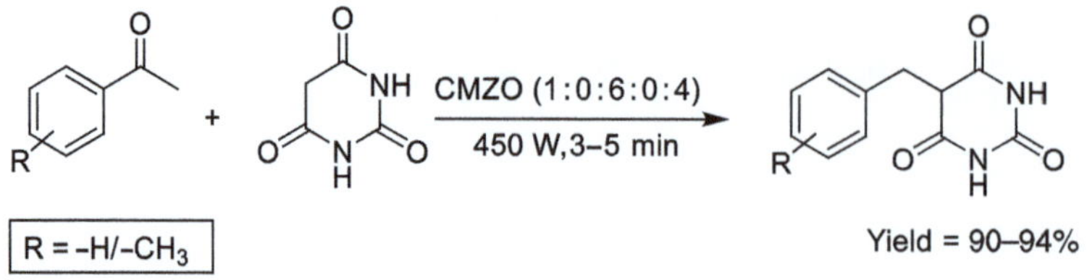

Scheme 5.8 Use of CMZO MMO for the synthesis of barbituric acid.

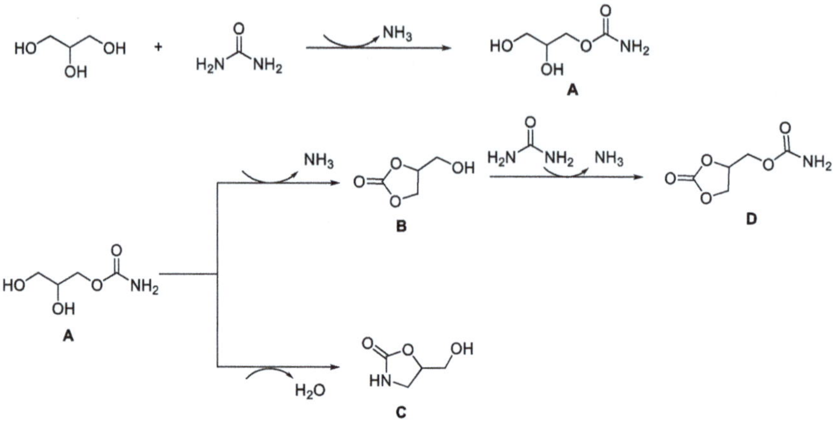

Scheme 5.9 **(2)**: The general route for producing GLC.

Scheme 5.10 represents the reaction cascade of isoniazid elimination in the presence of an MgZnO catalyst [33].

5.2.2 Zeolites

The utilization of zeolites in industries began in the 1960s with the substitution of catalysts established on indefinite aluminosilicates [34–37]. Zeolites are inorganic cation exchangers. Zeolites have many advantages such as the following:

(a) The Brϕnsted and Lewis acid centers in zeolite skeleton are altered in a guided manner throughout the synthesis by respective ion exchangers [38–49].
(b) The reactions catalyzed by protonic acids or Lewis acids can change a homogenous procedure to a heterogenous procedure in the presence of zeolites, making it easy for separation and disposal of the catalyst. The powerful effect of zeolite is that it offers a selection for the shape of zeolite over the composition of the product.
(c) Zeolites are capable of increasing the activity of the catalyst as well as the stability of the catalyst toward high temperatures by acting as carriers for active components.
(d) Zeolites, being thermally stable, offer an advantage to be utilized in reactions with thermodynamic equilibrium.
(e) Zeolite catalysts can also be reproduced in air by simply flaming the coke.

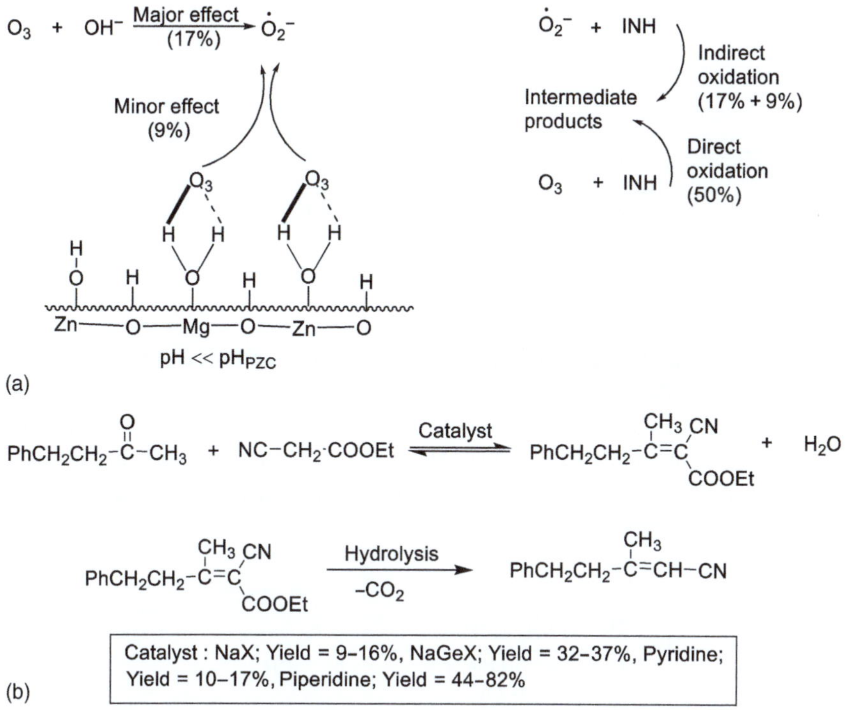

Scheme 5.10 Mechanism of isoniazid removal accompanied with MgZnO catalyst. Knoevenagel condensation reaction.

Zeolites have been abundantly used in carrying out organic synthesis for properties such as the following [50]:

(a) Isomerization of saturated compounds with acidic or basic zeolites [51], such as pinacol and skeletal rearrangement [52, 53]
(b) Electrophilic substitution reactions such as alkylations [54], acylations [55], nitration reactions, oxidations with oxygen [56], ammonia oxidation [57], oxidation with H_2O_2 [58], and oxidation with N_2O [59]
(c) Cyclization reactions, such as Diels–Alder reaction of cycloadditions [60, 61], and formation of anthracene [62].
(d) Addition of nucleophiles and substitution reactions such as ammonia and amines addition to olefins [63], production of anilines [64], and phosphines added to olefins [65]
(e) Multistage synthesis involves hydroformylation [66, 67], hydrolysis, and hydrogenation [68].

NaX zeolites with Si, when replaced by germanium throughout the synthesis, are used as base catalysts in the survey of Knoevenagel condensation between 2-nitrobenzaldehyde and 2-trifluoromethyl benzaldehyde with ethyl acetoacetate (Scheme 5.10) [69]. The use of zeolites in Knoevenagel condensation and Michael addition has been shown in Scheme 5.11 [69].

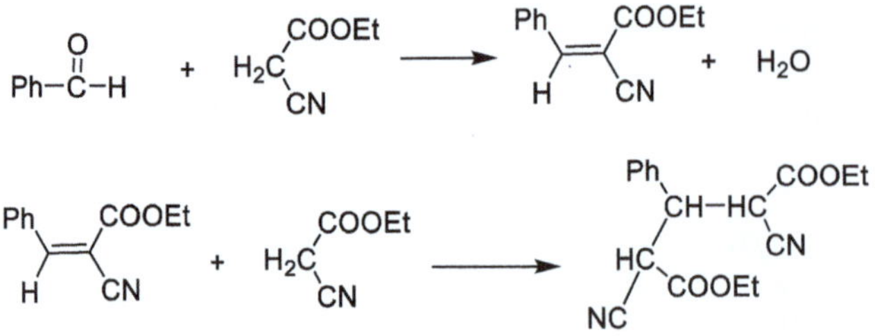

Scheme 5.11 Knoevenagel condensation followed by Michael addition accompanied with CsY zeolites.

Piyali Bhanja and Asim Bhaumik synthesized zeolites for eco-friendly organic transformations. The synthesized catalyst was used for Knoevenagel condensations with a variety of carbonyl compounds and active methylene group-containing compounds. The synthesis of NH_2-MCM-41 for base-catalyzed Knoevenagel condensation followed by Diels–Alder cyclization is shown in Scheme 5.12 [70].

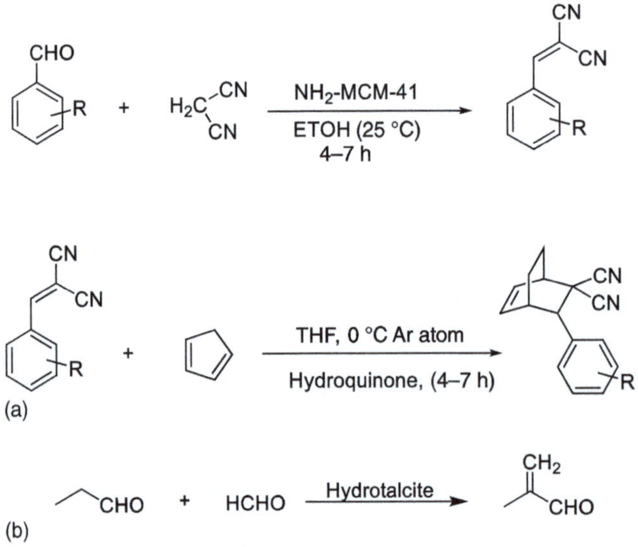

Scheme 5.12 Knoevenagel reaction using NH_2-MCM-41 catalyst. Propionaldehyde formaldehyde cross-condensation reaction.

5.2.3 Clays

Clay minerals consist of layered silicates with particle sizes ranging from 150 to less than 1 micron. The clay acidity present on their surfaces can be lowered by introducing silylpropylethylenediamine for base catalytic reactions such as Knoevenagel condensation [71]. Clays and modified clays are used in catalyzing a variety of organic reactions, such as Michael addition, hydrogenation, synthesis of heterocycles, aldol formation, allylation, insertion, pericyclic reactions, condensation reactions, acylation, carbene addition, dehydrogenation, imine synthesis, diazotization reactions, cyclization reactions, esterification reactions, oxidation of alcohols, rearrangement or isomerization reactions, and epoxidation, by functioning as a Brϕnsted and/or Lewis acids or bases. Clays are basically nanoparticles with layered structures, whose layers possess net negative charge nullified by cations, such as sodium ion, potassium ion, and calcium ion, occupying interlamellar space. The advantages of using clays in organic synthesis are as follows:

(a) Clays are safe to handle.
(b) Clays are reusable and inexpensive.
(c) Clays can prevent waste and promote the atom economy.
(d) Clays can lower the activation energy by stabilizing the transition state.
(e) Clays can act as general acids or bases during reaction.
(f) Clays are environmentally benign.
(g) Clays when used as catalysts are easier to recycle, hence increasing their economic efficiency.
(h) Clays may provide a surface for a reaction to occur.

Applications of hydrotalcites in organic synthesis are [72] (Schemes 5.12 and 5.13).

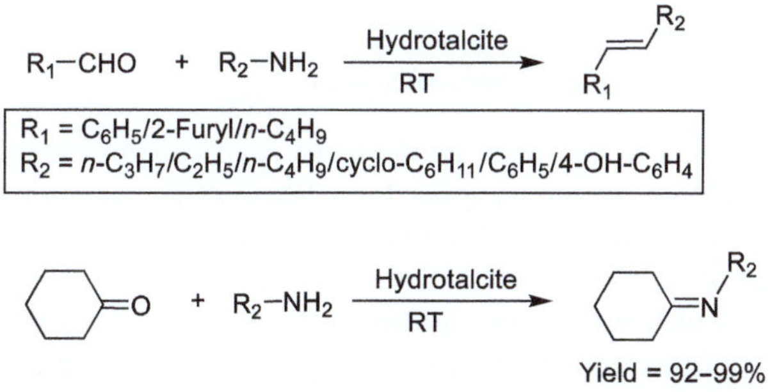

Scheme 5.13 Hydrotalcites for condensation of primary amines with aldehydes and ketones.

Deng et al. discovered a novel multiporous calcium aluminum hydrotalcite ($Ca_{12}Al_{14}O_{33}$) catalyst for transesterification reaction. The novel catalyst discovered had extraordinary properties such as stable catalytic activity during 4000 minutes coherent catalysis with 63% conversion and 100% selectivity, maintaining stability at even elevated temperatures of 900 °C, powerful alkali strength, and huge alkali amount. The catalyst was reported to be utilized in the transesterification reaction in the following way [73] (Scheme 5.14):

Organo-clay assemblies have been beneficial in organic transformations such as achieving thiocyanides, benzyl cyanides, and alcohols from responding benzyl halides and sodium cyanide, thiocyanate, and hydroxide. (Scheme 5.15) [74]

5.2.4 Solid-Supported Basic Catalysts

Alkali earth metals, alkali metals, and some basic organic groups are basically strong bases and nucleophiles and thus can perform catalysis of anionic

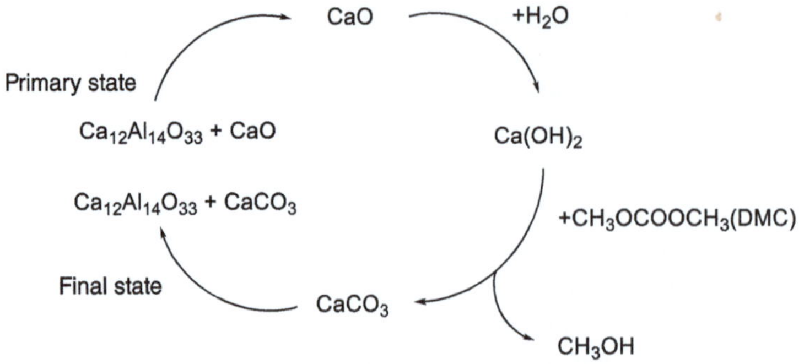

Scheme 5.14 The use of $Ca_{12}Al_{14}O_{33}$ for transesterification reaction.

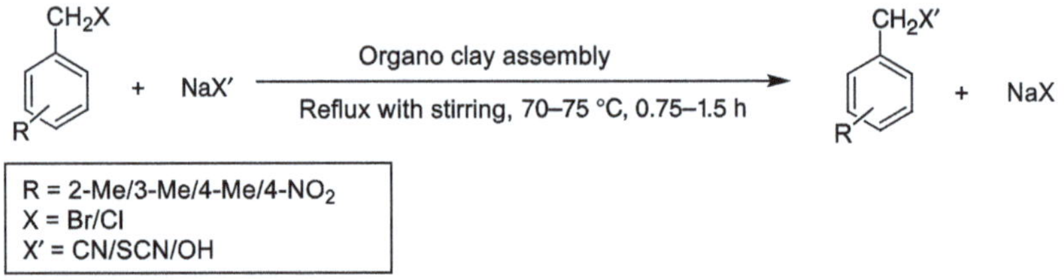

Scheme 5.15 Organic transformations using organic assemblies.

chain reactions as initiators [75–79]. The synthesis of tetraalkylammonium hydroxide-functionalized SBA-15 materials (SBA-15-pr-NR_3OH) by anchoring dimethyloctadecyl [3-(trimethoxysilyl)propyl] ammonium hydroxides (TMAOH) was carried out by Xie et al. [80]. The reaction was carried on the exterior of mesoporous SBA-15 silica which was further utilized as an efficient and stable SBC for transesterification of soybean oil with methanol [80]. The SBC used was found to provide an excellent yield of methyl esters of about 99.4% under the given reaction condition (methanol/oil molar ratio 12 : 1).

Another synthesis using a supported solid basic catalyst was performed by Bell et al. for the self-condensation of methyl ketones derived from biomass to the precursors of jet-diesel fuel by implanting secondary amines which are site-isolated on silica–alumina supports [81].

Huahua Zhao and colleagues presented the use of potassium oxide supported on mesoporous tetragonal zirconia (K_2O/M-ZrO_2-C) for the production of dimethyl carbonate. The use of a K_2O-anchored basic catalyst with carbon was found to exhibit better activity with a 60% higher yield [82] (Scheme 5.16).

Scheme 5.16 Transesterification of ethylene carbonate using K_2O/M-ZrO_2-C.

Michele Formica et al. used the bifunctional iminophospharane (BIMP) superbase for catalysis of various organic reactions such as 1,2-additions, 1,-4-additions, and polymer synthesis. The synthesized BIMP catalyst depicted can efficiently promote the enantioselectivity of addition reactions. The BIMP catalyst was synthesized according to the Staudinger reaction by reacting organoazide and phosphine at room temperature for 24 hours. The use of a BIMP catalyst for 1,2-additions is shown in Scheme 5.17a,b [83].

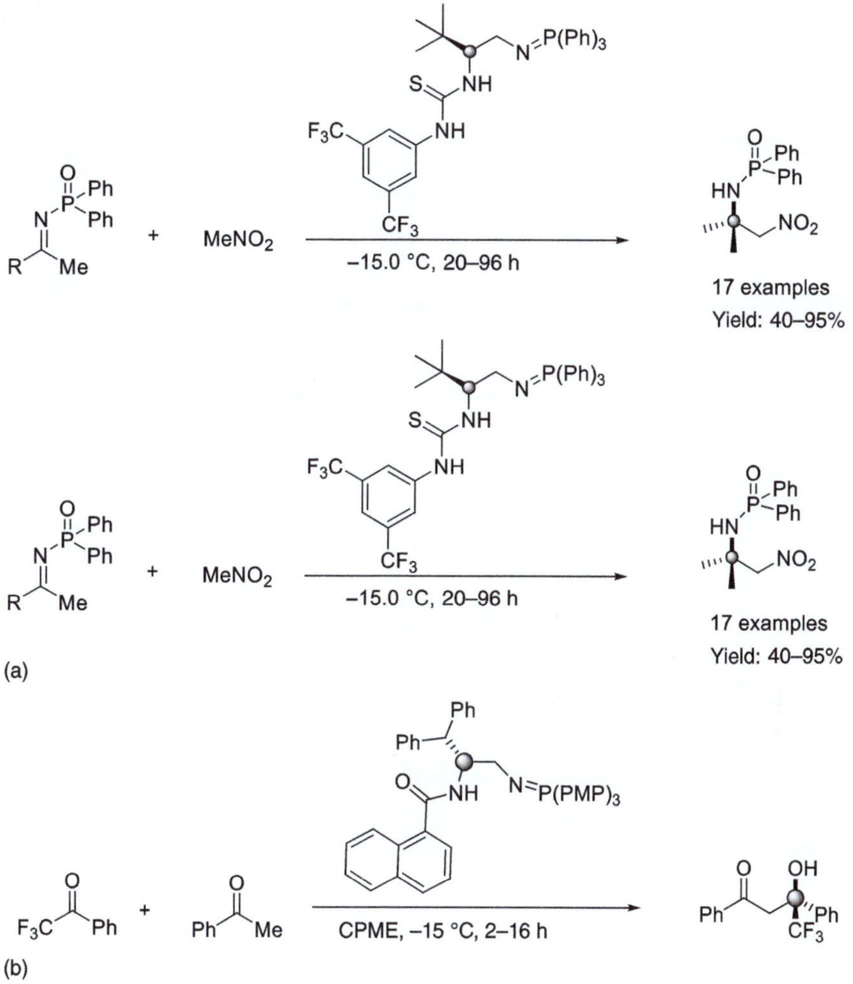

Scheme 5.17 (a) Use of BIMP catalyst for nitro-Mannich reaction of ketimines. (b) Use of BIMP catalyst for direct aldol addition to aryl ketones.

Saeed Askari and colleagues synthesized a Zr-based, UiO-66-NH_2-anchored heterogeneous catalyst for the production of xanthenes. The UiO-66-NH_2-based catalyst was synthesized according to Scheme 5.18a. The xanthenes were produced by reacting a variety of aldehydes with dimedone accompanied with the synthesized catalyst (Scheme 5.18b) [84].

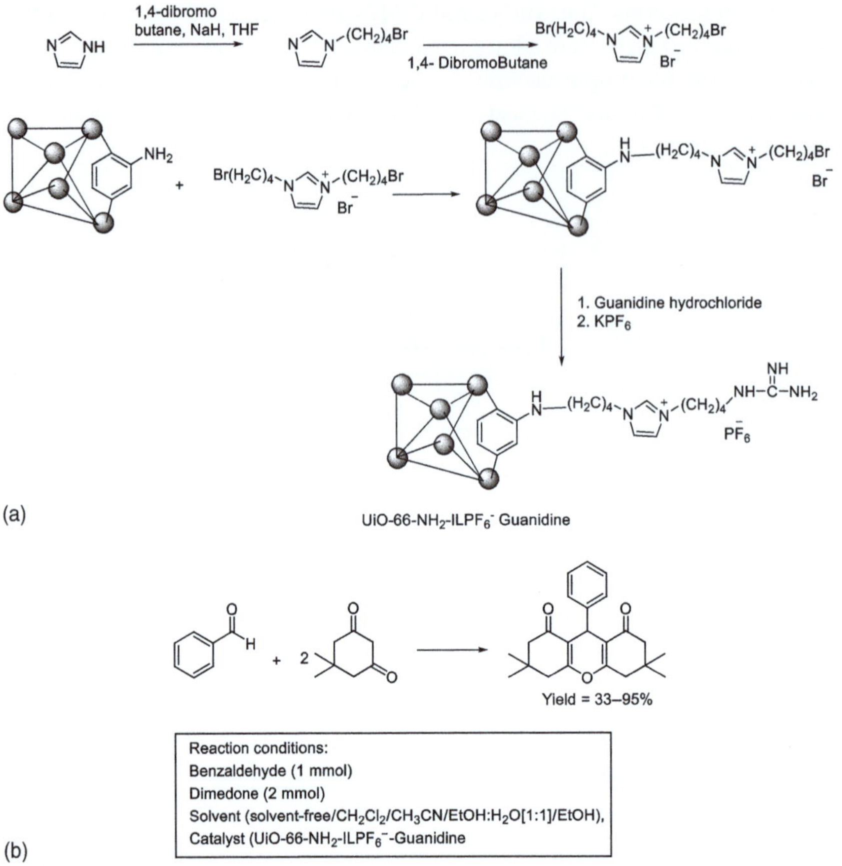

Scheme 5.18 (a) Synthesis of UiO-66-NH_2 catalyst (b) Zr-based catalyst for the synthesis of xanthenes.

Clacens et al. synthesized a CsF-supported catalyst for the reaction of transesterification of diethyl carbonate via a variety of alcohols and diols. The synthesized catalyst proved to be a strong base catalyst to synthesize Michael additions as well (Scheme 5.19) [85].

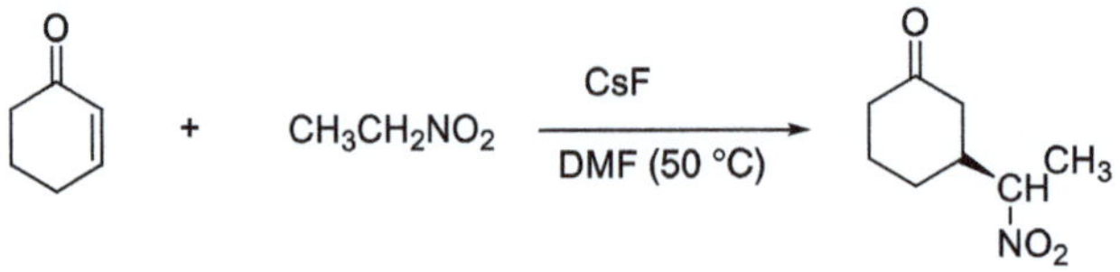

Scheme 5.19 CsF-supported basic catalyst for transesterification reactions.

Al-Maksoud et al. used a solid titanium-based catalyst for the oxidation of sulfides to corresponding sulfones by hydrogen peroxide. They performed oxidation of thianisole with substituted arylalkyl and diaryl sulfides and achieved 80–98%

yield of substituted sulfones. The reaction was carried out by reacting sulfide and o-dichlorobenzene in the presence of TiO_2 catalyst followed by the addition of 35% hydrogen peroxide (Scheme 5.20) [86].

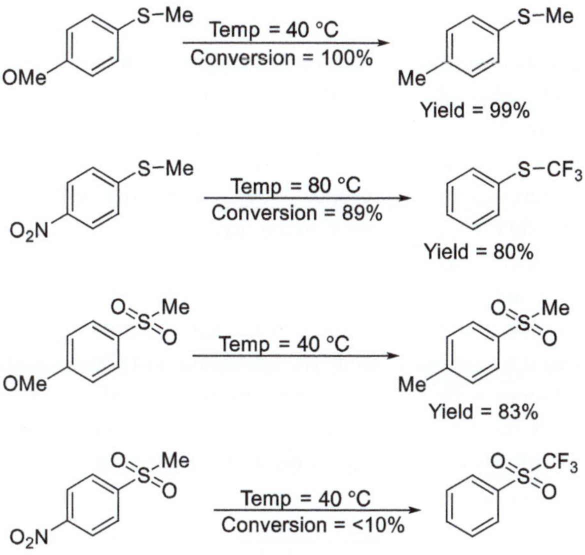

Scheme 5.20 Oxidation of sulfides by titanium-based catalyst.

Salam et al. synthesized mesoporous polytriallylamine (MPTA) and Cr-grafted Cr-MPTA and used them as base catalysts for oxidation of sulfides using hydrogen peroxide as an efficient oxidant. The base catalyst MPTA was prepared by aqueous-based polymerization of triallylamine accompanied in hydrothermal conditions. The oxidation reaction performed for direct aldol reactions of substituted aldehydes with acetone showed excellent catalytic activities in the presence of MPTA catalyst (Scheme 5.21) [87].

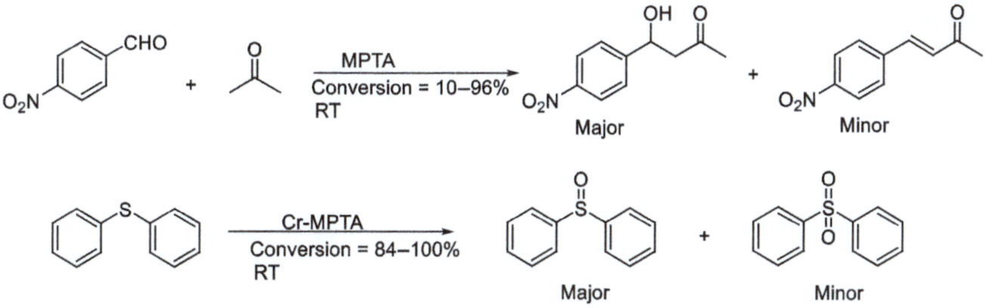

Scheme 5.21 Reaction pathways for sulfide oxidation using MPTA and Cr-MPTA.

Davydov et al. studied the catalytic activity of 20 metal oxides for Claus reactions and oxidation of H_2S. Their studies showed that vanadium oxide was the most active catalyst to be utilized in Claus reaction and for partial oxidation of H_2S.

The sequence of metal oxide-based catalysts in their decreasing order for their catalytic activity of partial oxidation of H_2S is as follows [88]:

$$V_2O_5 > Mn_2O_3 > CoO > TiO_2 > Fe_2O_3 > Bi_2O_3 > Sb_6O_{13} > CuO > Al_2O_3 = MgO = Cr_2O_3.$$

Dusi, Mallat, and Baiker briefly described epoxidation of various compounds such as allylic alcohols, homoallylic alcohols, unsaturated carbonyl compounds, and alkenes. A layered hydrotalcite, $Mg_{10}Al_2(OH)_{24}CO_3$, was identified for oxidizing a variety of linear as well as cyclic alkenes, cyclohexanol, and cyclohexanone accompanied with hydrogen peroxide in the presence of benzonitrile [89–92].

Fraile et al. described epoxidation of alkenes using hydrotalcite SBCs. The catalyst was prepared by precipitating solution containing a mixture metallic nitrates (Mg and Al) with aqueous solution of potassium hydroxide and potassium carbonate undergoing calcination and drying at the end to yield the final product. The formed hydrotalcite catalyst was tested to study the epoxidation of alkenes with hydrogen peroxide, chiral alkenes with TBHP and KF/alumina. Results showed that hydrotalcite catalyst with the ratio of Mg/Al = 3 showed highest catalytic activity as it has the highest number of basic sites present for reaction to occur (Scheme 5.22) [93].

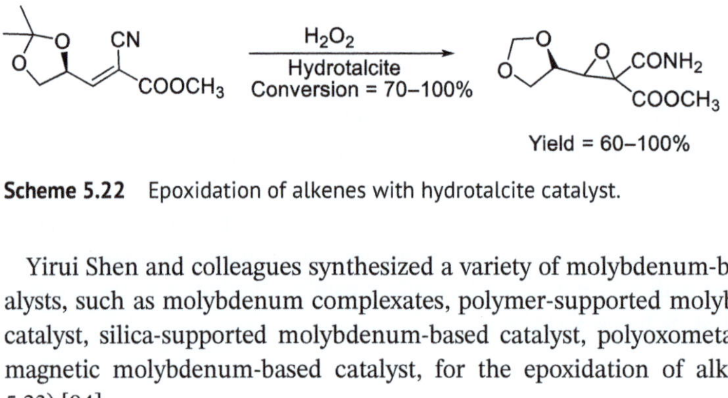

Scheme 5.22 Epoxidation of alkenes with hydrotalcite catalyst.

Yirui Shen and colleagues synthesized a variety of molybdenum-based solid catalysts, such as molybdenum complexates, polymer-supported molybdenum-based catalyst, silica-supported molybdenum-based catalyst, polyoxometalate catalysts, magnetic molybdenum-based catalyst, for the epoxidation of alkenes (Scheme 5.23) [94].

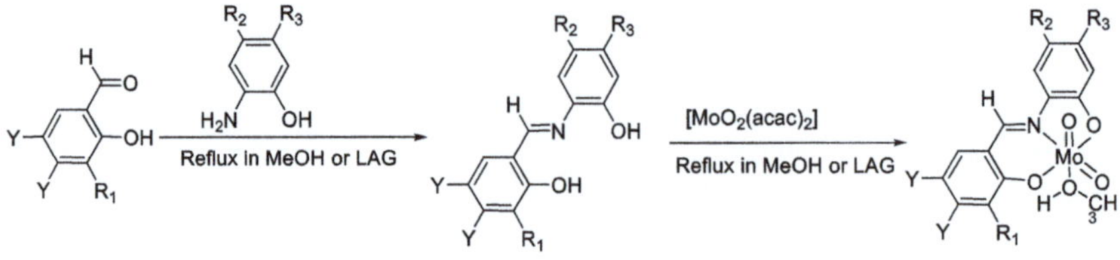

Scheme 5.23 Molybdenum-based catalyst for epoxidation of alkenes.

Tathod et al. synthesized platinum impregnated on γ-Al_2O_3 and hydrotalcite supported on a solid base and proved their importance in oxidation and hydrogenation reactions of C5 and C6 sugars. The synthesized solid base supported metal catalyst for the hydrogenation of glucose and was found to achieve 68% yield, while the C5 sugars achieved 82% yield (Scheme 5.24) [95].

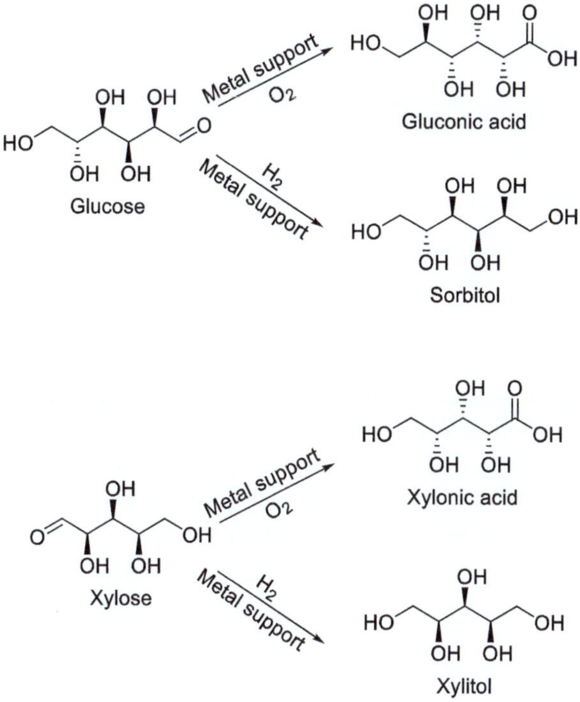

Scheme 5.24 Use of hydrotalcite base catalyst for oxidation of sugars.

Grant et al. synthesized and demonstrated the use of heterogenous metal oxide catalyst supported on VO_x/Al_2O_3 or silica (VO_x/silica) for aerobic partial oxidation of light alkanes following Mars–Van Krevelen reaction kinetic. The light alkanes having C_1–C_4 carbons are partially oxidized using solid metal oxides, MMO, and zeolites and are converted to form olefins, nitriles, aldehydes, carboxylic acids, and alcohols (Scheme 5.25) [96].

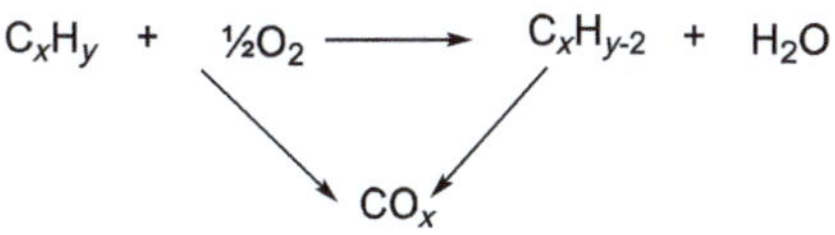

Scheme 5.25 General reaction of alkanes undergoing partial aerobic oxidations.

Cesar Jimenez Sachidrian and et al. prepared a variety of Mg/Al, Ma/Al/Zr, and Ma/Al/Sn hydrotalcite catalysts for Baeyer–Villiger oxidation of cyclohexanone in the presence of H_2O_2 or benzonitrile as solvent forming water and benzamide as by-products in the reaction. The hydrotalcite containing Zirconium promoted the decomposition of hydrogen peroxide and thereby affecting the oxidation reaction.

The hydrotalcite catalysts with of Mg/Al and Ma/Al/Sn were found to be the most promising catalyst for performing Baeyer–Villiger oxidation (Scheme 5.26) [97–102].

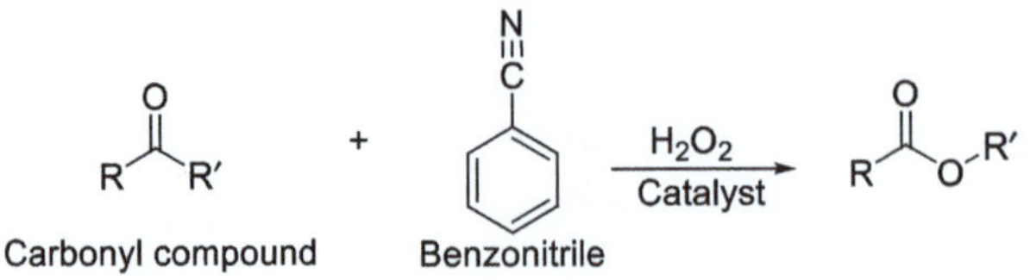

Scheme 5.26 Baeyer–Villiger oxidation.

5.3 Conclusion

The exploratory approach for organic synthesis using solid-based catalyst has been extraordinarily appealing in green catalytic procedures due to various precedence as mentioned in this chapter. Hence, development and study of solid basic catalysts have been important since the last decade in the fields of petrochemicals, academic, and industrial fields for their various organic synthetic applications.

List of Abbreviations

Al_2O_3	Aluminum oxide
CeO_2	Ceric oxide
CO_2	Carbon dioxide
CO_3O_4	Cobalt oxide
CuO	Copper(II) oxide
EtOH	Ethanol
Fe_2O_3	Ferric oxide
H_2O	Water
HCl	Hydrogen chloride
KF	Potassium fluoride
MgO	Magnesium oxide
O_2	Oxygen
RT	Room temperature
TBHP	*tert*-Butyl hydroperoxide
WO_3	Tungsten oxide
ZnO	Zinc oxide

Consent for Publication

All the authors agreed to publish the work as a book chapter.

Conflict of Interest

The authors declare no conflict of interest.

Acknowledgment

The author would like to express their sincere thanks to Eurofins Analytical Services, Bengaluru, Department of Chemistry, MMV, Banaras Hindu University, Varanasi, Uttar Pradesh, and Presidency University, Bangalore India, for providing the facilities to write and submit this article for publication.

References

1 Hattori, H. (2004). Solid base catalysts: generation, characterization, and catalytic behaviour of basic sites. *Journal of the Japan Petroleum Institute* 47 (2): 67–81. https://doi.org/10.1627/jpi.47.67.

2 Tomar, R., Kundra, P., Sharma, J. et al. (2024). An overview: synthesis of menthol using heterogeneous catalysis. *Letters in Organic Chemistry* 21 (1): 16–28.

3 Kumar, L., Verma, N., Tomar, R. et al. (2023). Development of bioactive 2-substituted benzimidazole derivatives using an MnO_x/HT nanocomposite catalyst. *Dalton Transactions* 52 (10): 3006–3015.

4 Tomar, R., Jain, S., Yadav, P. et al. (2022). Conversion of limonene over heterogeneous catalysis: an overview. *Current Organic Synthesis* 19 (3): 414–425.

5 Kumar, L., Verma, N., Sehrawat, H. et al. (2022). Successive oxidation–condensation reactions using a multifunctional gold-supported nanocomposite (Au/MgCe–HDO). *New Journal of Chemistry* 46 (7): 3472–3481.

6 Tomar, R., Singh, N., Kumar, N. et al. (2019). Base-free Suzuki–Miyaura coupling reaction using Palladium (II) supported catalyst in water. *Catalysis Letters* 149 (6): 1589–1594.

7 Tomar, R., Singh, N., Rathee, G. et al. (2017). Synthesis and characterization of hybrid $Mg(OH)_2$ and $CeCO_3OH$ composite with improved activity towards henry reaction. *Asian Journal of Organic Chemistry* 6 (12): 1728–1732.

8 Tomar, R., Sharma, J., Nishimura, S., and Ebitani, K. (2016). Aqueous oxidation of sugars into sugar acids using hydrotalcite-supported gold nanoparticle catalyst under atmospheric molecular oxygen. *Chemistry Letters* 45 (7): 843–845.

9 Arya, K. and Tomar, R. (2013). Microporous zeolite catalyst system: an eco-approach for regioselective synthesis of pyrimidobenzimidazoles. *Research on Chemical Intermediates* 41 (6): 3389–3400.

10 Tomar, R., Ebitani, K., and Chandra, R. (2019). Hydrotalcite-supported ceria nanoparticles as a heterogeneous catalyst for one-pot synthesis of imines under atmospheric air. *ChemistrySelect* 4 (12): 3577–3581.

11 Tomar, R., Rathee, G., Chandra, I. et al. (2018). Synthesis and characterization of magnesium hydroxide & cerium oxide composite. *ChemistrySelect* 3 (6): 1645–1649.

12 Molnár, Á. and Papp, A. (2017). Catalyst recycling—a survey of recent progress and current status. *Coordination Chemistry Reviews* 349: 1–65. https://doi.org/10.1016/j.ccr.2017.08.011.

13 Pines, H. and Haag, W.O. (1958). Communications—stereoselectivity in the carbanion-catalyzed isomerization of 1-butene. *The Journal of Organic Chemistry* 23 (2): 328–329. https://doi.org/10.1021/jo01096a627.

14 Hong, Z., Xiong, C., Zhao, G., and Zhu, Z. (2019). Side-chain alkylation of toluene with methanol to produce styrene: an overview. *Catalysis Science & Technology* 9 (24): 6828–6840. https://doi.org/10.1039/C9CY01581G.

15 Védrine, J.C. (2017). Heterogeneous catalysis on metal oxides. *Catalysts* 7 (11): 1–25. https://doi.org/10.3390/catal7110341.

16 Nakagawa, Y. and Tomishige, K. (2011). Heterogeneous catalysis of the glycerol hydrogenolysis. *Catalysis Science & Technology* 1 (2): 179–190. https://doi.org/10.1039/c0cy00054j.

17 Nakagawa, Y. and Tomishige, K. (2011). Catalyst development for the hydrogenolysis of 18 biomass-derived chemicals to value-added ones. *Catalysis Surveys from Asia* 15 (2): 111–116. https://doi.org/10.1007/s10563-011-9114-z.

18 Nakagawa, Y. and Tomishige, K. (2012). Production of 1,5-pentanediol from biomass via furfural and tetrahydrofurfuryl alcohol. *Catalysis Today* 195 (1): 136–143. https://doi.org/10.1016/j.cattod.2012.04.048.

19 Nakagawa, Y., Tamura, M., and Tomishige, K. (2013). Catalytic reduction of biomass-derived furanic compounds with hydrogen. *ACS Catalysis* 3 (12): 2655–2668. https://doi.org/10.1021/cs400616p.

20 Nakagawa, Y., Tamura, M., and Tomishige, K. (2014). Catalytic materials for the hydrogenolysis of glycerol to 1,3-propanediol. *Journal of Materials Chemistry A* 2 (19): 6688–6702. https://doi.org/10.1039/C3TA15384C.

21 Nakagawa, Y., Liu, S., Tamura, M., and Tomishige, K. (2015). Catalytic total hydrodeoxygenation of biomass-derived polyfunctionalized substrates to alkanes. *ChemSusChem* 8 (7): 1114–1132. https://doi.org/10.1002/cssc.201403330.

22 Nakagawa, Y., Tamura, M., and Tomishige, K. (2015). Catalytic conversions of furfural to pentanediols. *Catalysis Surveys from Asia* 19 (4): 249–256. https://doi.org/10.1007/s10563-015-9194-2.

23 Nakagawa, Y., Tamura, M., and Tomishige, K. (2018). Perspective on catalyst development for glycerol reduction to C3 chemicals with molecular hydrogen. *Research on Chemical Intermediates* 44 (7): 3879–3903. https://doi.org/10.1007/s11164-018-3481-2.

24 Nakagawa, Y., Tamura, M., and Tomishige, K. (2019). Recent development of production technology of diesel- and jet-fuel-range hydrocarbons from inedible biomass. *Fuel Processing Technology* 193: 404–422. https://doi.org/10.1016/j.fuproc.2019.05.028.

25 Lang, R., Du, X., Huang, Y. et al. (2020). Single-atom catalysts based on the metal-oxide interaction. *Chemical Reviews* 120 (21): 11986–12043. https://doi.org/10.1021/acs.chemrev.0c00797.

26 Mabate, T.P., Maqunga, N.P., Ntshibongo, S. et al. (2023). Metal oxides and their roles in heterogeneous catalysis: special emphasis on synthesis protocols, intrinsic properties, and their influence in transfer hydrogenation reactions. *SN Applied Sciences* 5 (7): 196. https://doi.org/10.1007/s42452-023-05416-6.

27 Pirkanniemi, K. and Sillanpää, M. (2002). Heterogeneous water phase catalysis as an environmental application: a review. *Chemosphere* 48 (10): 1047–1060. https://doi.org/10.1016/s0045-6535(02)00168-6.

28 Mogudi, B.M., Ncube, P., and Meijboom, R. (2016). Catalytic activity of mesoporous cobalt oxides with controlled porosity and crystallite sizes: evaluation using the reduction of 4-nitrophenol. *Applied Catalysis. B, Environmental* 198: 74–82. https://doi.org/10.1016/j.apcatb.2016.05.051.

29 Chapelle, A., Oudrhiri-Hassani, F., Presmanes, L. et al. (2010). CO_2 sensing properties of semiconducting copper oxide and spinel ferrite nanocomposite thin film. *Applied Surface Science* 256 (14): 4715–4719. https://doi.org/10.1016/j.apsusc.2010.02.079.

30 Wang, Y., Arandiyan, H., Scott, J. et al. (2017). Recent advances in ordered meso/macroporous metal oxides for heterogeneous catalysis: a review. *Journal of Materials Chemistry A* 5 (19): 8825–8846. https://doi.org/10.1039/C6TA10896B.

31 Gawande, M.B., Pandey, R.K., and Jayaram, R.V. (2012). Role of mixed metal oxides in catalysis science-versatile applications in organic synthesis. *Catalysis Science & Technology* 2 (6): 1113–1125. https://doi.org/10.1039/c2cy00490a.

32 Lü, Y., Jiang, Y., Zhou, Q. et al. (2017). Heterometallic metal-organic framework-templated synthesis of porous Co_3O_4/ZnO nanocage catalysts for the carbonylation of glycerol. *Journal of Solid State Chemistry* 256: 93–100. https://doi.org/10.1016/j.jssc.2017.08.035.

33 Shen, T., Zhang, X., Lin, K.A., and Tong, S. (2020). Solid base Mg-doped ZnO for heterogeneous catalytic ozonation of isoniazid: performance and mechanism. *Science of the Total Environment* 703: 134983. https://doi.org/10.1016/j.scitotenv.2019.134983.

34 Weisz, P.B., Frilette, V.J., Maatman, R.W., and Mower, E.B. (1962). *Catalysis Journal* 1: 307.

35 Plank, C.J., Rosinski, E.J., and Howthorne, M.P. (1964). *Industrial & Engineering Chemistry Product Design & Development* 3: 165.

36 Venuto, B.P. and Habib, E.T. Jr., (1979). *Chemical Industries*, vol. I. Marcel Dekker.

37 Holderich, W. and Gallei, E. (1984). *Chemie Ingenieur Technik* 56: 908.

38 Breck, D.W. (1974). *Zeolite Molecular Sieves*. Wiley.

39 Barrer, R.M. (1978). *Zeoliter and Cloy Minerals as Sorbents and Molecular Sieves*. Academic Press.

40 Rabo, J.A. (1976). *ACS Monograph* 171: 332.

41 Jacobs, P.A. (1977). *Carboniogenic Actiuity of Zeolites*. Elsevier.

42 Magee, J.S. and Blazek, J.J. (1976). *ACS Monograph* 171: 615.

43 Morrison, R.A. (1980). DBP 2848849; Mobile Oil Corp., May 22.

44 Butter, S.T., Keading, W.W. (1975). US-Pat. 3906054, September 16; Mobil Oil Corp.

45 Nayak, V.S. and Choudhary, V.R. (1983). *Journal of Corals* 8: 26.
46 Nakamoto, H. and Takahasi, H. (1982). *Zeolires* 2: 67.
47 Sherry, J.S. (1971). *Advanced Chemical Series* 101: 350.
48 Barthomeuf, D. (1977). Acidic and catalytic properties of zeolites. *ACS Symposium Series* 40: 453–472. https://doi.org/10.1021/bk-1977-0040.ch038.
49 Marosi, L., Stabenow, J., Schwarzmann, M. (1981). DBP 3006471; BASF, August 27.
50 Martin-Luengo, M.A.M. and Yates, M. (1995). Zeolite materials as catalysts for organic syntheses. *Journal of Materials Science* 30 (18): 4483–4491. https://doi.org/10.1007/BF01153053.
51 Martens, L.R., Grobet, P.J., Vermeiren, W.J., and Jacobs, P.A. (1986). New developments in zeolite science and technology. In: *Studies in Surface Science and Catalysis*, vol. 28 (ed. Y. Murakami, A. Iijima, and J.W. Ward), 935–941. Amsterdam: Elsevier.
52 Stöcker, M., Hemmersbach, P., Raeder, H., and Grepstad, J.K. (1986). Isomerization of n-butane over zeolites depending on catalyst structure. *Applied Catalysis* 25 (1–2): 223–230. https://doi.org/10.1016/S0166-9834(00)81240-3.
53 Molnar, A., Bucsi, S., and Bartok, H. (1982). Heterogeneous catalysis and fine chemicals". P6rto. In: *Studies in Surface Science and Catalysis*, vol. 41 (ed. M. Guisnet, J. Barrault, C. Bouchoule, et al.), 203. Elsevier.
54 Parton, R.F., Jacobs, J.M., Huybretchs, D.R. et al. (1989). Zeolites as Catalysts. Sorbents and Detergent Budden. In: *Studies in Surface Science and Catalysis*, vol. 46 (ed. H.G. Karge and J. Weitkamp), 163–192. Elsevier.
55 Chiche, B., Finiels, A., Gauthier, C., and Geneste, P. (1987). The effect of structure on reactivity in zeolite catalyzed acylation of aromatic compounds: A $\rho - \sigma+$ relationship. *Applied Catalysis* 30 (2): 365–369. https://doi.org/10.1016/S0166-9834(00)84126-3.
56 Tagiev, D.B. and Minachev, K.H.M. (1986). New developments in zeolites science and technology. In: *Studies in Surface Science and Catalysis*, vol. 28 (ed. Y. Murakami, A. Iijima, and J.W. Ward), 981–988. Elsevier.
57 Trifiro', F. and Jirů, P. (1988). About some possibilities and causes of changes in selectivity of vanadium containing zeolitic catalysts in oxidation reactions. *Catalysis Today* 3 (5): 519–524. https://doi.org/10.1016/0920-5861(88)87037-8.
58 Ferrini, C. and Kouwenhoven, H.W. (1990). New developments in selective oxidation. In: *Studies in Surface Science and Catalysis*, vol. 55 (ed. G. Centi and F. Trifiro), 53–62. Elsevier.
59 Ono, Y., Tohmori, K., Suzuki, S. et al. (1988). Heterogeneous catalysis and fine chemicals. In: *Studies in Surface Science and Catalysis*, vol. 4t (ed. M. Guisnet, J. Barrault, C. Bouchoule, et al.), 75. Elsevier.
60 Laszlo, P. and Lucchetti, J. (1984). Easy formation of Diels-Alder cycloadducts between furans and α,β-unsaturated aldehydes and ketones at normal pressure. *Tetrahedron Letters* 25 (39): 4387–4388. https://doi.org/10.1016/S0040-4039(01)81445-3.
61 Gosh, S. and Bauld, N.L. (1985). Hole-catalyzed and photoassisted hole-catalyzed pericyclic reactions on zeolites. *Journal of Catalysis* 95 (1): 300–304. https://doi.org/10.1016/0021-9517(85)90032-6.

62 Ganesan, K.; Pillai, C. N., ibid. 119 1990, 8.

63 Deeba, M. and Ford, M.E. (1988). Heterogeneous acid-catalyzed amination of isobutene to *tert*-butylamine. *The Journal of Organic Chemistry* 53 (19): 4594–4596. https://doi.org/10.1021/jo00254a039.

64 Warawdekar, M.G. and Rajadhyaksha, R.A. (1987). Aromatic nucleophilic substitutions on zeolites. I. Amination of anisole and phenol. *Zeolites* 7 (6): 574–578. https://doi.org/10.1016/0144-2449(87)90101-1.

65 Women's Fund of Hawaii, Elderich, M.H., and Sattler, E. (1988). *Proceedings of the 9th International Congress Catalysis, Vol. 1* (ed. M.J. Philips and M. Ternan), 316. Ottawa: The Chemical Institute of Canada.

66 Rode, E., Davis, M.E., and Hanson, B.E. (1985). *Journal of the Chemical Society, Chemical Communications* 11: 716.

67 Da Vis, M.E., Butler, P.M., Rossin, J.A., and Hanson, B.E. (1985). Hydroformylation of 1-hexene by soluble and zeolite-supported rhodium species. *Journal of Molecular Catalysis* 31 (3): 385–395. https://doi.org/10.1016/0304-5102(85)85120-8.

68 Maxwell, I.E., Downing, R.S., De Boer, J.J., and Van Langen, S.A. (1980). Copper-exchanged zeolite catalysts for the cyclodimerization of butadiene I. Catalyst stability and regenerability. *Journal of Catalysis* 61 (2): 485–492. https://doi.org/10.1016/0021-9517(80)90395-4.

69 Sheldon, R.A. and Van Bekkum, H. (2000). *Fine Chemicals Through Heterogenous Catalysis*. Wiley Publication https://doi.org/10.1002/9783527612963.

70 Bhanja, P. and Bhaumik, A. (2016). Functionalized porous nanomaterials as efficient heterogenous catalyst for eco-friendly organic transformations. *Journal of Nanoscience and Nanotechnology* 16 (9): 9050–9062. https://doi.org/10.1166/jnn.2016.12899.

71 Subba Rao, Y.V. and Choudary, B.M. (1991). Knoevenagel condensation catalysed by new montmorillonitesilylpropylethylenediamine. *Synthetic Communications* 21 (10–11): 1163–1166. https://doi.org/10.1080/00397919108021035.

72 Bhat, B.M. (2012). Synthesis and characterization of hydrotalcite and hydrotalcite compounds and their application as a base catalyst for aldol condensation reaction. *Oriental Journal of Chemistry* 28 (4): 1751–1760. https://doi.org/10.13005/ojc/280426.

73 Deng, W., Yao, J., and Shi, L. (2022). Highly thermally stable multi-porous calcium aluminium hydrotalcite catalyst for efficient carbonate synthesis. *Science Talks* 100026: 1–8. https://doi.org/10.1016/j.sctalk.2022.100026.

74 Nagendrappa, G. (2011). Organic synthesis using clay and clay–supported catalysts. *Applied Clay Science* 53 (2): 106–138. https://doi.org/10.1016/j.clay.2010.09.016.

75 Sun, G., Li, Y., Cai, Z. et al. (2017). K_2CO_3-loaded hydrotalcite: a promising heterogeneous solid base catalyst for bolubricant base oil production from waste cooking oils. *Applied Catalysis. B, Environmental* 209: 118–127. https://doi.org/10.1016/j.apcatb.2017.02.078.

76 Sano, N., Yamada, K., Tsunauchi, S., and Tamon, H. (2017). A novel solid base catalyst for transesterification of triglycerides toward biodiesel production:

carbon nanohorn dispersed with calcium ferrite. *Chemical Engineering Journal* 307: 135–142. https://doi.org/10.1016/j.cej.2016.08.010.

77 Zhang, P., Han, Q., Fan, M., and Jiang, P. (2014). Magnetic solid base catalyst $CaO/CoFe_2O_4$ for biodiesel production: influence of basicity and wettability of the catalyst in catalytic performance. *Applied Surgical Science* 317: 1125–1130.

78 Matsuhashi, H. and Kawamura, A. (2012). Preparation of a novel solid base catalyst of CaO covered with SiO_2. *Catalysis Today* 185 (1): 236–240. https://doi.org/10.1016/j.cattod.2011.08.015.

79 Zhu, F., Yang, D., Zhang, F., and Li, H. (2012). Amine-bridges periodic mesoporous organosilica nanospheres as an active reusable solid base-catalyst for water-medium and solvent-free organic reactions. *Journal of Molecular Catalysis* 363–364: 387–397. https://doi.org/10.1016/j.molcata.2012.07.015.

80 Xie, W. and Fan, M. (2014). Biodiesel production by transesterification using tetraalkylammonium hydroxides immobilized onto SBA-15 as a solid catalyst. *Chemical Engineering Journal* 239: 60–67. https://doi.org/10.1016/j.cej.2013.11.009.

81 Sankaranarayanapillai, S., Sreekumar, S., Gomes, J. et al. (2015). Catalytic upgrading of biomass-derived methyl ketones to liquid transportation fuel precursors by an organocatalytic approach. *Angewandte Chemie (International Edition in English)* 54 (15): 4673–4677. https://doi.org/10.1002/anie.201412470.

82 Zhao, H., Song, H., Wang, F. et al. (2020). Low-temperature fabrication of K_2O supported mesoporous tetragonal ZrO_2 solid base for synthesis of dimethyl carbonate. *Molecular Catalysis* 495: 111141. https://doi.org/10.1016/j.mcat.2020.111141.

83 Formica, M., Rozsar, D., Su, G. et al. (2020). Bifunctional iminophosphene superbase catalysis: applications in organic synthesis. *Accounts of Chemical Research* 53 (10): 2235–2247. https://doi.org/10.1021/acs.accounts.0c00369.

84 Askari, S., Khodaei, M.M., and Jafarzadeh, M. (2020). Basic ionic liquid anchored on UiO-66-NH_2 metal-organic framework: a stable and efficient heterogenous catalyst for the synthesis of xanthenes. *Research on Chemical Intermediates* 47: 2881–2899. https://doi.org/10.1007/s11164-021-04439-1.

85 Clacens, J.M., Genuit, D., Veldurthy, B. et al. (2004). CsF supported by A-alumina: an efficient basic catalyst. *Applied Catalysis* 53 (2): 95–100. https://doi.org/10.1016/j.apcatb.2004.05.006.

86 Maksoud, W., Daniele, S., and Sorokin, A.B. (2008). Practical oxidation of sulphides to sulfones by H_2O_2 catalyst by titanium catalyst. *Green Chemistry* 447–451: 10. https://doi.org/10.1039/b717696a.

87 Salam, N., Mondal, P., Mondal, J. et al. (2012). Highly efficient base catalysis and sulphide oxidation reactions over new functionalized mesoporous polymers. *RSC Advances* 2: 9494–6477. https://doi.org/10.1039/c2ra20630g.

88 Dayvdov, A.A., Marshneva, V., and Shepotko, M.L. (2003). Metal oxides in hydrogen sulphide oxidation by oxygen and sulfer dioxide I. The comparison study of the catalytic activity. Mechanism of the interactions between H_2S and SO_2 on some oxides. *Applied Catalysis, A: General* 93–100.

89 Cativiela, C., Figueras, F., Fraile, J.M. et al. (1995). Hydrotalcite-promoted epoxidation of electron-deficient alkenes with hydrogen peroxide. *Tetrahedron Letters* 36: 4125.

90 Choudary, B.M., Kantam, M.L., Bharathi, B., and Reddy, C.V. (1998). Superactive Mg-Al-O-t-Bu Hydrotalcite for Epoxidation of Olefins. *Synlett* 1203.

91 Ueno, S., Yamaguchi, K., Yoshida, K. et al. (1998). Hydrotalcite catalysis: heterogeneous epoxidation of olefins using hydrogen peroxide in the presence of nitriles. *Chemical Communications* 295.

92 Dusi, M., Mallat, T., and Baiker, A. Epoxidation of functionalized olefins over solid catalysts. *Catalysis Reviews: Science and Engineering* 42 (1–2): 213–278. https://doi.org/10.1081/CR-100100262.

93 Fraile, J.M., Garcia, J.I., Marco, D. et al. (2000). Epoxidation of electron-deficient alkenes using heterogenous basic catalysts. *Studies in Surface Science and Catalysis* 1673–1678. https://doi.org/10.1016/S0167-2991(00)80441-3.

94 Shen, Y., Jiang, P., Wai, P.T., and Zhang, W. (2018). Recent progress in application of molybdenum-based catalysts for epoxidation of alkenes. *Catalogue* 9: 31. https://doi.org/10.3390/catal9010031.

95 Tathod, A., Kane, T., Sanil, E.S., and Dhepe, P.L. (2013). Solid base supported metal catalysts for the oxidation of hydrogenation of sugars. *Journal of Molecular Catalysis* 1–10. https://doi.org/10.1016/j.molcata.2013.09.014.

96 Grant, J.T., Venegas, J.M., McDermott, P.W., and Hermans, I. (2018). Aerobic oxidations of light alkanes over solid metal oxide catalysts. *Chemical Reviews* 118: 2769–2815. https://doi.org/10.1021/acs.chemrev.7b00236.

97 Sanchidrian, C.S., Hidalgo, J.M., Llamas, R., and RuiZ, J.F. (2006). Baeyer–Villiger oxidation of cyclohexanone with hydrogen peroxide/benzonitrile over hydrotalcites as catalysts. *Applied Catalysis* 312: 86–94. https://doi.org/10.1016/j.apcata.2006.06.031.

98 Baeyer, A. and Villiger, V. (1899). *Berichte der Deutschen Chemischen Gesellschaft* 32: 3625.

99 Kaneda, K., Ueno, S., and Imanaka, T. (1994). Heterogeneous Baeyer–Villiger oxidation of ketones using an oxidant consisting of molecular oxygen and aldehydes in the presence of hydrotalcite catalysts. *Journal of the Chemical Society, Chemical Communications* 25: 797.

100 Kaneda, K. and Ueno, S. (1996). *ACS Symposium Series* 638: 300.

101 Ueno, S., Ebitani, K., Ookubo, A., and Kaneda, K. (1997). The active sites in the heterogeneous Baeyer-Villiger oxidation of cyclopentanone by hydrotalcite catalysts. *Applied Surface Science* 121 (122): 366.

102 Aramendia, M.A., Borau, V., Jiménez, C. et al. (2000). Epoxidation of limonene with layered-double hydroxides as catalysts. *Studies in Surface Science and Catalysis* 130: 1667.

6

Multicomponent Reactions for Eco-compatible Heterocyclic Synthesis Over Solid Base Catalysts

Amanpreet Singh and Jasdeep Kaur

Chandigarh University, Department of Chemistry, UIS, Mohali, Punjab 140413, India

6.1 Introduction

Heterocyclic compounds (HCs) form the core structure of various natural products, agrochemicals, and therapeutic agents [1]. These compounds are valuable synthetic frameworks in the areas of drug discovery, biomaterials, synthetic dyes, fragrances, polymers, and supramolecular chemistry. Owing to their spectacular pharmacological profiles, the production of these heterocycles is a significant area of interest [2]. Consequently, researchers are working on the development of sustainable, green, efficient, and eco-friendly synthetic approaches to prepare biologically important HCs [3].

In this concern, the multicomponent reactions (MCRs) are providing efficient route to synthesize biologically important heterocycles in a simple, economical, and selective manner [4]. The MCRs generally involve the condensation of carbonyl compounds with other chemical entities to synthesize bioactive compounds such as chromenes, pyrimidines, xanthene, oxiranes, naphtha-furan, benzodiazepines, pyran, pyrrole, and imidazole-based analogs [5]. In these MCRs, the probability of undesirable side reactions can lead to unwanted product formation. Therefore, the use of catalyst is precedent to enhance the selectivity of MCRs in order to avoid the possible side reactions [6].

Perusal of literature revealed that the traditional homogeneous catalysts are difficult to separate from the reaction mixture and require high catalytic loading, long reaction times, and tedious purification techniques [7]. In order to overcome the limitations associated with homogeneous catalysts, heterogeneous catalysts are more preferred [8]. The MCRs can be acid catalyzed or base catalyzed in which reactant molecules act as bases or acids, respectively. Solid acid-catalyzed MCRs are getting more attention in the past years as compared to solid base-catalyzed MCRs. Therefore, there is a necessity to explore solid base catalysts for MCRs. These solid base catalysts are more eco-friendly as they can provide solvent-free or solvent low conditions to reduce environment impact [9–18]. Furthermore, solid base catalysts are less corrosive, easy to handle; provide mild reaction conditions, easy separation,

Solid Base Catalysts: Synthesis, Characterization, and Applications, First Edition.
Edited by Ravi Tomar, K.K. Pant, and Ramesh Chandra.

more selectivity, chemical stability; and have broader substrate scope for MCRs [19]. Thus, this chapter includes introduction to MCRs, their types, solid base catalysts, the role of solid base catalysts in MCRs to prepare various heterocycles followed by conclusion and future prospective.

6.2 Multicomponent Reactions (MCRs)

MCRs are the types of organic reactions, which involve the one-pot simultaneous participation of three or more reactant molecules to produce complex and highly functionalized compound in one single step. In an ideal MCR, majority of the atoms present in the reactant molecule must be incorporated in the product structure. These MCRs have more advantages over stepwise linear synthesis (Figure 6.1) such as less waste generation, easy purification, atom economy, time and energy economy, and high convergence. Therefore, MCRs are much better than classical linear synthesis for rapid and efficient synthesis of complex heterocycles [20].

The concept of MCRs originated back to the middle of nineteenth century with the development of few seminal reactions that laid the foundation of this field. Recently, the area of MCRs has experienced exponential growth, and these MCRs have become an important tool for modern synthetic organic chemistry. Advantages of MCRs are not only limited to green and sustainable approach but also biological properties of the products synthesized via MCRs [21].

The intermediates formed in the MCRs are generally not isolated. It has been observed experimentally that mechanism of these MCRs is independent of the preferred reaction pathways. Since, all the possible reaction pathways are leading to the formation of same final product, and the overall path leading to the formation of

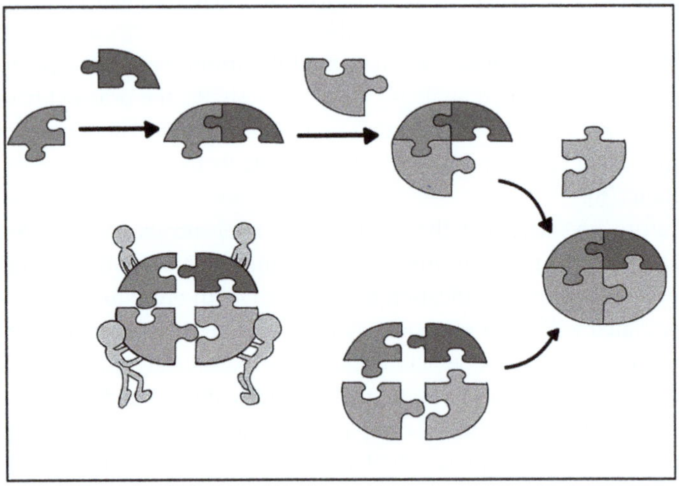

Figure 6.1 Diagrammatic representation of stepwise linear synthesis and multicomponent reaction approach.

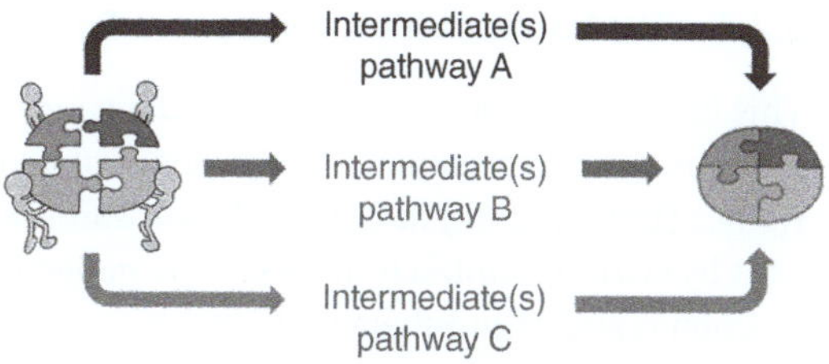

Figure 6.2 Graphical representation of possible reaction mechanisms associated with multicomponent organic transformations.

the final product is convergent (Figure 6.2). Among all MCRs, Biginelli, Hantzsch, Mannich, Passerini, Ugi, and Gewald reactions are most popular and well-studied and widely used in the synthesis of various bioactive compounds [22]. The various possible mechanisms proposed for these MCRs are supported by DFT and tandem electrospray mass spectrometry and depend upon the nature of catalyst, solvent system, and reaction conditions.

6.2.1 The Biginelli Multicomponent Reaction

It involves the condensation of an aldehyde, a β-ketoester, and urea or thiourea in one pot to give dihydropyrimidinones (DHPMs) or dihydropyrimidine thione derivatives [23]. This reaction was discovered by Pietro Biginelli in 1891. There are three possible mechanisms reported in the literature for the Biginelli reaction, namely iminium-, enamine-, or Knoevenagel mechanism. The general reaction scheme for the Biginelli reaction is shown in Scheme 6.1.

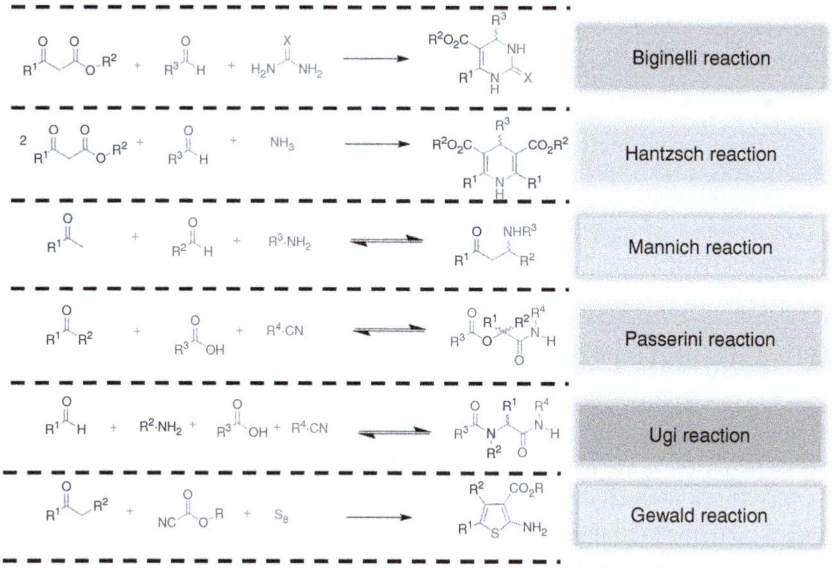

Scheme 6.1 General scheme for various types of multicomponent reactions (MCRs).

6.2.2 The Hantzsch Multicomponent Reaction

This is a classical MCR that involves the condensation of an aldehyde or a ketone, β-ketoester, and ammonia in the presence of a catalyst in one pot to give dihydropyridines (DHPs) [23]. The Hantzsch reaction is a three-component reaction discovered by Arthur Hantzsch in 1882. The general reaction scheme for the Hantzsch reaction is shown in Scheme 6.1.

6.2.3 The Mannich Multicomponent Reaction

The Mannich reaction is a three-component MCR, which involves the condensation reaction of aldehyde or ketone, amine, and compound containing acidic hydrogen most commonly α-carbonyl group to produce β-amino carbonyl compound, as shown in Scheme 6.1 [23]. This reaction was discovered by Franz Mannich in 1912.

6.2.4 The Passerini Multicomponent Reaction

This reaction involves the condensation of an aldehyde, an isocyanide, and a carboxylic acid leading to the formation of α-acyloxy amides (Scheme 6.1) [23]. This is a three-component reaction was discovered by Mario Passerini in 1921.

6.2.5 The Ugi Multicomponent Reaction

The Ugi reaction is an example of four-component MCR, which involves the one-pot synthesis of bis-amide via the condensation of an aldehyde or a ketone, an isocyanide, an amine, and a carboxylic acid (Scheme 6.1) [23]. This reaction was discovered by Ivar Karl Ugi in 1959.

6.2.6 The Gewald Multicomponent Reaction

It involves the condensation reaction of carbonyl compound with α-cyanoester in the presence of elemental sulfur to give substituted thiophene derivatives (Scheme 6.1) [23].

6.3 Solid Base Catalysts for Organic Reactions

The role of solid base catalyst in the organic transformations can be traced back to 1958 based on series of reactions studied by Pines and Haag. In this investigation, they have observed that alkali metal-supported alumina displayed excellent catalytic activity in the double bond isomerization of alkenes, which proceed through carbanion formation. One of the advantages of these solid base catalysts for isomerization of alkenes over solid acid catalysts is that these catalysts are not poisoned by the substrates (allyl amines and allyl ethers) containing heteroatoms such as N and O. In 1970, a number of catalysts were investigated for base-catalyzed

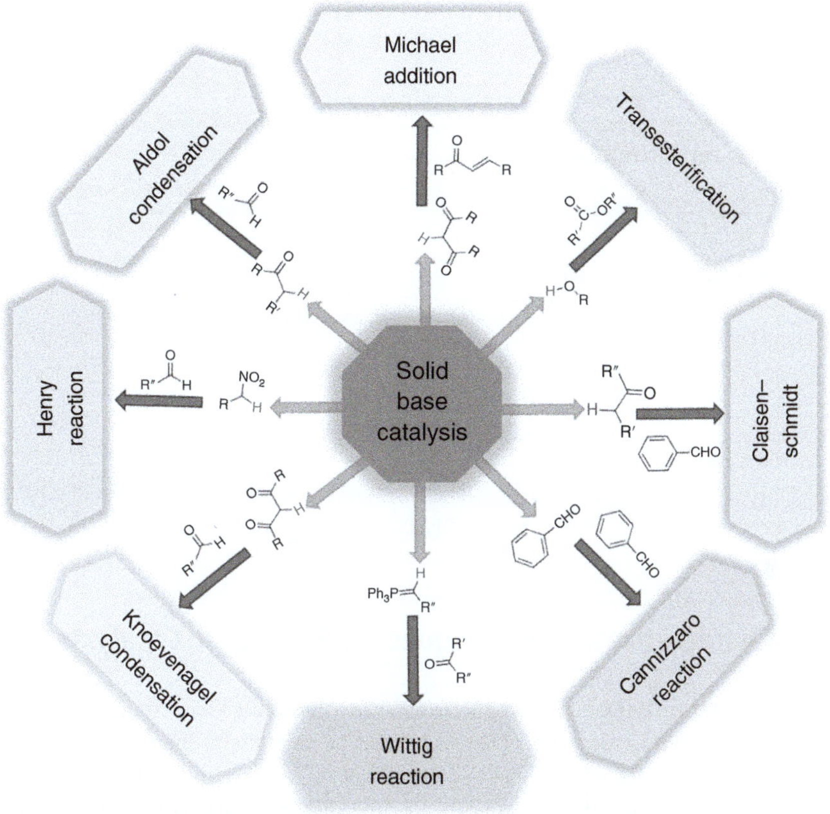

Figure 6.3 Typical solid base-catalyzed organic reactions.

organic reactions. In general, these solid base catalysts involve the abstraction of H^+ ion from the substrate by basic sites. With the recent advances in catalysis and material chemistry, wide range of solid base catalysts are known for extensive synthesis of fine chemicals in organic chemistry via various reaction routes, which includes double bond migration, Michael addition, transesterification, aldol condensation, transesterification, Knoevenagel, Claisen–Schmidt, Henry reaction, and Meerwein–Ponndorf reaction. (Figure 6.3) [24, 25].

Generally, the solid base catalysts employed for organic reactions can be classified into two subgroups (Figure 6.4). The first classification includes solid base catalysts with intrinsic basic sites. For example, solid metal oxides (MgO, CaO, SrO, BaO, ZnO, La_2O_3, Sm_2O_3, Mg-Al and Mg-Zr mixed oxides, MgO-TiO_2, ZnO-Al_2O_3, etc.) and hydroxides [($Mg(OH)_2$, $CaSn(OH)_6$, hydrotalcites (HTs), rehydrated HTs, etc.] come under this category. The second classification includes the basic species supported on nonbasic substrates. For example, supported alkali metal/compound (Na/Al_2O_3, Na/SiO_2, Na_2O/Al_2O_3, CsF/Al_2O_3, Na_2O/MgO) and grafted organic bases (amines and guanidine) on support (silica, alumina, or carbon-based materials) [21, 24].

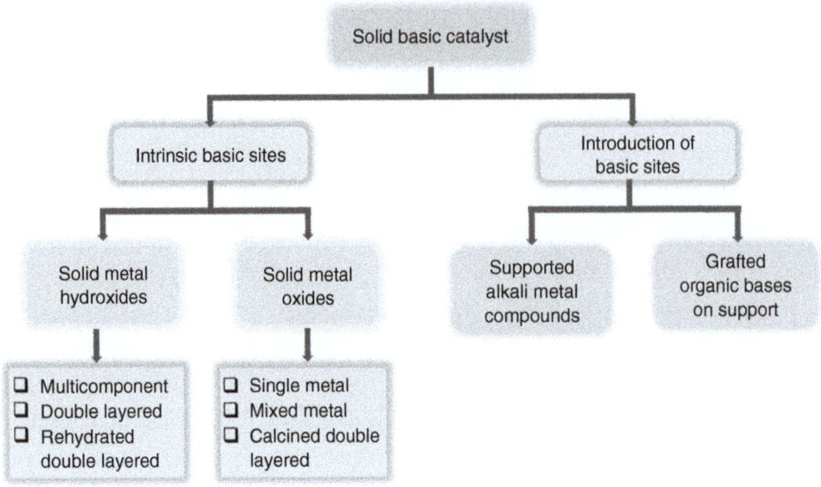

Figure 6.4 Classification of solid base catalysts used in organic synthesis.

6.4 Characterization Techniques for Solid Base Catalysts

In order to optimize the catalytic processes, it is required to have in-depth understanding of catalytic properties of the catalyst. The catalytic properties of the solid base catalyst are affected by basic properties that include the number of basic sites, their location, and basic strength. In the literature, various techniques such as acid–base titrations, solid-state NMR spectroscopy, temperature-programmed desorption (TPD), X-ray photoelectron spectroscopy (XPS), electron paramagnetic resonance (EPR) spectroscopy, atomic force microscopy (AFM), scanning tunneling microscopy (STM) have been employed for the characterization of solid base catalysts [19]. In acid–base titrations, a known amount of an acid is added to the catalyst and the number of basic sites is quantified that provides the information regarding the basic strength. XPS can identify the binding energy of the basic sites to provide better insights into the basic strength of the catalyst [24]. Furthermore, TPD technique is used to study the desorption process of the acidic probes adsorbed on the basic sites of the catalyst. This method basically involves the heating of the catalyst in inert gas atmosphere to study the desorption. The temperature at which the desorption of adsorbed species takes place can provide the information about the nature as well as strength of the basic sites present on the catalyst. IR spectroscopy is also employed to characterize the various basic functional groups by studying various vibrational modes in the IR spectrum. The effect of adsorbed molecules on the catalyst can be studied using the change in vibrational frequency of the basic sites before and after the interaction with the adsorbed species [19]. To study the local environment of the particular nucleus, NMR spectroscopy is a reliable tool. NMR technique provides the information about the nature and strength of the basic sites by studying the interaction of basic sites with the adsorbed

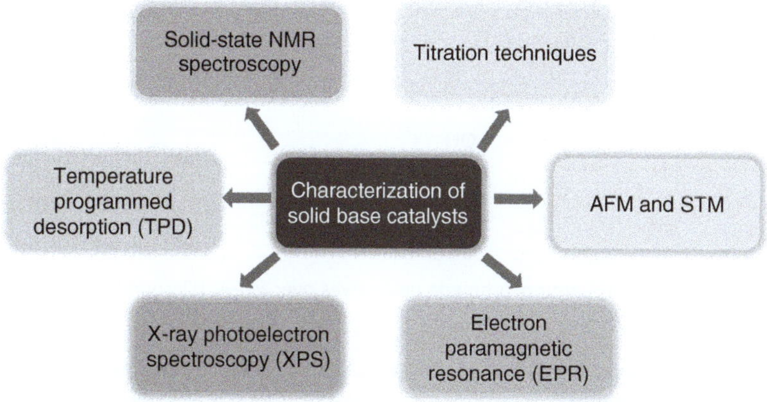

Figure 6.5 Characterization of basic sites of solid base catalysts.

species. Additionally, the morphology and the distribution of the basic sites on the catalyst surface can be analyzed using STM and AFM analysis of the catalyst surface [24] (Figure 6.5).

6.5 Heterocycle Synthesis Using Solid Base-Catalyzed MCRs

Perusal of literature revealed that variety of solid base catalysts are used for MCRs. The most of MCRs for HC synthesis uses carbonyl compounds as one of the components. This type of MCRs mainly involves general organic reactions such as imine formation, Knoevenagel condensation, Michael addition, and intramolecular cyclization reactions. [26] In base-catalyzed MCRs, HTs are emerging as excellent candidates over traditional base catalysts used for organic transformations. These materials can be used for the synthesis of various types of HCs, such as pyrans, chromenes, pyrans, pyrazoles, and triazoles, via one-pot MCR strategy. HTs are basically anionic clays having double-layered hydroxides (DLHs) structure. For example, in Mg-Al HT, i.e. $Mg_6Al_2(OH)_{16}CO_3.4H_2O$, the divalent Mg^{2+} ions are replaced by trivalent Al^{3+} ions, which results in the formation of layered structure, and the charge is balanced via the carbonate and H_2O molecules present inside the layers [27]. Surpur et al. have reported the microwave-assisted solvent-free synthesis of 2-amino-4H-chromene synthesis via one-pot MCR of aromatic aldehydes, malononitrile, and α-naphthol using Mg-Al HT as a heterogeneous catalyst. The reaction mechanism involves the Knoevenagel condensation of an aldehyde with malononitrile catalyzed by HT. The resulting intermediate adds to α-naphthol at third position followed by cyclization to afford required chromene heterocycle. The basic properties of HT can be varied by changing the Mg/Al ratio. It was observed that HT that have Mg/Al ratio equal to 3 showed the excellent efficiency with 84% yield of product. Furthermore, the calcined HTs exhibited significantly lower activity for the model reaction between benzaldehyde, malononitrile, and α-naphthol due to release of CO_2 and water from HTs via cratering mechanism at high temperatures [28] (Scheme 6.2).

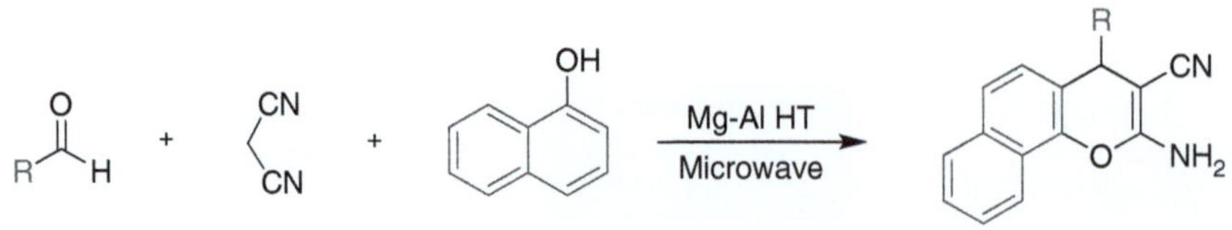

Scheme 6.2 Synthesis of 2-aminochromene derivatives over Mg-Al HT solid base catalyst.

Antonyraj et al. reported the first synthesis of DHP derivatives using HTs as a solid base catalyst via condensation of ethyl acetoacetate (EAA), ammonium acetate, and aldehyde. In this report, they have employed different HT-based catalysts, and among all the screened catalysts $MgAl_2$-HT (Mg/Al atomic ratio = 2) showed the maximum yield of the products. The reaction was optimized under different conditions, and the proposed mechanism is shown in Scheme 6.3 [29].

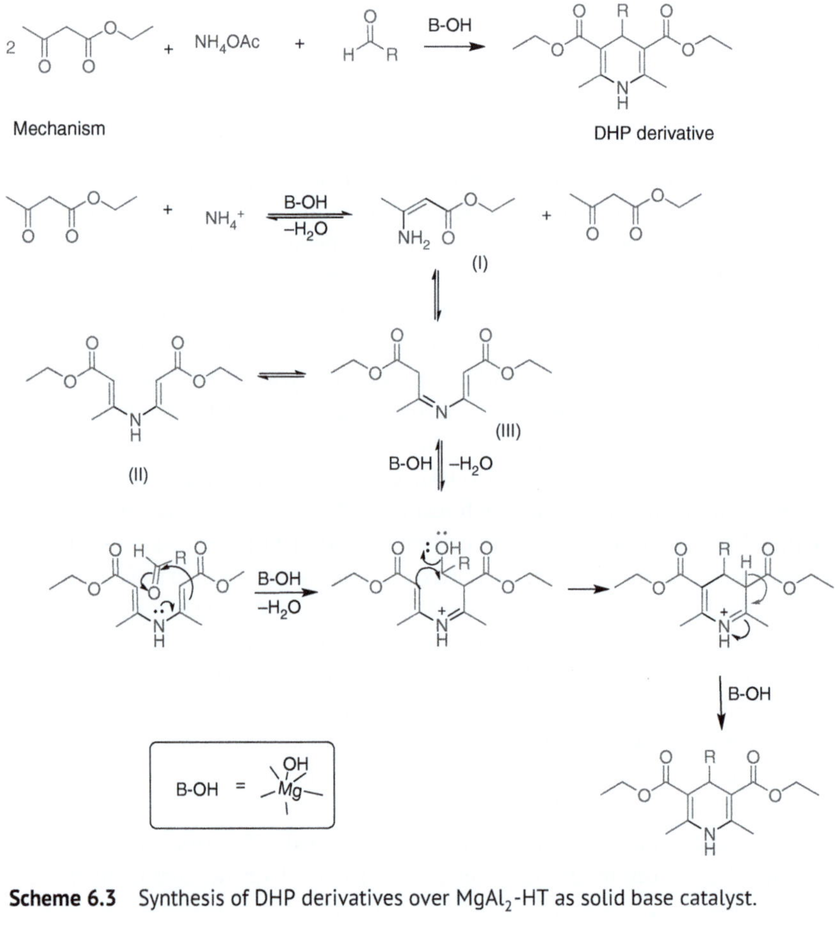

Scheme 6.3 Synthesis of DHP derivatives over $MgAl_2$-HT as solid base catalyst.

The first step involves the abstraction of acidic proton from the EAA by the basic site of the catalyst followed by reaction with ammonium acetate to give Intermediate I. This intermediate undergoes further reaction with the second molecule of EAA to form imine Intermediate II, which undergoes tautomerism to form an Enamine III followed by condensation of aldehyde to afford DHP derivative [29].

Solid base catalysts derived from metal and mixed metal-based oxide precursors are also emerging as promising candidates for MCRs. Babu et al. reported Mg/La mixed oxide-based heterogeneous catalyst for one-pot MCR of an aldehyde, malononitrile, and a keto ester to afford pyran derivatives. The presence of La increases the activity of the catalyst and the prepared catalyst can be reused even after four cycles without any significant lose in activity. The strong basic sites present in the catalyst promotes the abstraction of proton from malononitrile and generates the carbanion, which undergoes Knoevenagel condensation with an aldehyde followed by Michael addition with keto ester to give functionalized 4H-pyran derivatives [25] (Scheme 6.4).

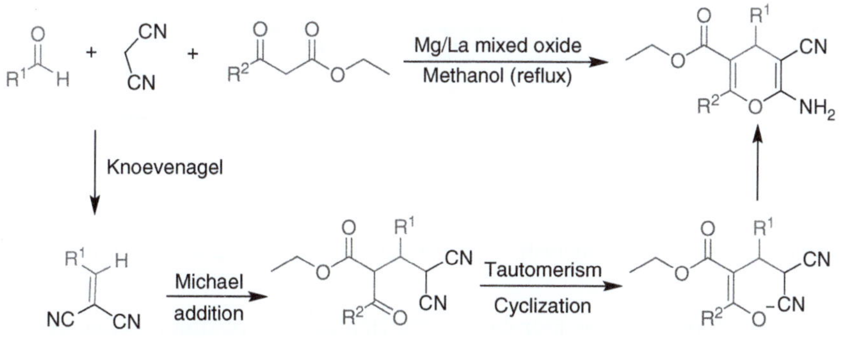

Scheme 6.4 Synthesis of 4H-pyran derivatives using Mg/La mixed oxide as solid base catalyst.

Similarly, nano-sized CaO has been used as solid base catalyst for the preparation of substituted pyridines via one-pot MCR of aldehyde, thiol, and malononitrile. The yield of the product obtained from the nano-sized CaO is significantly higher than the bulk CaO catalyst. Furthermore, the yield of the pyridine derivatives varies with the choice of the catalyst as follows: $SiO_2 < Mn_3O_4 < CuO < NiO < CaO$ [30] (Scheme 6.5).

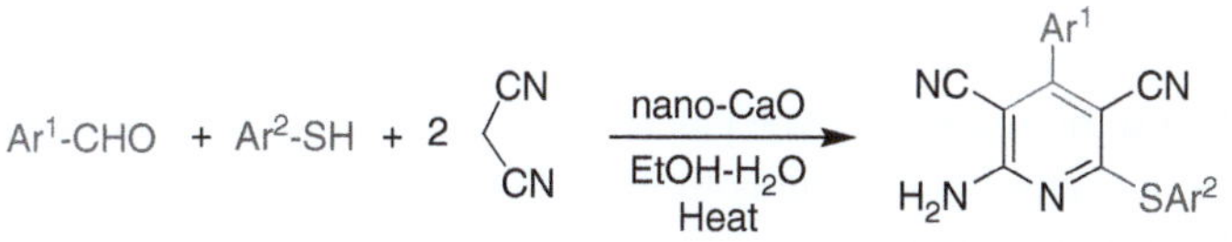

Scheme 6.5 Synthesis of pyridine derivatives using nano-sized calcium oxide as solid base catalyst.

Tamaddon et al. reported selectivity in Biginelli versus Hantzsch reactions using ZnO-based solid base catalyst. There are two possibilities associated with the urea, either it can participate as component for MCR or dissociate into ammonia, which increases the competition between Biginelli versus Hantzsch reaction. It has been observed that with ZnO the Biginelli reaction occurs at comparatively lower temperature under solvent-free conditions, whereas the Hantzsch reaction precedes in water at high temperature or under microwave conditions [8] (Scheme 6.6).

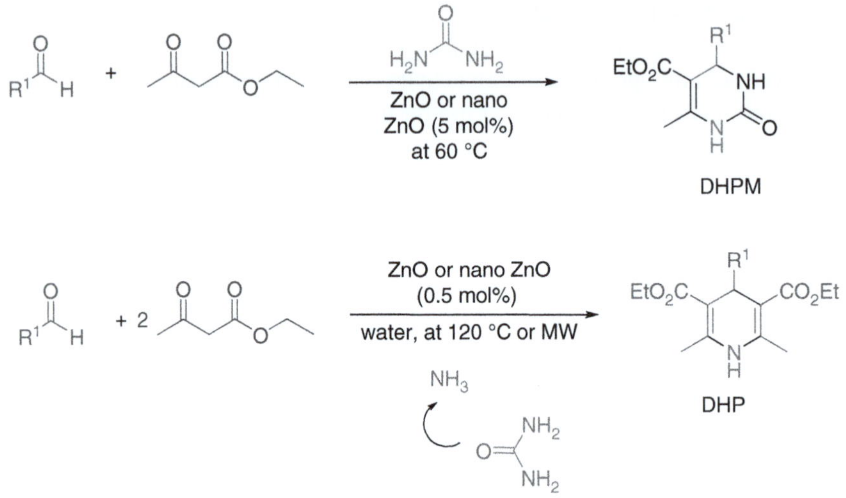

Scheme 6.6 Selectivity in Biginelli versus Hantzsch reactions using ZnO as catalyst.

Tekale et al. reported the four-component Hantzsch polyhydroquinoline derivative synthesis using heterogeneous La_2O_3 as an effective catalyst in trifluoroethanol in short reaction time. The reaction involves one-pot condensation of an aldehyde, dimedone, alkyl acetoacetate, and ammonium acetate [31] (Scheme 6.7).

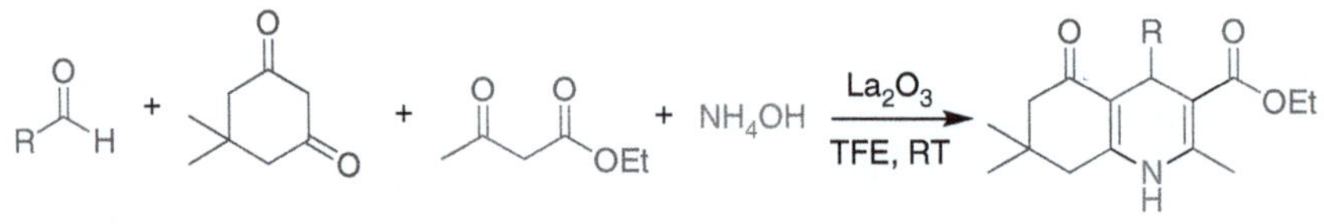

Scheme 6.7 Synthesis of polyhydroquinoline derivatives using La_2O_3 as solid base catalyst.

Ansari et al. described the one-pot MCR for the production of steroidal pyridines derivatives with the use of nano-sized MgO as heterogeneous, reusable, and mild catalyst. The prepared catalyst is bearing both Mg^{2+} (Lewis acid) and O^{2-} (Lewis base) sites. In addition to this, there are various cationic and anionic vacancies in the crystal lattice. Perusal of literature revealed that the MgO nitrophenol (NPs) are extremely effective as basic and heterogeneous catalyst for Michael addition and Knoevenagel condensation reactions. In the catalytic cycle, the active sites present on the reactant molecules are well exposed to the reaction components and stabilized the intermediates formed to afford the desired products in excellent yields [26] (Scheme 6.8).

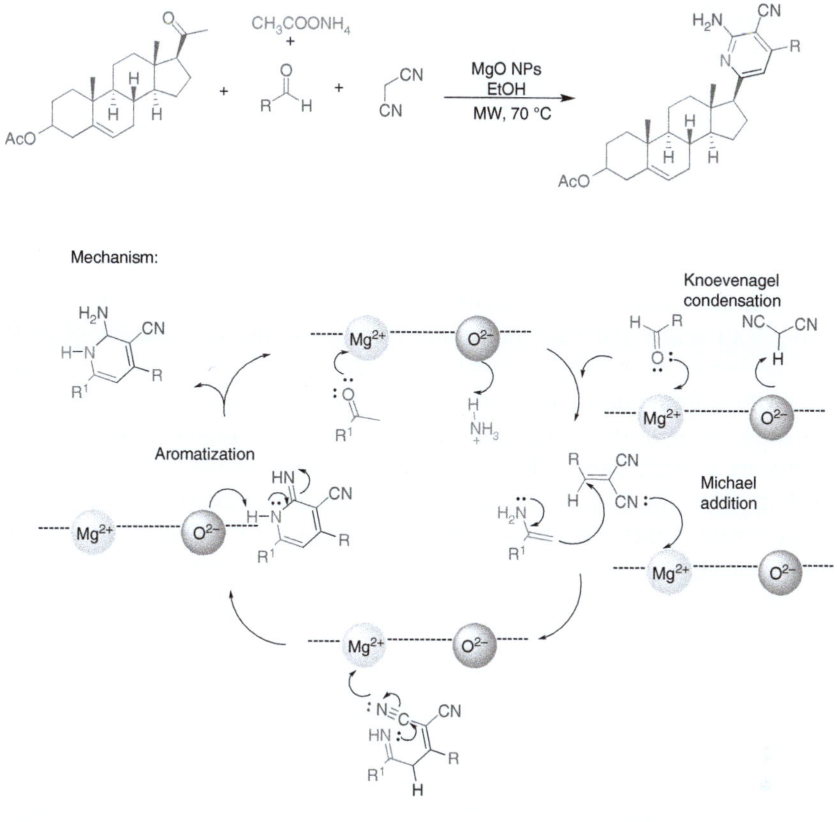

Scheme 6.8 Synthesis of steroidal pyridine derivatives using MgO NPs as solid base catalyst.

Aydin et al. reported the use of KF/Al_2O_3 as a solid base catalyst for the preparation of Hantzsch DHP derivatives via one-pot MCR of an aldehyde, β-ketoester, and ammonia. The ammonia used for the synthesis can be obtained either by thermal decomposition of ammonium acetate or by bubbling ammonia through the reaction medium [32] (Scheme 6.9).

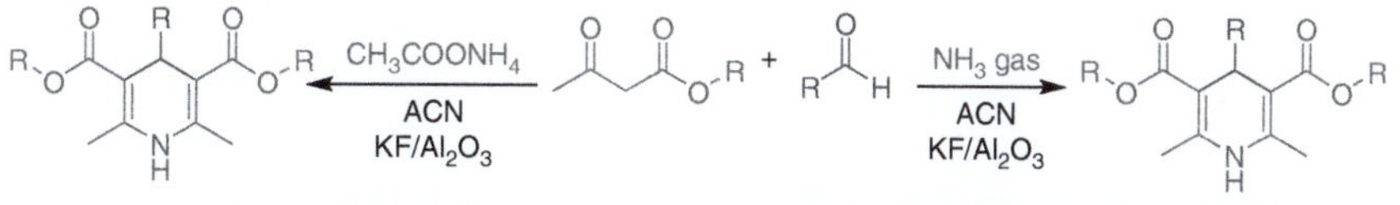

Scheme 6.9 KF/Al_2O_3 as a solid base catalyst for the preparation of Hantzsch dihydropyridine derivatives.

Nimkar et al. reported the use of CsOH supported on Al_2O_3 as a reusable, heterogeneous basic catalyst for the one-pot MCR of acetophenone, aromatic aldehydes, and guanidine to synthesize 2-amino-4,6-diaryl pyrimidine derivatives via green and efficient approach. The developed catalyst can be reused up to five catalytic cycles, and the catalytic system generates the final products in short reaction time and excellent yields [33].

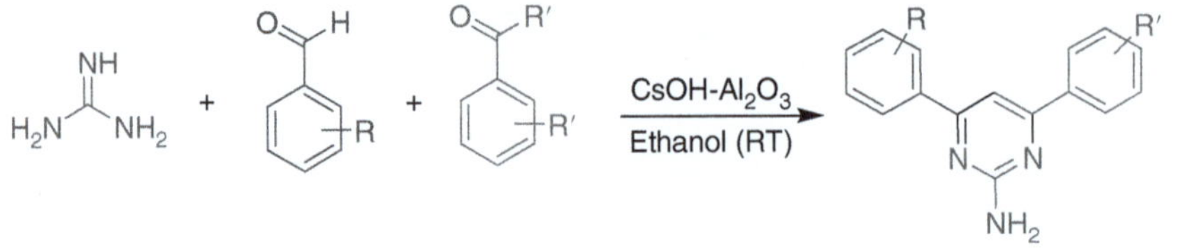

Singh et al. reported carbon quantum dots (CQDs)-based bifunctional catalyst bearing both acidic and basic functional groups. During the synthesis of CQDs from the lemon peels, ethylene amine was added to functionalize the prepared CQDs with amino functional groups on their surface. Furthermore, the functionalization of some of the amino groups was done using sulfonic acid derivatives, which were initially derived from the reaction of sulfone derivative with corresponding aldehyde. The developed catalyst possesses both H-bond donor and H-bond acceptor sites. This bipolar catalyst can be used for efficient synthesis of various benzo[g]chromene and pyrimidine derivatives in pure aqueous medium [1] (Scheme 6.10).

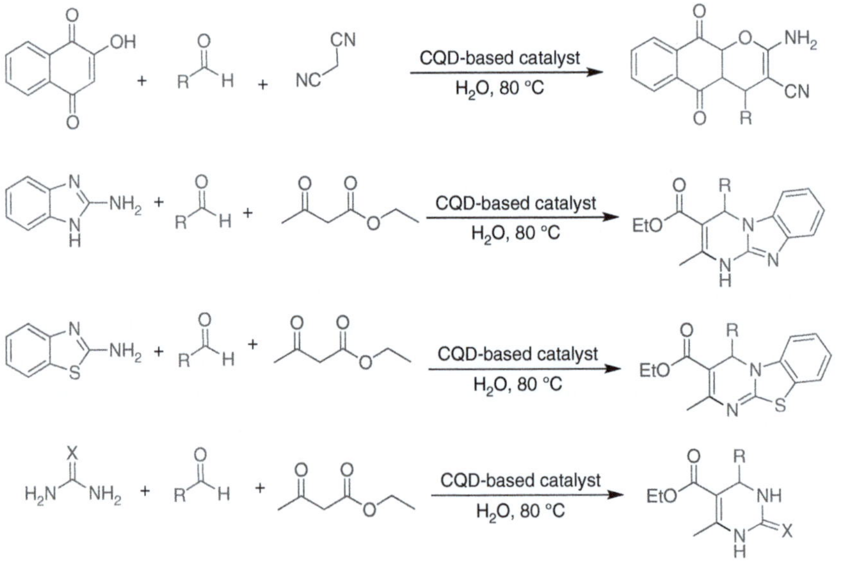

Scheme 6.10 Synthesis of various benzo[*g*]chromene and pyrimidine derivatives in pure aqueous medium using CQD-based bifunctional catalyst.

6.6 Conclusion and Future Trends

In the synthesis of heterocycles via MCRs, solid base catalysts have significant benefits over conventional homogeneous catalysts. Researchers are continuously investigating the synthesis and use of different solid base catalysts for different applications in the field of catalysis. Furthermore, research is needed to synthesize more solid base catalysts for eco-friendly and sustainable approach to synthesize various heterocycles in a short time interval. The forthcoming research on the role of solid base catalyst for MCRs may focus on the asymmetric synthesis of various heterocycles using specific catalysts. This application of solid base catalysts in MCRs can replace the expensive chiral reagents used in asymmetric synthesis.

Acknowledgment

The authors are thankful to Chandigarh University for the research facilities.

References

1 Singh, A., Singh, G., Sharma, S. et al. (2022). Metal-free, biomass-derived nano-architectured carbon quantum dots as an efficient acid-base bifunctional catalyst for facile synthesis of benzo[g]chromene and pyrimidine analogs. *ChemistrySelect* 7.

2 Hülsey, M.J., Yang, H., and Yan, N. (2018). Sustainable routes for the synthesis of renewable heteroatom-containing chemicals. *ACS Sustainable Chemistry & Engineering* 6: 5694–5707.

3 El-Nassan, H.B. (2021). Amberlyst 15®: an efficient green catalyst for the synthesis of heterocyclic compounds. *Russian Journal of Organic Chemistry* 57: 1109–1134.

4 Rotstein, B.H., Zaretsky, S., Rai, V., and Yudin, A.K. (2014). Small heterocycles in multicomponent reactions. *Chemical Reviews* 114: 8323–8359.

5 Yu, J.T. and Pan, C. (2016). Radical C-H functionalization to construct heterocyclic compounds. *Chemical Communications* 52: 2220–2236.

6 Bosica, G. and Abdilla, R. (2022). Recent advances in multicomponent reactions catalysed under operationally heterogeneous conditions. *Catalysts* 12: 725.

7 Elinson, M.N., Dorofeev, A.S., Miloserdov, F.M. et al. (2008). Catalysis of salicylaldehydes and two different C-H acids with electricity: first example of an efficient multicomponent approach to the design of functionalized medicinally privileged 2-amino-4H-chromene scaffold. *Advanced Synthesis and Catalysis* 350: 591–601.

8 Tamaddon, F. and Moradi, S. (2013). Controllable selectivity in Biginelli and Hantzsch reactions using nanoZnO as a structure base catalyst. *Journal of Molecular Catalysis A: Chemical* 370: 117–122.

9 Tomar, R., Kundra, P., Sharma, J. et al. (2024). An overview: synthesis of menthol using heterogeneous catalysis. *Letters in Organic Chemistry* 21 (1): 16–28.

10 Kumar, L., Verma, N., Tomar, R. et al. (2023). Development of bioactive 2-substituted benzimidazole derivatives using an MnO_x/HT nanocomposite catalyst. *Dalton Transactions* 52 (10): 3006–3015.

11 Tomar, R., Jain, S., Yadav, P. et al. (2022). Conversion of limonene over heterogeneous catalysis: an overview. *Current Organic Synthesis* 19 (3): 414–425.

12 Kumar, L., Verma, N., Sehrawat, H. et al. (2022). Successive oxidation–condensation reactions using a multifunctional gold-supported nanocomposite (Au/MgCe–HDO). *New Journal of Chemistry* 46 (7): 3472–3481.

13 Tomar, R., Singh, N., Kumar, N. et al. (2019). Base-free Suzuki–Miyaura coupling reaction using Palladium (II) supported catalyst in water. *Catalysis Letters* 149 (6): 1589–1594.

14 Tomar, R., Singh, N., Rathee, G. et al. (2017). Synthesis and characterization of hybrid $Mg(OH)_2$ and $CeCO_3OH$ composite with improved activity towards henry reaction. *Asian Journal of Organic Chemistry* 6 (12): 1728–1732.

15 Tomar, R., Sharma, J., Nishimura, S., and Ebitani, K. (2016). Aqueous oxidation of sugars into sugar acids using hydrotalcite-supported gold nanoparticle catalyst under atmospheric molecular oxygen. *Chemistry Letters* 45 (7): 843–845.

16 Arya, K. and Tomar, R. (2013). Microporous zeolite catalyst system: an eco-approach for regioselective synthesis of pyrimidobenzimidazoles. *Research on Chemical Intermediates* 41 (6): 3389–3400.

17 Tomar, R., Ebitani, K., and Chandra, R. (2019). Hydrotalcite-supported ceria nanoparticles as a heterogeneous catalyst for one-pot synthesis of imines under atmospheric air. *ChemistrySelect* 4 (12): 3577–3581.

18 Tomar, R., Rathee, G., Chandra, I. et al. (2018). Synthesis and characterization of magnesium hydroxide & cerium oxide composite: application in organic transformation. *ChemistrySelect* 3 (6): 1645–1649.

19 Hattori, H. (2015). Applied catalysis A: general solid base catalysts: fundamentals and their applications in organic reactions. *Applied Catalysis A: General* 504: 103–109.

20 John, S.E., Gulati, S., and Shankaraiah, N. (2021). Recent advances in multicomponent reactions and their mechanistic insights: a triennium review. *Organic Chemistry Frontiers* 8: 4237–4287.

21 El-khateeb, A.Y., Hamed, S.E., and Elattar, K.M. (2022). Recent advancements in the multicomponent synthesis of heterocycles integrated with a pyrano. 11808–11842. https://doi.org/10.1039/d2ra00927g.

22 Mcrs, U. and Neto, B.A.D. (2014). RSC advances what do we know about multicomponent reactions? *Mechanisms and Trends for the Biginelli* 54282–54299. https://doi.org/10.1039/c4ra10651b.

23 Cioc, R.C., Ruijter, E., and Orru, R.V.A. (2014). Multicomponent reactions: advanced tools for sustainable organic synthesis. *Green Chemistry* 16: 2958–2975.

24 Bing, W. and Wei, M. (2019). Recent advances for solid basic catalysts: structure design and catalytic performance. *Journal of Solid State Chemistry* 269: 184–194.

25 Seshu Babu, N., Pasha, N., Venkateswara Rao, K.T. et al. (2008). A heterogeneous strong basic Mg/La mixed oxide catalyst for efficient synthesis of polyfunctionalized pyrans. *Tetrahedron Letters* 49: 2730–2733.

26 Ansari, A., Ali, A., Asif, M., and Shamsuzzaman, V. (2018). Microwave-assisted MgO NP catalyzed one-pot multicomponent synthesis of polysubstituted steroidal pyridines. *New Journal of Chemistry* 42: 184–197.

27 Burange, A.S. and Gopinath, C.S. (2021). Catalytic applications of hydrotalcite and related materials in multicomponent reactions: concepts, challenges and future scope. *Sustainable Chemistry and Pharmacy* 22: 100458.

28 Surpur, M.P., Kshirsagar, S., and Samant, S.D. (2009). Exploitation of the catalytic efficacy of Mg/Al hydrotalcite for the rapid synthesis of 2-aminochromene derivatives via a multicomponent strategy in the presence of microwaves. *Tetrahedron Lett.* 50: 719–722.

29 Antonyraj, C.A. and Kannan, S. (2008). Hantzsch pyridine synthesis using hydrotalcites or hydrotalcite-like materials as solid base catalysts. *Applied Catalysis A General* 338: 121–129.

30 Fedorova, O.V., Titova, Y.A., and Ovchinnikova, I.G. (2021). Metal oxides in multicomponent synthesis of heterocycles. *Chemistry of Heterocyclic Compounds* 57: 900–904.

31 Tekale, S.U., Pagore, V.P., Kauthale, S.S., and Pawar, R.P. (2014). La_2O_3/TFE: an efficient system for room temperature synthesis of Hantzsch polyhydroquinolines. *Chinese Chemical Letters* 25: 1149–1152.

32 Aydin, F. and Ozen, R. (2004). Potassium fluoride/alumina mediated simple synthesis of Hantzsch 1,4-dihydropyridines. *Journal of Chemical Research* 486–487: https://doi.org/10.3184/0308234042037284.

33 Amey Nimkar, M.M.V.R., Betkar, R., Ranade, P., and Mundhe, B. (2016). CsOH/c-Al_2O_3: a heterogeneous reusable basic catalyst for one-pot synthesis of 2-amino-4,6-diaryl pyrimidines. *New Journal of Chemistry* 40: 2541–2546.

7

Industrial Applications of Solid Base Catalysis

Navdeep Kaur and Nibedita Banik

Chandigarh University, UIS, Department of Chemistry, Gharuan, NH-05, Ludhiana Highway, Mohali, Punjab, India

7.1 Introduction to Solid Base Catalysis

7.1.1 Definition and Characteristics of Solid Base Catalysts

A sort of catalytic reaction known as "solid base catalysis" uses a solid substance, also called a "solid base," to act as a base and promote a chemical reaction. In contrast to conventional catalysis, which often uses soluble catalysts for liquid or uniform catalysis, solid base catalysis uses a solid catalyst that has the ability to both take and contribute essential locations in order to facilitate reactions.

Ono and Hattori [1] conducted the first work on solid base catalysts, demonstrating that sodium metal dissolved on alumina acted as a potent initiator for the double-bond isomerization of alkenes [1]. The advancement of reactions catalyzed by solid bases has persisted. The thorough reviews that have been published recently make it abundantly evident that this advancement is accelerating as solid bases are acknowledged to be harmless or nontoxic catalysts. Solid base catalysis has several traits, as shown in Figure 7.1:

Additional benefits of solid-base catalysis include:

- For instance, stoichiometric amounts of liquid bases are frequently needed for certain organic processes in order to produce carbanions [1] as well as metal salts as by-products. In this case, under phase transfer conditions, the methylation of phenyl acetonitrile ($PhCH_2CN$) with methyl iodide (CH_3I) [1] occurs in the presence of bases.

$$PhCH_2CN + CH_3I + NaOH \rightarrow PhCH(CH_3)CN + NaI + H_2O$$

In this particular case, neutralizing the hydrogen iodide (HI) created and maintaining the basicity of the solution requires more sodium hydroxide [1, 2] than a stoichiometric amount. Furthermore, inevitably, a stoichiometric amount of sodium iodide (NaI) [3] will form, and it must be disposed of properly. As a result, there is no catalytic reaction. Consequently, there is a great need for catalytic reactions to prevent these issues [1–3].

Solid Base Catalysts: Synthesis, Characterization, and Applications, First Edition.
Edited by Ravi Tomar, K.K. Pant, and Ramesh Chandra.

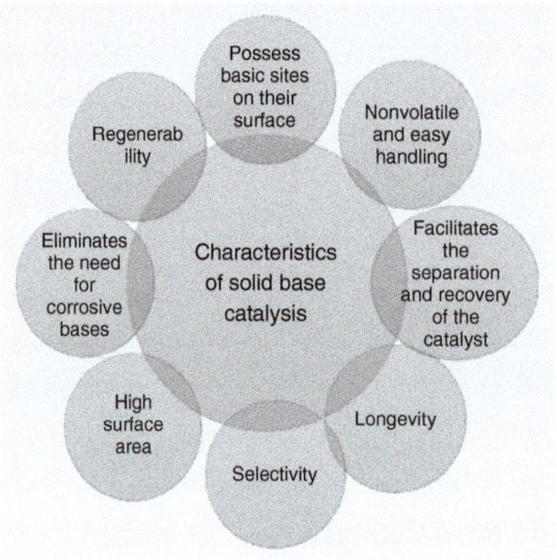

Figure 7.1 Characteristics of solid base catalysis. Source: Ono and Hattori [1], 2012/with permission of Springer Nature.

Alkali-exchanged zeolites can be used as catalysts in conjunction with methanol (CH_3OH) or dimethyl carbonate ($C_3H_6O_3$) as the methylating agent to accomplish the methylation of phenyl acetonitrile. The process can be carried out in the vapor phase [1].

$$PhCH_2CN + CH_3OH \rightarrow PhCH(CH_3)CN + H_2O$$

$$PhCH_2CN + CH_3OCOOCH_3 \rightarrow PbCH(CH_3)CN + CH_3OH + CO_2$$

There are no inorganic salts produced here, such as sodium iodide, because the reactions are catalytic. A stoichiometric quantity of base compounds is not required. As a result, the reactions may be significantly less wasteful and hence more atomically efficient.

- Solid bases have a significant advantage over homogeneous bases in terms of the solvent. Solvents in reactions catalyzed by solid bases are only required for the dissolution of reactants (as well as products). This greatly broadens the selection of solvents. It is established that the outcomes of solvents in solid-base catalyzed reactions contrast with those in homogeneous corrosive reactions. For instance, water (H_2O) and hydrocarbons (C_nH_{2n+2}) like pentane work well as solvents in the Knoevenagel condensation of 2-furaldehyde ($C_5H_4O_2$) with ethyl cyanoacetate ($C_5H_7NO_2$) on xonotlite [4] (Ca_2O_4Si). Additionally, there is no need for a solvent to conduct the reactions [5, 6]. Reactions can occur even in the vapor phase.
  In these situations, it is evident that separating the solvents once the reactions are finished substantially simplifies the procedure. There is no waste produced from solvents like organic solvents having alkaline water [1] or water-containing organics. Less-complicated separation means less energy is needed. Additionally, solid bases can be separated and used again. This aids in the development of high atom-economy catalyst [1]. More significantly, having more opportunities to discover novel reactions is made possible by being freed from solvent selection constraints.

7.1.2 Importance in Industrial Catalysis

Enhanced selectivity along with specificity in chemical processes can frequently be achieved with solid base catalysts. This implies that they can assist in ensuring that the intended product is generated without any unwanted by-products. Chemical, energy sources, and drug manufacturing are just a few of the industrial processes [7] that employ solid base catalysts. Alkylation, the process of hydrogenation, esterification, and transesterification are among the processes [7, 8] in which they are engaged. The utilization of solid-base catalysts in industries was surveyed as in Table 7.1.

Solid base catalysts frequently make it possible to conduct operations under gentler circumstances, which lowers the energy needed and the environmental effect of industrial reactions. This is consistent with industry's increased focus on environmentally friendly and sustainable chemistry practices. Also, solid base catalyst allows better control [24] over the reaction and simplifies the downstream processing of the final products.

In conclusion, because of their adaptability, advantages for the environment, increased selectivity, ability to resist poisoning, thermal endurance, rapid separation, security, control over reaction kinetics, decreased risk of corrosion, and flexibility, solid base catalysts are an essential component of industrial catalysis. Their use in a variety of industrial processes [25] promotes innovation across several sectors, increases efficiency [25], and lessens environmental effects.

7.1.3 Comparison with Solid Acid Catalysts

Solid base and solid acid catalysts are two types of heterogeneous catalysts that play essential roles in various chemical reactions. They differ in their mode of action and the types of reactions they can catalyze. They are in a different phase than the reactants and products in a chemical reaction. They are typically used in the form of powders or beads [26–36], which can be easily separated from the reaction mixture after the reaction is complete. This makes them attractive for industrial applications [25, 26], as they can be reused multiple times. Solid acid and solid base catalysts are contrasted here as in Table 7.2.

Because of this, solid base and solid acid catalysts are selected according to the particular needs of the reaction and have varying levels of reactivity. Choosing the right catalyst for a particular chemical transformation requires a grasp of the characteristics of the catalyst and the nature of the reactions.

7.2 Biodiesel Production

Transesterification [37] is a process that turns vegetable [37] or animal fats [37] into biodiesel—an environmentally [37] benign and renewable fuel substitute for conventional diesel. Triglycerides ($C_6H_8O_6$), which are present in oils and fats, react with methanol (CH_3OH) or ethanol [38] (C_2H_5OH) to form biodiesel (RCO_2CH_3 or $C_6H_{13}N_3O_2$) and glycerol [38] ($C_3H_8O_3$). This process is frequently aided by a catalyst. Catalysts with solid bases are essential to this process.

Table 7.1 The specific industrial catalysts and methods.

Process	Catalyst	Company	Reference
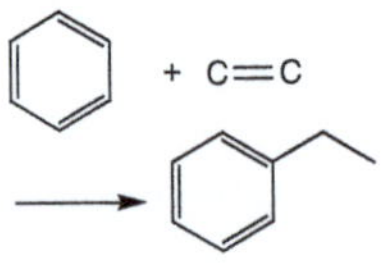 *1.*	H-ZSM-5 vapor phase [8]	Mobil ± Badge [8]	[8–10]
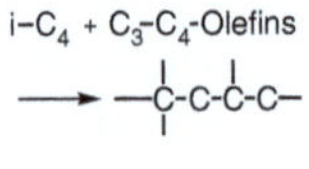 *2.* *Alkylation processes*	CF_3SO_3H/silicon dioxide [8] Antimony pentafluoride/SiO_2 Sulfated zirconium dioxide	Kellog FBA-process/ Haldor Topsoe CDTECH [8] Orient Catalyst [8]	[8, 11, 12] [8, 13, 14] [8, 15]
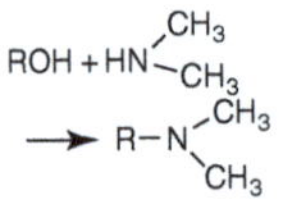 *Amination process*	Copper, nickel/ SiO_2-Al_2O_3	Kao [8]	[8, 16]
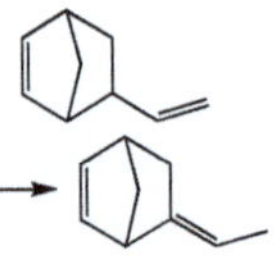 *Isomerization process*	Sodium/sodium hydroxide/Al_2O_3	Sumitomo chemical [8]	[8, 17]
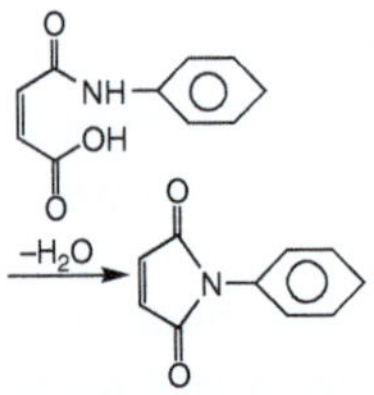 *Dehydration and condensation process*	SiO_2/H_3PO_4-aniline salt	Nippon Shokubai [8]	[8, 18]
Deep gas oil cracking in vacuum *Cracking process*	$Na_nAl_nSi_{96}$–nO_{192}·$16H_2O$ $(0 < n < 27)$ (pentasil zeolite)	China Petro [8]	[8, 19]
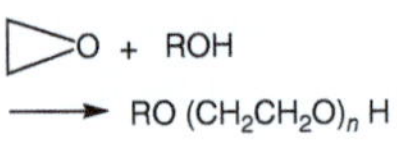 *Etherification process*	Hydrotalcite $Mg_6Al_2O_8(OH)_2$	Henkel [8]	[8, 20]

(Continued)

Table 7.1 (Continued)

Process	Catalyst	Company	Reference
Naphtha → Aromatics	Pt L-zeolite	Chevron Chem. AROMAX process [8]	[8, 20]
Aromatization process	R-132 catalyst	UOP LLR-Platforming [8]	[8, 21]
Hydrocracking of gas oils wax + Hydrogen → gasoline *Hydrocracking process*	Amorphous silicon dioxide–Al_2O_3 with zeolite		[8, 9, 22]
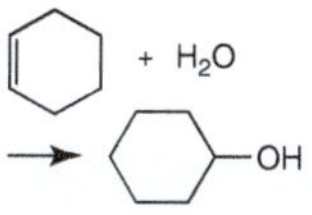 *Hydration process*	Ultrasiliceous H-ZSM-5 [8], less than 1 mm	Asahi Chemicals	[8, 16]
$CO + H_2$ → gasoline *Hydrogenation process*	Zeolite	BP	[8, 23]

Source: Adapted from Tanabe and Hölderich [8].

Table 7.2 Difference in characteristics of solid acid catalysts and solid base catalysts.

Characteristics	Solid base catalyst	Solid acid catalyst
Nature of active sites	Basic active sites [27] that can take and donate electrons, such as hydroxide (OH–) or amine (NH_2–) groups, are found in solid base catalysts. By taking in protons or reacting with acids, these sites [27] help reactions happen more quickly	Acidic active sites found in solid acid catalysts, such as sulfonic acid groups (RSO_2OH) or acidic oxides [28], can donate protons (H^+ ions) and facilitate proton [28] transfer, which in turn can drive processes
Types of reactions	Employed mostly in nucleophilic substitution processes, like Knoevenagel condensations [29], aldol condensations [30], and the transesterification of esters [31]. Additionally, they are capable of catalyzing processes in which acidic protons are removed	Usually used in reactions like esterification [32], dehydration, and hydrolysis reactions [33] where protonation is an essential step. They may also encourage Fischer esterification and other acid-catalyzed processes
Mechanisms	Solid base catalysts work by either taking up or removing protons [27-28] from molecules of the reactants. They accomplish this by giving reactants a surface to interact with, frequently due to the surface's inclusion of basic sites [28] or functional groups	In contrast, protons [34] are donated to the molecules of the reactant by solid acid catalysts. Usually, they have acidic functional groups [34] or locations that facilitate protonation-induced bond breakage
Strength of catalysis	Solid base catalysts are often stronger and more efficient catalysts compared to solid acid catalysts in basic catalyzed reactions	Solid acid catalysts are typically stronger and more efficient catalysts in acid catalyzed reactions
Reactivity	More reactive in the presence of acidic compounds, as they can neutralize the acidity by accepting protons [27, 28]	More reactive in presence of basic compounds, as they can donate protons to basic species [28, 34]
Selectivity	When certain functional groups are deprotonated [35], solid base catalysts can demonstrate high selectivity [7], which can result in regional selectivity as well as stereo selectivity	Solid acid catalysts can be selective in promoting specific reactions, such as the fonnation of specific carbocation [36] intermediates, leading to regioselectivity as well as stereoselectivity
Sensitivity to moisture and impurities	Less sensitive to moisture and impurities [37] because the basic sites can tolerate some level of contamination	More sensitive to moisture and impmities, as the acidic sites can be deactivated by moisture and other impurities [37]
Examples	Metal oxides (MgO, CaO), anion exchange resins, alkaline earth metal hydroxides, zeolites [8, 38] with basic sites etc.	Sulfated metal oxides, solid acids like silica alumina, zeolites [8] with acidic sites etc.
Applications	Used in reactions like aldol condensation [30], Knoevenagel condensation [29], and various organic transformations. They are employed in biodiesel production [37] and some types of hydrocarbon cracking processes	Catalyse reactions such as esterification [32], cracking of large hydrocarbons [39] into smaller ones. They are also used in the production of high- octane fuels and petrochemicals [40]
Regeneration and stability	Solid base catalysts can be prone to deactivation due to poisoning by acidic impurities or reaction byproducts. They may require regeneration or replacement	Solid acid catalysts can also undergo deactivation over time, especially when exposed to highly basic or acidic reactants, and may need regeneration or reactivation

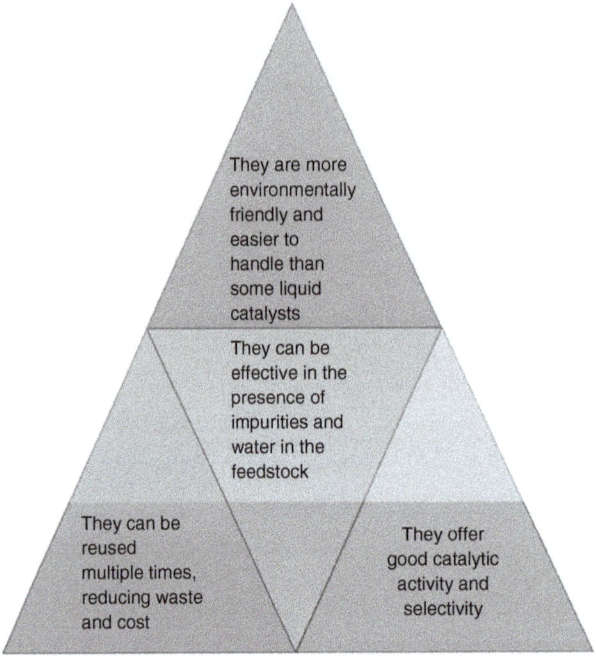

Figure 7.2 Advantages of using solid base catalysts in biodiesel production.

Solid base catalysts are typically made from alkaline materials, such as metal oxides (e.g., calcium oxide (CaO), magnesium oxide (MgO) [8, 39], or sodium hydroxide (NaOH)) supported on solid materials like silica or alumina. They act as a base to neutralize the acidic components present in the feedstock, such as free fatty acids (FFA) and moisture [40], which can hinder the transesterification reaction [37, 38, 40]. By neutralizing these acidic impurities, solid base catalysts facilitate the reaction and prevent side reactions, resulting in higher biodiesel yields and purity. When producing biodiesel, solid base catalysts have the following benefits as shown in Figure 7.2.

7.2.1 Transesterification Reactions

The key chemical reaction in biodiesel production is transesterification [37]. During this reaction, triglycerides (which consist of three fatty acid chains linked to a glycerol molecule) are converted into biodiesel and glycerol [38]. The reaction involves the replacement of the glycerol's ester bonds with alcohol (methanol or ethanol) to form biodiesel molecules and glycerol [38].

Many heterogeneous catalysts [40], including alkaline earth metal oxides and other alkali metal complexes based on aluminum or zeolite [40–44], have been investigated recently for the transesterification of vegetable oils into biodiesel. Nevertheless, the majority of supported alkali catalysts [45] have limited catalyst lifetimes due to the active components' easy methanol corrosion. Since CaO, MgO, Barium Oxide, and SrO has basic sites that are stronger than H = 26.5 [45–47],

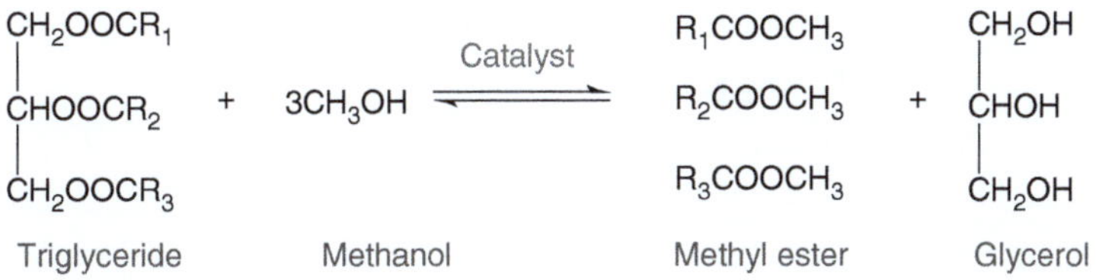

Figure 7.3 Transesterification reaction.

numerous researchers have looked into it as a solid strong base catalyst. Using calcium oxide (CaO), MgO, Barium Oxide, and SrO [48] as a solid base catalyst to produce biodiesel has several benefits, including increased activity, tolerant reaction conditions, extended catalytic lifetimes, cheap catalyst costs, and more.

7.2.2 Catalysts and Mechanisms

The oil needs to be pretreated by heating it above 100 °C [49] for one hour to eliminate water and other contaminants. Using a conventional titrimetric technique, the oil's FFA content [49] was ascertained; the equation for the transesterification reaction [49] is shown in Figure 7.3.

The following is a description of the mechanism [49] involved in the conversion of triglyceride ($C_6H_8O_6$) to biodiesel when a metal oxide catalyst is present.

Step I: Initially, H^+ is extracted from CH_3OH by surface O^{2-}, resulting in CH_3O^-—a highly basic compound with catalytic properties in the transesterification reaction [50].

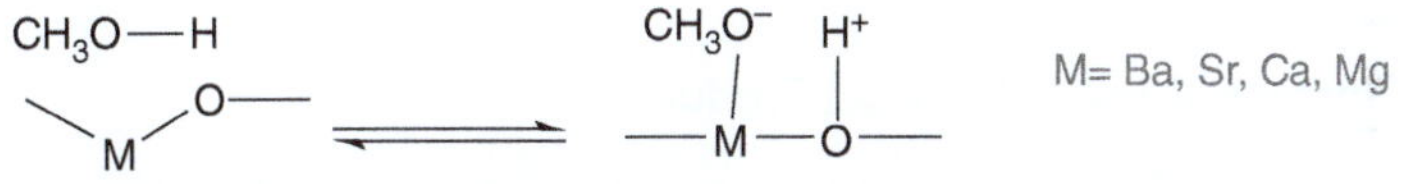

Step II: In the second stage, a methoxide anion is drawn to the carbonyl carbon [49] atom of the triglyceride molecule from MO's surface to create a tetrahedral intermediate, where R_1 is the alkyl group's long chain.

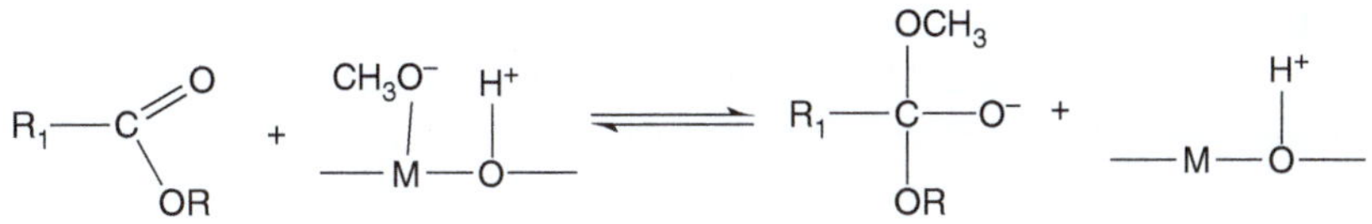

Step III: The tetrahedral intermediate removes H^+ from the MO surface in the third phase. Methoxide anions [49] can be produced by the reaction between the tetrahedral intermediate and methanol [51].

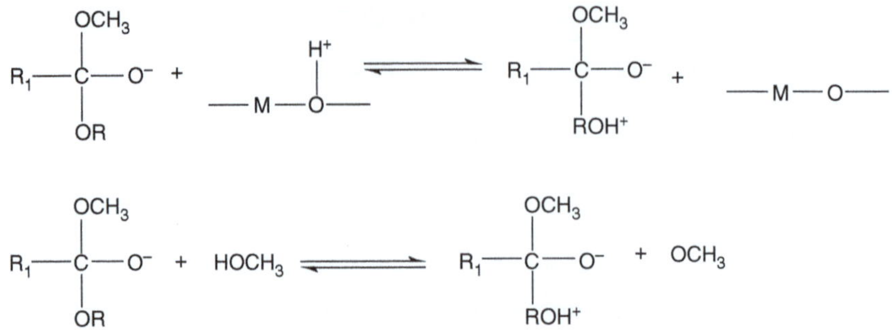

Step IV: Biodiesel is created in the final phase when the tetrahedral intermediate [52] is rearranged.

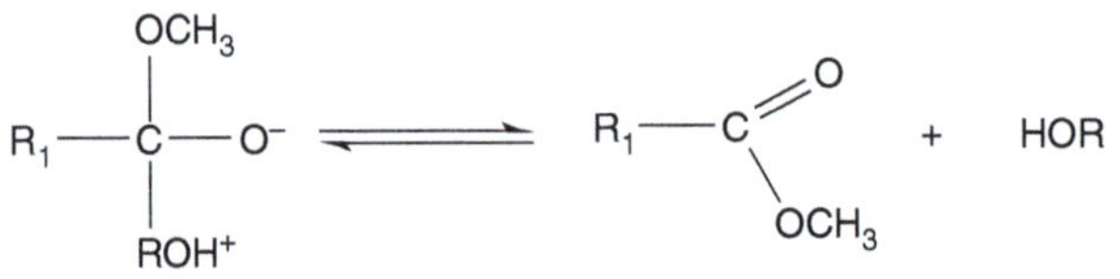

It is crucial to remember that the economics and overall efficiency of producing biodiesel can be impacted by the catalyst selection and process parameters. Solid base catalysts are just one of the various strategies used in the sector. It is imperative to meticulously evaluate multiple variables, including feedstock quality, catalyst loading, reaction temperature, and pressure, to optimize the transesterification process and yield high-grade biodiesel with optimal efficiency.

7.2.3 Industrial-scale Biodiesel Production

Large-scale procedures are used in industrial-scale biodiesel [53] manufacturing to produce biodiesel from feedstocks such as animal fats, vegetable oils, or other materials [53, 54] that contain lipids. From the preparation of the feedstock to the storage of the finished product, the production process usually comprises multiple steps. A summary of the essential procedures and factors to be considered while producing biodiesel on an industrial scale is given as under:

Feedstock preparation: The feedstock must be prepared to remove impurities [55, 56], such as water, FFA, and phospholipids, which can interfere with the transesterification [55] reaction. Impurities can be removed by a variety of methods, such as heating, filtration, and chemical treatment [56]. The fatty acid content of the feedstock may also need to be adjusted to meet the requirements of the transesterification process. This can be done by blending different feedstocks or by adding or removing fatty acids [56].

Transesterification: The transesterification reaction is the most important step in biodiesel production [37]. It is a chemical reaction that converts triglycerides (the main component of vegetable oils and animal fats) into fatty acid methyl esters [37, 38] (FAMEs), which are the main components of biodiesel. The transesterification reaction is typically carried out in a reactor at an elevated temperature (50–65 °C) in the presence of a catalyst [57]. The catalyst can be either acidic or basic. Acidic catalysts, such as sulfuric acid, are more effective but can also produce more impurities. Basic catalysts, such as sodium hydroxide or potassium hydroxide, are less effective but produce fewer impurities.

Separation: Once the transesterification reaction is complete, the biodiesel must be separated from the glycerol and other impurities [58]. This can be done by a variety of methods, such as gravity settling, centrifugation, and washing. The most common method is gravity settling. In this method, the biodiesel and glycerol are allowed to settle in a separation tank [59]. The biodiesel, which has a lower density than glycerol, floats to the top of the tank. The glycerol and other impurities, which have a higher density than biodiesel [58, 59], sink to the bottom of the tank. The biodiesel can then be removed from the top of the tank.

Purification: The biodiesel must be purified [60] to meet quality standards [60]. This typically involves washing the biodiesel with water to remove any remaining glycerol and other impurities. The biodiesel may also be filtered to remove any solid particles.

Storage: The biodiesel can then be stored in tanks [58–60] until it is ready to be used. Biodiesel can be stored for several months without losing quality [60].

The following system (Figure 7.4) is used by commercial and large-scale biodiesel production enterprises to produce biodiesel on a huge scale.

Industrial biodiesel production necessitates a large infrastructure, specialized knowledge, and regulatory compliance. Based on local conditions and consumer demand, several choices may be made regarding feedstock, production techniques, and process optimization. With constant advancements in technology and environmentally friendly practices, it is a dynamic industry.

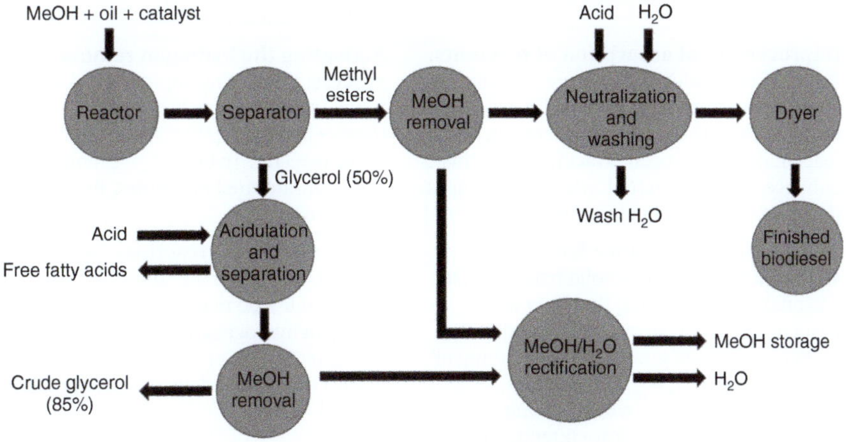

Figure 7.4 Process flow diagram for the manufacturing of biodiesel.

7.3 Hydrogenation and Dehydrogenation Reactions

7.3.1 Role of Solid Base Catalysts

By promoting these chemical changes at the surface of solid materials, solid base catalysts [1] are essential to hydrogenation as well as dehydrogenation reactions. Let us examine how solid base catalysts function in each of these reactions in Table 7.3.

Some specific examples of hydrogenation and dehydrogenation reactions that are performed in catalytic membranes [65] are shown in Figure 7.5.

Additional functions of solid base catalysts in processes involving hydrogenation and dehydrogenation can be:

- Because of their stability and reusability [1], solid base catalysts are frequently chosen. These catalysts can be recycled and utilized again in later reactions due to their solid form, which facilitates simpler separation from the reaction mixture [1].
- Maintaining catalytic activity across several reaction cycles and minimizing the requirement for frequent catalyst replacement depends on the durability [66] of the solid base catalyst.

Table 7.3 Role of solid base catalyst in hydrogenation and dehydrogenation reactions.

Hydrogenation reaction	Dehydrogenation reaction
Energizing (activation of) hydrogen: By allowing hydrogen molecules to adhere to their surfaces [61], solid base catalysts can activate them. In hydrogenation processes, when hydrogen is added to unsaturated molecules [61] (such as C_nH_{2n} or C—O compounds), this activation is a crucial step	**The reactant's activation:** Solid base catalysts are involved in dehydrogenation [62] processes, which remove hydrogen from molecules by adsorbing onto the surfaces of the reactant molecules and activating [62] them. The reactant is ready for the next step of hydrogen elimination after this activation
Improvement of adsorption of reactants: Reactant molecules can be more effectively adsorbed onto the surfaces of solid base catalysts [63]. The interaction between the reactants and hydrogen is facilitated by this enhanced adsorption [61], which encourages the creation of reaction intermediates [61–63]	**Activating the hydrogen removal process:** By offering appropriate reaction routes, solid base catalysts can help remove hydrogen [63] from the molecules of the reactant. To break C—H bonds and create unsaturated molecules, this is necessary
Surface-based assistance for reaction steps: On their surfaces, solid base catalysts can offer active sites [64] that help with particular reaction steps, including the reduction of C—O groups or the breaking of double bonds in C_nH_{2n}. The result is saturated compounds, which is what hydrogenation processes are intended to produce	**Avoiding adverse reactions:** Solid base catalysts can aid in the suppression of undesirable side reactions that arise during dehydrogenation [61–64], guaranteeing the process's selectivity toward the intended unsaturated product [62, 63]

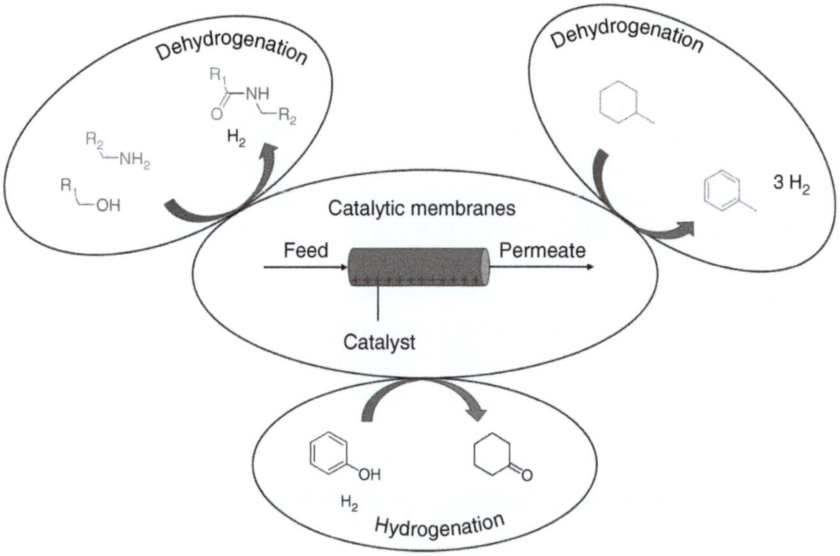

Figure 7.5 Examples of dehydrogenation as well as hydrogenation performed in catalytic membrane. Source: Izák et al. [65], 2018/with permission of John Wiley & Sons.

- Solid base catalysts can have their designs customized for certain hydrogenation and dehydrogenation processes [61]. This entails maximizing the catalyst's composition, acidity/basicity, and surface characteristics to improve its performance in a given reaction [61–64].

Additionally, novel technologies for hydrogen storage and carbon capture and utilization [67] (CCU) are being developed using solid base catalysts. For instance, reversible hydrogenation and dehydrogenation of aromatic hydrocarbons can be facilitated by solid base catalysts, providing a secure and effective means of storing hydrogen. CO_2 can also be converted to methanol (CH_3OH) and other valuable compounds with the use of solid base catalysts [68].

All things considered, solid base catalysts are crucial to several hydrogenation and dehydrogenation processes. These catalysts are being used to develop novel technologies for CCU and hydrogen storage because they have several benefits over conventional liquid or gaseous bases.

7.3.2 Case Studies: Hydrogenation of Oils and Dehydrogenation of Hydrocarbons

Case Study 7.1

Hydrogenation of Oils

Vegetable oils are chemically changed into saturated fats through a process called hydrogenation [69]. This method is frequently used to make margarine

(Continued)

Case Study 7.1 (Continued)

[70], butter, and other cooking oils. Solid base catalysts are essential for the hydrogenation [61] of oils because they act as a surface for the process.

Ni/MgO catalysts [71, 72], which are solid base catalysts supported on magnesium oxide, are among the most widely utilized for the hydrogenation of oils. For this reaction, Ni/MgO is a very active and selective catalyst [71, 72] that can be utilized to create hydrogenated oils with a variety of melting points.

Copper chromite [73] (Cu/Cr) is another solid base catalyst that is frequently used to hydrogenate oils. Compared to Ni/MgO, Cu/Cr is a less active catalyst, but it is more selective in producing fully hydrogenated oils. Cu/Cr [73] is a suitable option for hydrogenating low-quality oils since it is also more resistant to poisoning from contaminants in the oil.

A detailed example of a case study showing vegetable oil hydrogenation in the food industry including the catalyst used, the process, and the objective of case study and possible outcomes is displayed in Figure 7.6.

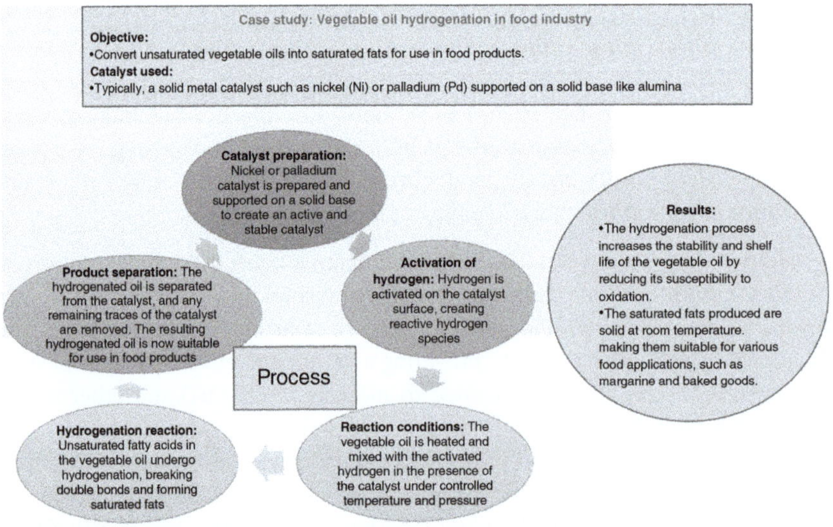

Figure 7.6 Case study: Hydrogenation of oils.

Case Study 7.2

Dehydrogenation of hydrocarbons

Hydrocarbons undergo a chemical process called dehydrogenation that transforms them into more unsaturated hydrocarbons [74]. The petrochemical industry uses alkenes as major feedstocks, and this technique is frequently employed to generate them. Solid base catalysts are essential for the dehydrogenation [62, 63] of hydrocarbons because they act as a surface for the reaction.

One of the most often used solid base catalysts for the dehydrogenation of hydrocarbons is potassium-promoted nickel supported on alumina [75, 76] (K-Ni/Al_2O_3). For this reaction, K-Ni/Al_2O_3 is a very active and selective catalyst that may be utilized to create a range of C_nH_{2n}, such as C_4H_6, C_3H_6, and C_2H_4.

Zinc-promoted chromium oxide (Zn/Cr_2O_3) is another widely used solid base catalyst for the dehydrogenation of hydrocarbons. While Zn/Cr_2O_3 is not as active as K-Ni/Al_2O_3, it is more resilient to sulfur compounds in the hydrocarbon feedstock that might poison it. Because of this, it is a suitable option for dehydrogenating heavy hydrocarbons, which frequently have significant sulfur content.

A detailed example of case study showing catalytic dehydrogenation of propane in petrochemical industry [77] including the catalyst used, the process, and the objective of case study and possible outcomes is displayed in Figure 7.7.

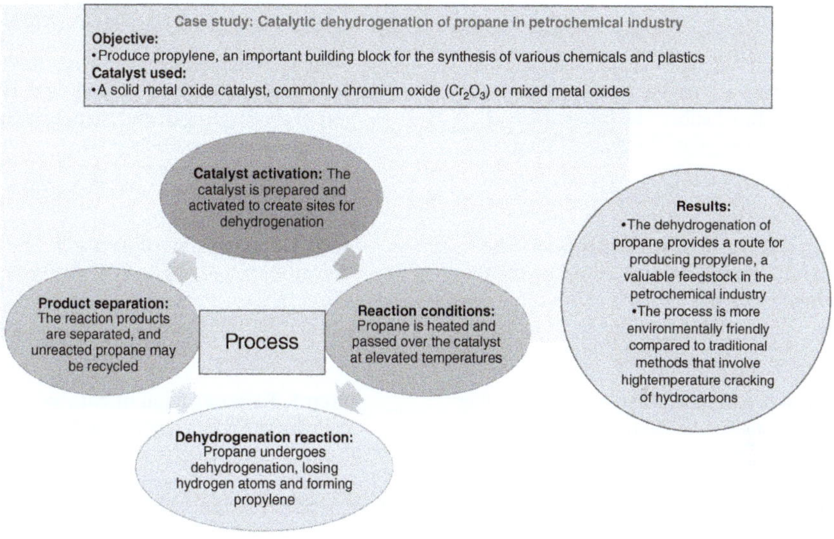

Figure 7.7 Case study: Dehydrogenation of hydrocarbons.

These case examples demonstrate the significance of dehydrogenation in converting hydrocarbons into useful compounds and the role of hydrogenation in altering the characteristics of oils for certain uses. Because they enable the intended reactions and raise overall process efficiency, solid catalysts are essential to both processes.

7.3.3 Catalytic Mechanisms

Both hydrogenation and dehydrogenation processes can be catalyzed by solid bases, and the mechanisms underlying this catalysis can change based on the particular reactants and circumstances. An outline of solid base catalytic processes for reactions involving hydrogenation along with dehydrogenation is provided in Table 7.4.

Table 7.4 General catalytic mechanism involved in hydrogenation as well as dehydrogenation.

Hydrogenation	Dehydrogenation
Step 1. Adsorption of reactants: Hydrogen and unsaturated substrates like alkenes are adsorbed onto the surface of the solid base catalyst $C_nH_{2n} + H_2 \xrightarrow{\text{Adsorption}} C_nH_{2n}$ $H_2 \xrightarrow{\text{Adsorption}} 2H^*$	**Step 1. Adsorption of reactants:** The hydrocarbon substrate is adsorbed onto the surface of the solid base catalyst $C_nH_{2n+2} \xrightarrow{\text{Adsorption}} C_nH_{2n+2}$
Step 2. Activation of hydrogen: Adsorbed hydrogen molecules dissociate into reactive hydrogen atoms on the catalyst surface $2H^* \xrightarrow{\text{Adsorption}} 2H$	**Step 2. Activation of substrate:** The adsorbed hydrocarbon undergoes activation on the catalyst surface $C_nH_{2n+2} \xrightarrow{\text{Activation}} C_nH^+{}_{2n+2} + e^-$
Step 3. Hydrogen transfer: Activated hydrogen atoms react with the adsorbed unsaturated substrate, causing the hydrogenation of the double bond $C_nH_{2n} + 2H \rightarrow C_nH_{2n+2}$	**Step 3. Dehydrogenation reaction:** Dehydrogenation occurs, leading to the removal of hydrogen and the formation of unsaturated products $C_nH^+{}_{2n+2} \xrightarrow{\text{Dehydrogenation}} C_nH_{2n} + H^*$
Step 4. Desorption of product: The hydrogenated product desorbs from the catalyst surface $C_nH_{2n+2} \xrightarrow{\text{Desorption}} C_nH_{2n+2}$	**Step 4. Desorption of product:** The dehydrogenated products desorb from the catalyst surface $C_nH_{2n} \xrightarrow{\text{Desorption}} C_nH_{2n}$
Step 5. Regeneration of catalyst: The catalyst surface, having released the product, is available for the next cycle of reaction	**Step 5. Regeneration of catalyst:** The catalyst surface is regenerated for subsequent cycles

These mechanisms provide a simplified overview of the dehydrogenation and hydrogenation using solid base catalyst. The actual mechanism can depend on the specific conditions and catalyst used, which may lead to variations in the detailed steps of these processes.

7.4 Bimolecular Reactions

A bimolecular reaction [64] catalyzed by a solid base catalyst involves a reaction in which two molecules come together to form a product, and the catalyst facilitating [64] this process is a solid base. Solid base catalysts [1] are materials that can act as bases and promote chemical reactions without undergoing significant dissolution or change in their physical state. These catalysts are often used in various industrial processes and are of interest in green chemistry [78] due to their potential to reduce environmental impact.

7.4.1 Dialkyl Carbonate Synthesis

The synthesis of dialkyl carbonates [79] using a solid base catalyst is an important and environmentally friendly [80] process. Dialkyl carbonates are compounds that contain two alkyl groups (hydrocarbon chains) bonded to a carbonate group ($O{=}C(OR)_2$, where R [81] is an alkyl group). They have applications as solvents, fuel additives, and intermediates in the production of pharmaceuticals and agrochemicals. Using solid base catalysts for the synthesis of dialkyl carbonates [79–81] has advantages such as mild reaction conditions, reduced waste generation, and the potential for catalyst recycling.

Here is a general outline of the synthesis of dialkyl carbonates through a solid base catalyst.

Reaction scheme:

$$\text{Alcohol} + \text{diethyl carbonate} \xrightarrow{\text{Solid base catalyst}} \text{Dialkyl carbonate} + \text{ethanol}$$

Reaction mechanism: The general reaction involves the transesterification of an alcohol with diethyl carbonate in the presence of a solid base catalyst. The solid base catalyst facilitates the reaction by abstracting a proton [82] from the alcohol, generating the alkoxide ion [82]. This alkoxide ion then reacts with diethyl carbonate [82, 83] to form the dialkyl carbonate and ethanol.

Example: Consider the synthesis of dimethyl carbonate (DMC) [84] from methanol [84, 85] and diethyl carbonate using a solid base catalyst.

$$CH_3OH + (C_2H_5O)_2CO \xrightarrow{\text{Solid base catalyst}} (CH_3O)_2CO + C_2H_5OH$$

Reaction conditions: The reaction is typically carried out at moderate temperatures, usually between 100 and 150 °C, and may require a slight excess of one of the reactants to drive the equilibrium toward the formation of the desired dialkyl carbonate.

Advantages:

- **Mild reaction conditions**: The use of solid base catalysts allows for the synthesis to occur at moderate temperatures [63].
- **Reduced waste**: The process is more environmentally friendly [24, 80] compared to traditional methods that may use toxic or hazardous reagents.
- **Catalyst recyclability**: Solid base catalysts can often be separated [26, 58] and reused in subsequent reactions.

Challenges:

- **Catalyst stability**: Ensuring the stability and activity of the solid base catalyst over multiple reaction cycles.
- **Selectivity**: Achieving high selectivity [25, 61–64] for the desired dialkyl carbonate product.

7.4.2 Catalyst Selection and Reaction Pathways

The selection of a catalyst and the reaction pathway for the synthesis of dialkyl carbonates through solid base catalysis involves careful consideration of factors such as catalyst activity, stability, and selectivity. Here is an overview of catalyst selection and potential reaction pathways.

Catalyst selection:

a) **Metal oxides: Magnesium oxide (MgO):** MgO is a common solid base catalyst [8, 39, 48] for the synthesis of dialkyl carbonates. It exhibits good catalytic activity and stability.
b) **Supported alkali metal catalysts: Sodium hydroxide (NaOH) on solid support:** NaOH supported on a solid substrate can serve as an effective catalyst. The solid support enhances the stability and recyclability [1] of the catalyst.

Reaction pathways:

a) **Transesterification reaction:** The primary reaction pathway for the synthesis of dialkyl carbonates involves the transesterification of an alcohol with a dialkyl carbonate. The general reaction scheme is as follows:

$$\text{Alcohol} + \text{diethyl carbonate} \xrightarrow{\text{Solid base catalyst}} \text{Dialkyl carbonate} + \text{alkoxide ion}$$

b) **Alkoxide ion formation:** The solid base catalyst initiates the reaction by abstracting a proton from the alcohol, forming an alkoxide [86] ion.

$$\text{Alcohol} \xrightarrow{\text{Solid base catalyst}} \text{Alkoxide ion} + \text{proton}$$

c) **Reaction with diethyl carbonate:** The alkoxide ion [86] then reacts with diethyl carbonate, leading to the formation of the dialkyl carbonate and regeneration of the solid base catalyst [86]:

$$\text{Alkoxide ion} + \text{diethyl carbonate} \xrightarrow{\text{Solid base catalyst}} \text{Dialkyl carbonate} + \text{alcohol}$$

Reaction conditions:

- The reaction is typically carried out at moderate temperatures (100–150 °C) to achieve a balance between reaction rate and catalyst stability. The use of a slight excess of one of the reactants may be necessary to drive the equilibrium toward the formation of the desired dialkyl carbonate.

Challenges and optimization:

- Challenges may include maintaining catalyst stability over multiple reaction cycles and optimizing reaction conditions for high selectivity toward the desired dialkyl carbonate product.

7.4.3 Applications and Industrial Scale-Up

Bimolecular reactions play a crucial role in various industrial processes, and their applications are diverse across different sectors. In the pharmaceutical industry [87], bimolecular reactions are integral to the synthesis of complex

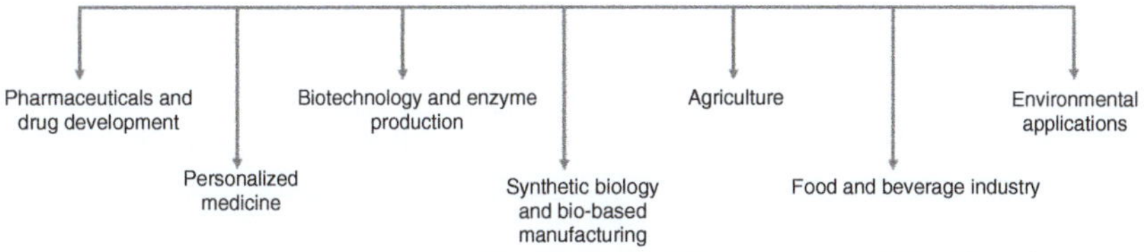

Figure 7.8 Industrial scale-up of bimolecular reactions in various fields.

organic compounds, enabling the production of drugs [87] and pharmaceutical intermediates. In the field of polymers and materials [88], bimolecular reactions are employed for the creation of advanced materials with tailored properties. In the petrochemical industry, processes like alkylation [89] and polymerization, which involve bimolecular reactions, are used to produce fuels and polymers [89]. The scale-up of bimolecular reactions to an industrial level involves optimizing reaction conditions, improving yields, and ensuring cost-effectiveness. The development of efficient catalytic systems and reactor designs is crucial for achieving large-scale production. Additionally, advancements in process intensification and sustainable practices contribute to the successful industrial scale-up of bimolecular reactions, aligning with the principles of green chemistry [78] and reducing environmental impact.

The following (Figure 7.8) are some important uses and things to think about when scaling up biomolecular reactions to an industrial scale.

In an industrial setting, scaling up biomolecular reactions necessitates a multidisciplinary strategy incorporating knowledge of chemical engineering, molecular biology, biochemistry, and process optimization. Reproducibility maintenance, risk reduction from contamination, reaction condition optimization, and cost-effectiveness assurance are possible obstacles. Technological developments in automation, high-throughput screening, and process monitoring help make biomolecular reactions a successful industrial application.

7.5 Methanol and DME Synthesis

Methanol and dimethyl ether (DME) [90] are versatile chemicals with significant importance in various industrial applications, contributing to energy production, fuel alternatives [91], and chemical manufacturing [91, 92].

7.5.1 Importance of Methanol and DME

Methanol is a valuable fuel source used in fuel cells as a fuel additive and in the production of biodiesel [93]. It has the potential to serve as a green alternative to traditional fossil fuels [91]. It is a crucial building block [94] for the synthesis of various chemicals and materials, including formaldehyde, acetic acid, and olefins [94], contributing to the production of plastics, resins, and pharmaceuticals. It can be used as a carrier of hydrogen, providing a means for energy storage and transportation, particularly in fuel cell applications.

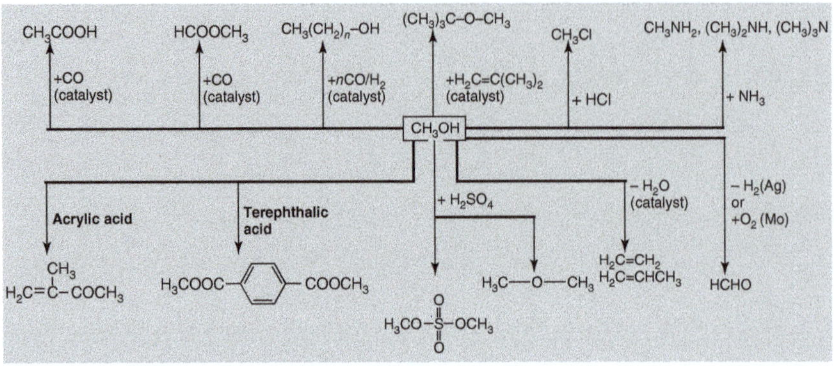

Figure 7.9 Methanol reactions with industrial [95] significance.

Some industrial important reactions of methanol [95] are shown in Figure 7.9.

DME is considered a clean-burning alternative fuel, with properties similar to liquefied petroleum gas [96] (LPG). It can be used as a substitute for diesel in engines and as a propellant in aerosol products. It is widely used as a propellant in aerosol products such as deodorants [97], hairsprays, and spray paints, providing a safe and effective alternative to traditional hydrocarbon propellants and a valuable intermediate for the production of chemicals and polymers, contributing to the synthesis of various industrial products.

The raw components, production process, and applications of DME [98] are displayed in Figure 7.10:

The adaptability of methanol and DME synthesis as fuels, chemical feedstocks, and energy carriers accounts for their significance. Because of its potential to lower greenhouse gas emissions and provide cleaner energy sources, the production of methanol and DME [91] is garnering more attention, as the globe looks for more environmentally friendly and sustainable alternatives. Catalytic techniques are used

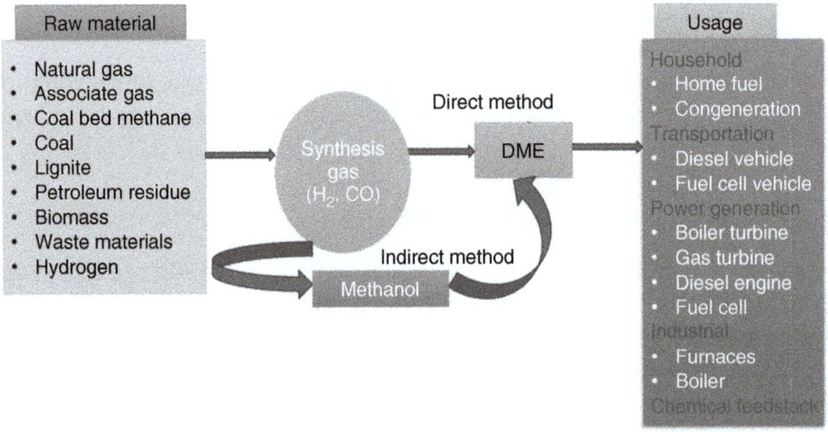

Figure 7.10 DME applications, manufacturing process, and raw materials.

Table 7.5 Distinguished features of methanol as well as DME.

Property	Methanol synthesis	Dimethyl ether synthesis
Catalyst type	$Cu/ZnO/Al_2O_3$	Acidic zeolites
Reaction type	Hydrogenation of CO and CO_2	Dehydrogenation of methanol
Reaction equation	$CO + 2H_2 \rightarrow CH_3OH$	$2CH_3OH \rightarrow CH_3OCH_3 + H_2O$
Operating conditions	High pressure, elevated temperature	Moderate temperature, low pressure
Applications	Fuel, chemical feedstock	Alternative fuel, aerosol propellent, chemical feedstock
Importance	Key building block, fuel source	Clean burning fuel, Aerosol propellent, chemical feedstock

in the synthesis of these chemicals, and current research aims to increase the production's selectivity, efficiency, and environmental sustainability.

7.5.2 Catalysts and Reaction Conditions

Here is a tabular (Table 7.5) summary of catalyst and reaction selection for the synthesis of methanol and dimethyl ether.

These catalysts and reactions highlight the diverse applications of methanol as a fuel and chemical feedstock and dimethyl ether as a clean-burning fuel and versatile chemical intermediate. The catalysts and reaction conditions are tailored to optimize the synthesis processes for each compound based on their intended applications.

7.5.3 Technological Advancements

Technological advancements in the synthesis of methanol and DME have been pivotal in enhancing the efficiency, sustainability, and adaptability of these chemical processes. In methanol synthesis, catalyst innovations have played a significant role, with the development of advanced catalysts like copper-based [99] systems, such as $Cu/ZnO/Al_2O_3$, that exhibit improved activity, selectivity, and stability. The use of nanostructured catalysts [100] has also been explored to increase the surface area and enhance methanol yields. Additionally, there has been a notable shift toward renewable feedstocks, with ongoing research in biological methanol production where microorganisms are engineered to convert carbon dioxide or methane directly into methanol. CCU [101] technologies are being integrated into methanol synthesis processes to harness carbon dioxide emissions and contribute to environmental sustainability. Moreover, novel reaction pathways, such as electrochemical methanol synthesis, show promise in utilizing renewable electricity for methanol production.

In the realm of dimethyl ether synthesis, advancements have been achieved through catalyst engineering and reactor design. Tailored zeolite catalysts, especially acidic zeolites like HZSM-5 [102], have been developed to enhance the efficiency of DME synthesis from methanol, while bifunctional catalysts incorporating both acidic and basic sites aim to improve reaction kinetics and DME [90, 98] selectivity. Reactor design innovations include the incorporation of reactive distillation processes and membrane reactors. Reactive distillation [103] allows for simultaneous methanol dehydration and DME synthesis, streamlining [103] the production process. Membrane reactors selectively remove water during DME synthesis, favoring thermodynamic equilibrium toward DME formation. Alternative feedstocks, such as biomass-derived syngas, are gaining attention, broadening the sustainability of DME synthesis. Furthermore, process intensification strategies, including the utilization of microreactor technology, contribute to better heat and mass transfer, offering improved control over reaction conditions in DME synthesis.

These technological advancements collectively demonstrate a commitment to making methanol and DME synthesis more sustainable, economically viable, and versatile in accommodating various feedstocks. Ongoing research and innovation in these areas continue to drive the chemical industry toward greener and more energy-efficient practices, aligning with global efforts to address environmental challenges and promote sustainable development.

7.6 Transesterification of Esters

Solid base catalysts are essential for enabling the interchange of alkyl groups during the transesterification of esters [104], which transforms one ester into another. This procedure is especially important for synthesizing different esters for use in industry, such as making biodiesel. Solid base catalysts play the following primary functions (shown in Figure 7.11) in ester-based transesterification reactions.

7.6.1 Role in Chemical and Petrochemical Industries

In the context of chemical and petrochemical industries [105], transesterification plays a significant role, especially in the production of biodiesel [105] and other specialty chemicals.

Role of solid base catalyst in transesterification of esters:

- Can activate the ester molecules which makes the ester more susceptible to nucleophilic attack, facilitating the transfer of the alkyl group from one ester to another
- Enhance the nucleophilicity of the alkoxide ion, promoting efficient substitution reactions
- Can help shift the equilibrium toward the desired product by favoring the forward reaction which is particularly important in industrial processes where high conversion rates are desired
- Can offer better selectivity in terms of the desired transesterification reaction, minimizing the formation of unwanted byproducts, crucial for obtaining a high-quality final product with the desired chemical properties
- Allow transesterification reactions to occur under milder conditions (lower temperatures and pressures), reducing the energy requirements and making the process more economically feasible
- Allowing for easy separation from the reaction mixture which enables the catalyst to be reused in subsequent reactions, contributing to cost-effectiveness and sustainability
- Can help minimize side reactions, such as saponification (hydrolysis of esters into carboxylic acids and alcohols)
- Important for maintaining the integrity of the reaction and improving the overall yield of the desired transesterification product

Figure 7.11 Role plays of solid base catalyst in transesterification of ester.

In the production of biodiesel, transesterification is used to convert triglycerides into FAMEs [106] (Figure 7.3), also known as biodiesel. Triglycerides are the main components of vegetable oils, such as soybean oil [107] and canola oil. Methanol [108] is typically used as the alcohol in the transesterification reaction. The FAMEs produced by transesterification are a renewable alternative to petroleum-based diesel fuel.

Transesterification is also used in the production of pharmaceuticals [109]. For example, it is used to synthesize methyl salicylate [110], which is an ingredient in pain relief medications. It is also used to produce various esters of propylene glycol [111], which are used as solvents and plasticizers.

In the petrochemical industry, transesterification is used to produce a variety of plastics, such as polyethylene terephthalate (PET) [112] and polybutylene terephthalate [113] (PBT). PET is used to make bottles and other containers, while PBT is used to make fibers and engineering plastics [112, 113].

The following are some particular instances of the application of transesterification in the petrochemical and chemical industries.

7.6.1.1 Producing Biodiesel

One essential stage in the creation of biodiesel [37, 38] is transesterification. This method involves the reaction of triglycerides [106], or fats and oils, with alcohol (often methanol or ethanol) in the presence of a catalyst, usually potassium hydroxide or sodium hydroxide.

The primary ingredients of biodiesel, fatty acid methyl or ethyl esters [106] (FAME or FAEE), are produced from the triglycerides in the feedstock. Through this procedure, the feedstock's fuel qualities are enhanced, making it a viable alternative to diesel.

7.6.1.2 Specialized Chemical Production

Many specialized compounds are synthesized by the process of transesterification. It can be used, for instance, to produce high-value chemicals such as flavors [114], perfumes, and medicines. Modifying the chemical structure of molecules through selective ester group exchange permits the production of a wide range of intricate compounds.

7.6.1.3 Procedures for Polymerization

Certain polymerization techniques involve transesterification. Transesterification reactions, for example, are commonly involved in the interaction between dicarboxylic acids (HOOC $(CH_2)_n$ COOH) and glycols ($C_2H_6O_2$) during the synthesis of polyester resins [115]. Films, fibers, and packaging materials are all produced using polyester resins.

7.6.1.4 Engineering Reactions and Catalysis

In transesterification processes, the selection of catalysts and reaction conditions is essential for maximizing reaction rates and product yields.

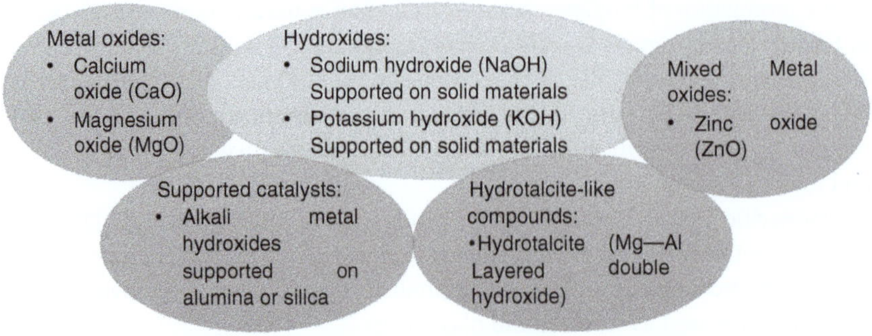

Figure 7.12 Solid base catalysts for the transesterification of esters [106–115].

In an effort to increase the effectiveness and selectivity of transesterification reactions, scientists and engineers in the chemical and petrochemical sectors investigate a variety of catalysts, both homogeneous and heterogeneous.

7.6.1.5 Resource Efficiency and Waste Reduction

Through the utilization of waste streams rich in esters or triglycerides [107], transesterification technologies can help reduce waste. Transesterification helps the chemical and petrochemical industries [74–77] adhere to the concepts of resource efficiency and sustainability by turning waste resources into useful products.

7.6.2 Catalysts for Transesterification

For transesterification processes in the synthesis of biodiesel, solid base catalysts are favored because of their high activity, simplicity of separation, and possibility for catalytic reuse [25, 61–64]. The particular ester feedstock, the reaction circumstances, and the intended product specifications all play a role in the catalyst selection process. In order to increase the effectiveness and sustainability of transesterification processes, researchers are still looking into and creating novel solid base catalysts. Typical solid base catalysts for the transesterification of esters include the following examples (displayed in Figure 7.12):

7.7 Alkylation and Isomerization Reactions

7.7.1 Solid Base Catalysis in Petrochemical Processes

In a number of petrochemical processes, solid base catalysis is used to speed up reactions that are essential to the creation of valuable products. Here are some instances of these procedures and the corresponding chemical reactions (Table 7.6):

Other applications of solid base catalysis in petrochemical processes:

1) **Epoxidation of olefins**: Epoxidation involves the addition of an oxygen atom across a carbon–carbon double bond in an olefin (alkene) molecule.

Table 7.6 Reactions using solid base catalysts in petrochemical industries.

Name of reaction	Illustration	Reaction	Catalyst	Equation
Isomerization of alkanes	Solid base catalysts are often used in the isomerization of hydrocarbons [61], particularly, in the production of high-octane gasoline. Isomerization reactions convert straight-chain hydrocarbons [61] into their branched isomers, improving the octane rating of the fuel	Isomerization of *n*-butane to isobutane [116]	Solid base catalysts like zeolites [116] or alkaline earth metal oxides	$n\text{-}C_4H_{10} \longrightarrow i\text{-}C_4H_{10}$
Hydrocracking	In hydrocracking processes [117], solid base catalysts are employed to promote the cracking of heavy hydrocarbons into lighter fractions [117]. This is an essential step in the production of high-value products like diesel and jet fuels	Hydrocracking of heavy hydrocarbons to produce lighter fractions	Solid base catalysts such as zeolites or mixed metal oxides	$C_{12}H_{26} \longrightarrow C_8H_{18} + C_4H_{10}$
Aldol condensation and dehydration reaction	Solid base catalysts are utilized in aldol condensation [61] and dehydration reactions, which are important for the synthesis of petrochemical intermediates and fine chemicals. These reactions involve the formation of carbon–carbon bonds and the removal of water from reactants	Aldol condensation of acetone [118] to form unsaturated compounds	Solid base catalysts like metal oxides or hydrotalcite-like materials	$2CH_3COCH_3 \longrightarrow CH_3COCH = CH_2 + CH_3COCH_2CH_3$
		Dehydration of alcohols or diols to form olefins [119]	Solid base catalysts such as metal oxides or zeolites	$CH_3CH_2CH_2OH \longrightarrow CH_2 = CHCH_2 + H_2O$
Alkylation reactions	Solid base catalysts are employed in alkylation reactions [86], where hydrocarbons are combined to produce higher-value products. For instance, the alkylation of benzene with olefins is a common petrochemical process used to produce valuable aromatic compounds	Alkylation of benzene [120] with olefins to produce alkylated aromatics	Solid base catalysts like zeolites or supported alkali metal hydroxides	$C_6H_6 + CH_3CH = CH_2 \longrightarrow C_6H_5CH_2CH_3$

(Continued)

Table 7.6 (Continued)

Name of reaction	Illustration	Reaction	Catalyst	Equation
Fischer–Tropsch synthesis:	In Fischer–Tropsch synthesis [121], which converts syngas (a mixture of carbon monoxide and hydrogen) into hydrocarbons, solid base catalysts may be used to promote the desired reactions. These catalysts can influence the selectivity [61–64] and efficiency of the process	Synthesis of hydrocarbons [122] from syngas ($CO + H_2$)	Solid base catalysts such as alkali-promoted iron or cobalt catalysts	$CO + 2H_2 \longrightarrow CH_3OH$
Desulfurization and denitrogenation:	Solid base catalysts are employed in processes aimed at removing sulfur [123] and nitrogen compounds from petroleum products. Desulfurization and denitrogenation [124] are essential steps to meet environmental regulations and produce cleaner fuels	Removal of sulfur compounds from petroleum feedstocks	Solid base catalysts, often oxides or hydrotalcite-like materials	$R{-}S{-}R' + 2H_2 \longrightarrow 2RH + H_2S$
Methanol and dimethyl ether (DME) synthesis:	Solid base catalysts are involved in the synthesis of methanol and dimethyl ether [125], both of which are valuable chemicals with applications in fuel production and as chemical intermediates	Synthesis of methanol from syngas [122]	Solid base catalysts, such as copper–zinc–alumina	$CO + 2H_2 \longrightarrow CH_3OH$

Solid base catalysts are employed in this reaction to produce epoxides [126], which serve as valuable intermediates in the synthesis of various chemicals and pharmaceuticals.

2) **Knoevenagel condensation**: The Knoevenagel condensation involves the reaction of an aldehyde or ketone with a malonic [127] or cyanoacetic ester in the presence of a base catalyst to produce α, β-unsaturated [127] carbonyl compounds. Solid base catalysts [13–26, 37–140] are preferred for this reaction due to their high selectivity [25] and ability to suppress side reactions.
3) **Michael addition**: Michael addition involves the addition of a nucleophile to an α, β-unsaturated carbonyl compound [141]. Solid base catalysts [141, 142] are often used to promote this reaction, particularly for the synthesis of fine chemicals and pharmaceuticals.
4) **Si—C bond formation**: Solid base catalysts can effectively promote the formation of Si—C bonds [143], which are essential in the synthesis of various silicones and organosilicon [143, 144] compounds.
5) **Halogenation of aromatic compounds**: Solid base catalysts can facilitate the halogenation of aromatic compounds, such as chlorination, bromination, and iodination [145]. These reactions are important for the synthesis of various pharmaceuticals and agrochemicals.
6) **Oxidation reactions**: Solid base catalysts can be used in various oxidation reactions, such as the oxidation of alcohols to aldehydes or ketones and the oxidation of alkenes to epoxides [146] or carbonyl compounds.
7) **Synthesis of fine chemicals and pharmaceuticals**: Solid base catalysts play a significant role in the synthesis of fine chemicals and pharmaceuticals [147], providing an efficient and environmentally friendly alternative to homogeneous catalysts.

In general, solid base catalysis has become a potent instrument in petrochemical operations, providing a host of benefits above conventional homogeneous catalysts. For a variety of significant reactions, its ease of separation, decreased soap formation, improved product quality, and environmental friendliness makes it the favored option. Solid base catalysis is anticipated to become even more important in petrochemical processes in the future, as research and development continue.

7.7.2 Environmental and Economic Implications

Ester transesterification is a chemical process [148] with several uses that is adaptable and safe for the environment. The process's advantages for the environment and the economy can be increased by using heterogeneous catalysts [40]. This method is frequently employed in organic synthesis.

Traditionally, transesterification has been carried out using homogeneous catalysts, such as sodium hydroxide [1, 2] or potassium hydroxide. These catalysts are not only effective, but they also have several environmental drawbacks. For example, they are corrosive and can generate large amounts of wastewater.

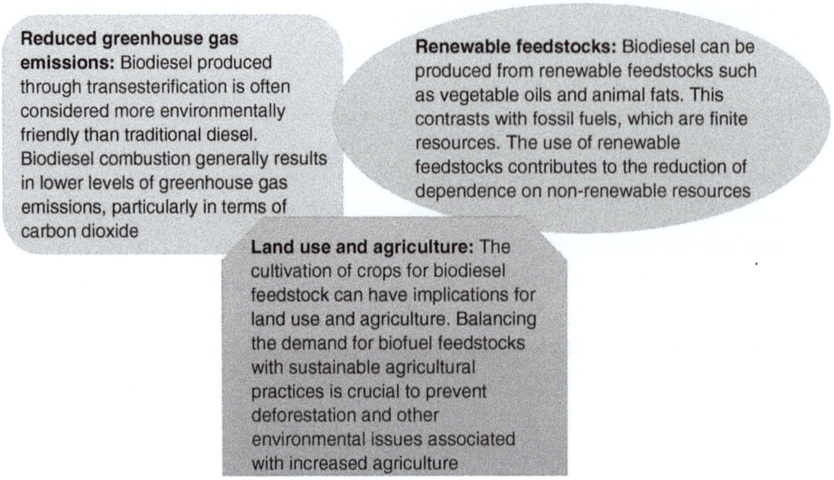

Figure 7.13 Environmental implications.

In addition, homogeneous catalysts [1, 2] can be difficult to separate from the product, which can lead to contamination.

In recent years, there has been a growing interest in the development of heterogeneous catalysts for transesterification. The use of heterogeneous catalysts [40] can significantly reduce the environmental impact of transesterification. The following (Figure 7.13) are some important things to remember.

7.7.2.1 Economic Implications

The use of heterogeneous catalysts can also have significant economic [149] implications. Heterogeneous catalysts are typically more expensive than homogeneous catalysts [150], but they can be reused many times. This can offset the initial cost of the catalyst and make the process more cost-effective overall. Other key considerations may be shown as under:

- Production of biodiesel broadens the sources of fuel used in transportation, enhancing energy security [151]. Nations with the ability to generate biodiesel [37] from domestic feedstocks can potentially improve energy security by lowering their need on imported petroleum.
- From the production of feedstocks to the distribution and processing of biodiesel, the biodiesel business can help create jobs. This can benefit the economy [149] on a number of levels, particularly in rural areas where feedstocks are frequently cultivated.
- Feedstock prices, which might fluctuate, have an impact on the biodiesel production process's economic sustainability. The profitability of producing biodiesel may be impacted by changes in the cost of animal or vegetable fats.
- Government policies and subsidies [152] frequently have an impact on the biodiesel production industry's economic viability. The competitiveness of biodiesel in relation to conventional fuel can be affected by mandates, tax credits, and subsidies.

- Infrastructure expenditures [153], such as distribution networks and biodiesel production facilities, may be necessary to make the switch to biodiesel. These expenditures may affect the overall cost-effectiveness of biodiesel and have an impact on the economy.

7.8 Environmental Applications

In recent decades, the escalating concerns over environmental degradation and the imperative to develop sustainable technologies have spurred a surge of interest in green chemistry and environmentally benign processes. Among the pivotal advancements in this realm, solid base catalysts have emerged as indispensable tools in various environmental applications. Solid base catalysts, characterized by their robust nature and enhanced reusability, play a pivotal role in addressing the challenges associated with traditional liquid or homogeneous catalysis systems. This shift toward solid base catalysts aligns with the broader global agenda of mitigating pollution, minimizing waste, and promoting cleaner production methods.

Solid base catalysts are distinct from their liquid counterparts in that they are characterized by a heterogeneous phase, often supported on solid materials like oxides, zeolites, or mesoporous structures. This unique structural configuration not only facilitates efficient catalytic reactions but also enables facile recovery and recycling, minimizing the environmental footprint associated with catalyst production and disposal. The applications of solid base catalysts span a wide spectrum, ranging from industrial processes to remediation strategies, showcasing their versatility in addressing diverse environmental challenges.

This introduction explores the burgeoning field of environmental applications of solid base catalysts, shedding light on their significance in catalyzing sustainable transformations, reducing energy consumption, and promoting ecofriendly practices. Through a comprehensive examination of their role in various processes, from pollutant degradation to biomass conversion, this discussion aims to underscore the pivotal role of solid base catalysts in forging a path toward a greener and more sustainable future.

7.8.1 Sulfur Removal from Flue Gas

The quest for cleaner and more sustainable energy sources has led to an increased focus on mitigating the environmental impact of fuel combustion processes. One of the paramount challenges in this context is the removal of sulfur [123] compounds from fuel gases – a task that holds significant implications for both environmental conservation and human health. Solid base catalysts have emerged as key players in this endeavor, offering an innovative and efficient solution to reduce sulfur content in fuel gases. This application of solid base catalysts represents a pivotal step toward enhancing the environmental performance [154] of combustion processes and aligning with global efforts to curb air pollution [155].

Sulfur-containing compounds, such as hydrogen sulfide (H_2S) and mercaptans [156], are commonly found in various fuel sources, including natural gas and petroleum derivatives. Their combustion releases sulfur dioxide (SO_2), a major contributor to air pollution and acid rain, with detrimental effects on air quality and ecosystems. Solid base catalysts, with their heterogeneous [40] nature and tailored active sites, present an effective means to selectively remove sulfur compounds from fuel gases, mitigating the environmental impact associated with sulfur emissions.

This exploration delves into the environmental application of solid base catalysts in sulfur removal from fuel gas. By examining the catalytic mechanisms, performance advantages, and broader implications for cleaner combustion, this discussion aims to highlight the pivotal role of solid base catalysts in promoting sustainable energy [157] practices and addressing the pressing challenges of sulfur-induced environmental degradation.

7.8.2 NO_x Reduction in Catalytic Converters

The automotive industry, a vital driver of modern mobility, faces a dual challenge of meeting increasing energy demands while concurrently minimizing its environmental footprint [158]. A significant aspect of this challenge lies in curbing the emissions of nitrogen oxides [159] (NO_x) from internal combustion engines, particularly in the context of stringent environmental regulations. In this pursuit, catalytic converters play a pivotal role, and solid base catalysts have emerged as a key player in the effective reduction of nitrogen oxides [159, 160]. This application represents a critical advancement in addressing air pollution and promoting cleaner automotive technologies.

Nitrogen oxides, primarily in the form of nitric oxide (NO) and nitrogen dioxide (NO_2), are notorious contributors to air pollution [159], posing serious health and environmental risks. Catalytic converters, integral components in vehicle exhaust systems, serve as catalysts for the reduction of NO_x [160] to harmless nitrogen (N_2) [159–161] and water (H_2O). Solid base catalysts, with their unique characteristics, have proven to be highly effective in this conversion process, offering advantages in terms of stability, durability, and performance over traditional catalysts.

7.8.3 Waste Remediation and Pollution Control

Waste remediation and pollution control [162] have become pressing imperatives in the face of escalating environmental challenges worldwide. Addressing the detrimental impacts of industrial processes and waste disposal on ecosystems and human health necessitates innovative solutions. In recent years, solid base catalysts have emerged as powerful tools in the realm of pollution control and waste remediation. These catalysts, characterized by their heterogeneous [40] nature and versatile catalytic properties, offer a sustainable approach to mitigating pollutants and transforming waste into environmentally benign substances. This introduction delves into the pivotal role of solid base catalysts in waste remediation and pollution control, highlighting their potential to revolutionize sustainable practices across various industries.

Solid base catalysts, often composed of metal oxides [39, 48] or supported basic materials, exhibit unique catalytic characteristics that make them well-suited for tackling environmental challenges. Their heterogeneous nature facilitates facile recovery and recycling, minimizing the generation of secondary waste. Moreover, the tailored active sites of solid base catalysts enable selective and efficient transformation of pollutants, promoting cleaner and more sustainable processes.

The applications of solid base catalysts in waste remediation encompass a broad spectrum, ranging from the degradation of organic pollutants [163] to the neutralization of acidic [163, 164] emissions. These catalysts play a crucial role in the conversion of harmful substances into less toxic or inert forms, contributing to the overall reduction of pollutants released into air, water, and soil.

7.9 Dehydration Reactions

7.9.1 Dehydration of Alcohols to Olefins

The dehydration of alcohols to olefins [165] using solid base catalysts constitutes a noteworthy advancement, aligning with the principles of green chemistry [78] and sustainable manufacturing.

Alcohol dehydration is a chemical reaction wherein water is removed from alcohol molecules, leading to the formation of olefins [166], commonly known as alkenes. Olefins are versatile compounds with applications [167] ranging from the production of plastics and polymers to the synthesis of specialty chemicals. The use of solid base catalysts in this context provides several advantages, including the ability to operate under milder reaction conditions, reduced corrosion of equipment, and the potential for catalyst recovery and reuse.

The dehydration of alcohols to olefins using solid base catalysts represents a cutting-edge application in catalysis [168], fostering more sustainable and efficient routes for the production of valuable chemicals and biofuels. Solid base catalysts offer distinct advantages over traditional liquid acid catalysts [1], such as improved selectivity [1, 2], enhanced stability, and the potential for catalyst recovery and reuse [25]. This innovation aligns with the principles of green chemistry, emphasizing the importance of minimizing environmental impact and optimizing resource utilization in chemical processes.

Examples:

1) **Isobutanol dehydration to isobutene:** One illustrative example of alcohol dehydration involves the conversion of isobutanol to isobutene [169].

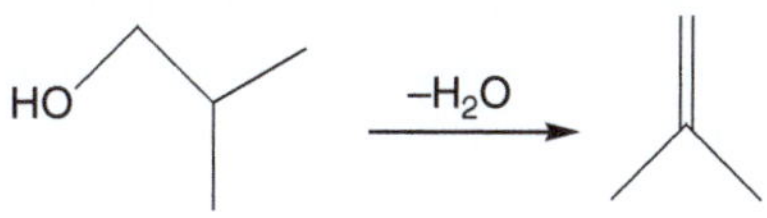

Solid base catalysts, often comprising metal oxides or supported bases, facilitate the removal of water from isobutanol, yielding isobutene [170]—a valuable building block to produce synthetic rubber and high-octane fuels. The use of solid base catalysts in this process not only enhances selectivity but also allows for the recyclability of the catalyst, reducing waste and energy consumption [170, 171].

2) **Ethanol dehydration to ethylene:** Another significant application is the dehydration of ethanol to ethylene [172], a crucial industrial intermediate.

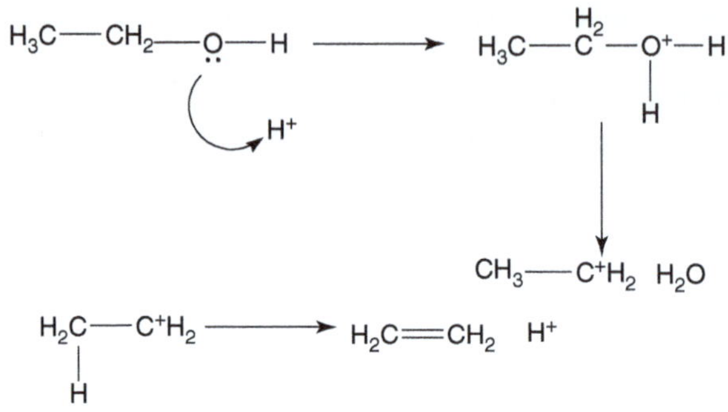

Solid base catalysts, such as alkaline earth metal oxides or hydrotalcite-like materials, prove effective in promoting the elimination of water from ethanol, generating ethylene [173]—a key component in the production of plastics, resins [115], and chemicals. The utilization of solid base catalysts in this context contributes to the sustainability of ethylene production by minimizing the environmental footprint associated with traditional methods.

3) **Glycerol dehydration to acrolein:** Glycerol, a by-product of biodiesel production, can be converted into acrolein [174] through dehydration.

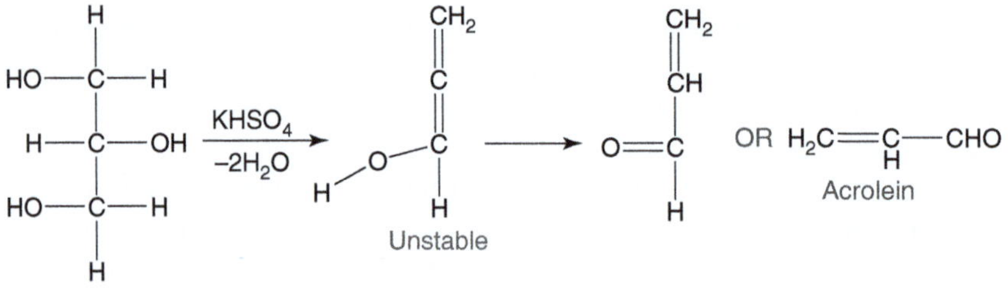

Solid base catalysts, including metal oxides like magnesium oxide [8, 39] or basic zeolites [116], play a crucial role in this transformation. Acrolein is a versatile compound used in the synthesis of various chemicals, and the use of solid base catalysts ensures efficient conversion and offers the potential for catalyst regeneration, supporting the ecofriendly utilization of glycerol [38] by-products.

These examples underscore the versatility and environmental benefits of employing solid base catalysts in alcohol dehydration processes, showcasing their role in advancing sustainable practices within the realm of chemical synthesis.

7.9.2 Dehydration of Alkanes

The dehydration of alkanes—a process involving the removal of hydrogen and subsequent formation of olefins [175]—is a transformative catalytic reaction with implications for the production of valuable hydrocarbons and petrochemicals. In recent advancements, solid acid catalysts have emerged as essential tools in facilitating alkane [176] dehydration, offering advantages such as increased selectivity, stability, and the potential for catalyst recycling. This catalytic approach aligns with the principles of green chemistry, promoting more sustainable and efficient pathways for hydrocarbon synthesis while minimizing environmental impact.

Examples:

1) ***n*-Butane dehydration to butenes:** The dehydration of *n*-butane to butenes [177] exemplifies the application of solid acid catalysts in alkane dehydration. Solid acid materials, including zeolites [116] and sulfated metal oxides [39, 40], play a crucial role in selectively removing hydrogen atoms from *n*-butane, leading to the formation of butenes.

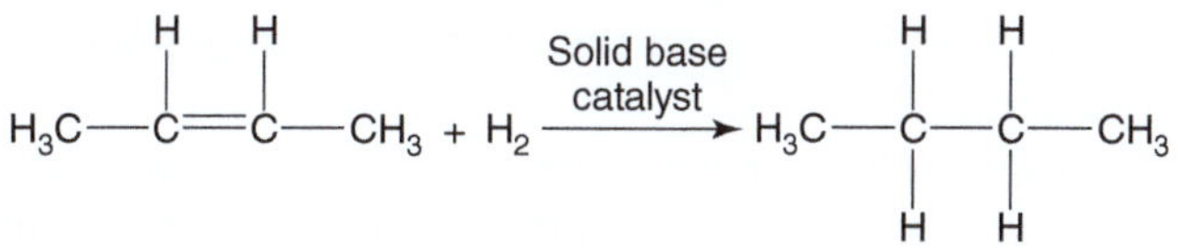

These olefins serve as essential building blocks in the production of polymers [178], plastics, and synthetic rubber. The use of solid acid catalysts enhances the efficiency of the process and allows for the reusability of the catalyst, contributing to sustainability in hydrocarbon processing.

2) **Isobutane dehydration to isobutene:** Isobutene [179], a valuable precursor in the synthesis of chemicals and fuels, can be obtained through the dehydration of isobutane. Solid acid catalysts, such as solid super acids or zeolites, facilitate this transformation by selectively abstracting hydrogen atoms from isobutane. The utilization of solid acid catalysts in this context not only improves the selectivity of isobutene [180] production but also enables the recovery and reuse of the catalyst, reducing waste and optimizing resource utilization [180, 181].
3) **Cyclohexane dehydration to cyclohexene:** The dehydration of cyclohexane to cyclohexene [182] is a notable example in the conversion of cycloalkanes to olefins.

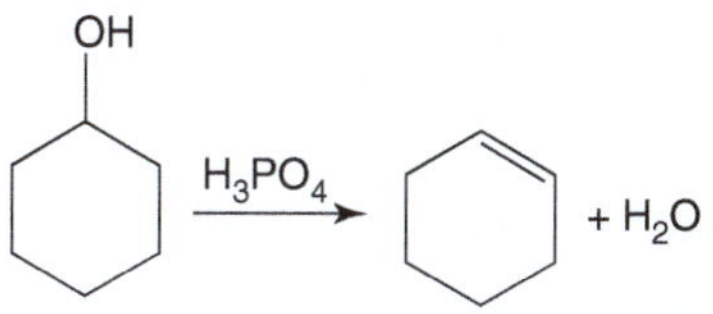

Solid acid catalysts, including aluminum-containing zeolites or silica–alumina materials [183], serve as effective promoters for the elimination of hydrogen from cyclohexane, yielding cyclohexene [182, 183]—a key intermediate in the production of nylon and other chemicals. The use of solid acid catalysts enhances the sustainability of cyclohexene synthesis by promoting more controlled and environmentally friendly reaction pathways.

7.9.3 Industrial Significance and Process Optimization

The industrial landscape heavily relies on dehydration reactions facilitated by solid base catalysts for the production of key chemical building blocks. From the synthesis of petrochemicals to the generation of platform chemicals, the importance of this catalytic process spans various sectors. For instance, the conversion of alcohols to olefins using solid base catalysts is integral to the production of polymers, plastics, and specialty chemicals, contributing to the development of diverse materials and industrial applications.

Moreover, in the realm of biofuels and renewable energy, solid base-catalyzed dehydration reactions play a vital role in the conversion of biomass-derived alcohols into biofuels, providing a sustainable alternative to traditional fossil fuels. This aligns with the global pursuit of greener energy sources and reduced environmental impact.

Solid base catalysts offer distinct advantages in the optimization of dehydration reactions. Their heterogeneous nature allows for efficient catalyst recovery and recycling, reducing operational costs and minimizing waste. The ability to operate under milder reaction conditions compared to traditional liquid catalysts further contributes to energy efficiency and process sustainability. Additionally, the tailored active sites of solid base catalysts enable selective transformations, enhancing the overall yield of desired products and reducing the formation of by-products.

Furthermore, the optimization of dehydration processes using solid base catalysts is closely tied to environmental considerations. By minimizing the use of hazardous or corrosive liquid reagents and reducing energy consumption, these catalysts contribute to the development of cleaner and more sustainable manufacturing practices.

7.10 Sulfur Removal in Fuel Refining

The industrial application of solid base catalysts for sulfur removal [123] in fuel refineries represents a cornerstone in the ongoing efforts to produce cleaner and more environmentally friendly transportation fuels. Sulfur, inherently present in crude oil and natural gas, possesses significant challenges, including environmental pollution and degradation of catalytic converters. In response, fuel refineries [184] worldwide are adopting advanced desulfurization processes, with solid base catalysts playing a pivotal role. These catalysts, often composed of metal oxides or supported basic materials, operate within hydrodesulfurization units to selectively eliminate sulfur compounds from fuel feedstocks. This process involves the catalytic promotion of hydrogenation reactions, breaking sulfur bonds [185] and resulting in cleaner, low-sulfur fuels. The industrial significance of solid base catalysts lies in their ability to enhance the efficiency and selectivity of desulfurization [123, 184–186], ensuring compliance with stringent environmental regulations. Their robust nature contributes to operational stability, minimizing downtime and maintenance costs. Moreover, the heterogeneous nature of solid base catalysts facilitates easy recovery and recycling, aligning with principles of green chemistry and promoting sustainability in the refining industry. As the global energy landscape evolves toward greater environmental responsibility, the industrial application of solid base catalysts [185] emerges as a crucial element in the production of fuels that meets stringent emissions standards and contributes to a cleaner and more sustainable energy future.

7.10.1 Hydrodesulfurization (HDS) Catalysts

Hydrodesulfurization (HDS) catalysis stands at the forefront of the global effort to produce cleaner and environmentally [187] responsible fuels. As the demand for energy continues to rise, the refining industry faces the challenge of mitigating the environmental impact of sulfur-containing compounds present in crude oil and natural gas. Hydrodesulfurization [188]—a catalytic process integral to fuel refinement—addresses this challenge by selectively removing sulfur compounds to meet stringent environmental standards.

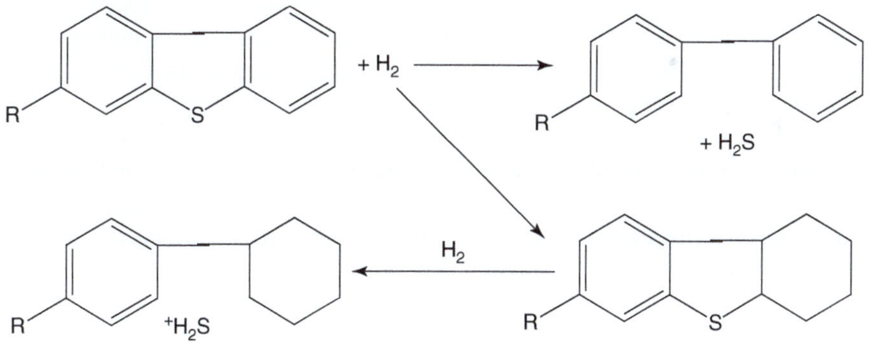

This catalytic transformation, employing solid base or metal-based catalysts, involves the reaction of sulfur-containing feedstocks [189] with hydrogen, resulting in the cleavage of sulfur bonds and the formation of hydrogen sulfide. The significance of hydrodesulfurization catalysis lies in its pivotal role in enhancing fuel quality, reducing emissions of sulfur dioxide (SO_2), and extending the life and efficiency of emission control systems in vehicles. This introduction explores the fundamental principles and industrial importance of hydrodesulfurization catalysis, emphasizing its crucial role in shaping the landscape of cleaner energy production and advancing sustainable practices within the refining sector.

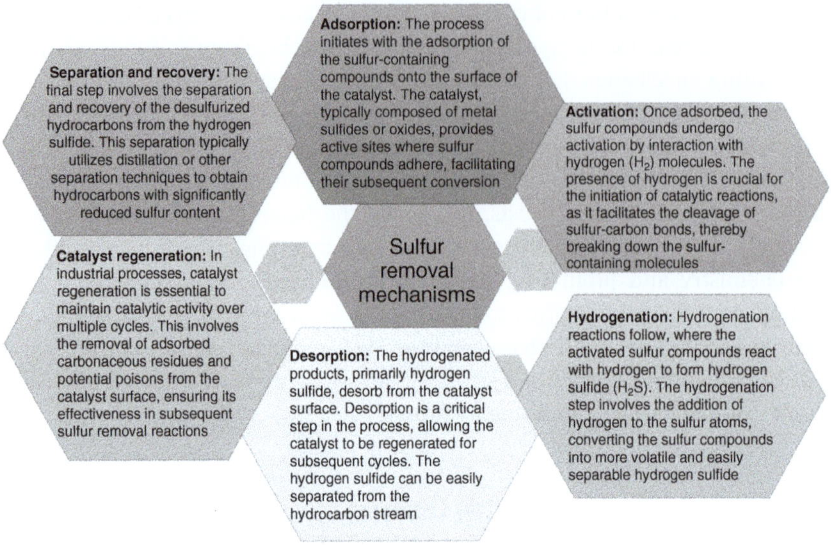

Figure 7.14 Sulfur removal mechanism during HDS [187, 190–192].

7.10.2 Sulfur Removal Mechanisms

Sulfur removal mechanisms in catalytic processes, particularly in hydrodesulfurization (HDS), involve intricate chemical transformations designed to selectively eliminate sulfur-containing compounds from hydrocarbon feedstocks [187, 190]. The predominant sulfur compounds targeted in these processes include thiols, sulfides, and thiophenes [191], all of which contribute to environmental pollution and adversely affect downstream catalysts. Here, we outline (Figure 7.14) the fundamental steps involved in the sulfur removal mechanism [192] during HDS.

The hydrodesulfurization mechanism is a well-established and efficient process in the refining industry [193], contributing significantly to the production of low-sulfur fuels that adhere to stringent environmental regulations. Advances in catalyst design and reaction engineering continue to refine these mechanisms, driving the industry toward cleaner and more sustainable [194] fuel production.

7.10.3 Impact on Clean Fuel Production

The impact of sulfur removal mechanisms, particularly through catalytic processes like HDS on clean fuel production is profound and far-reaching [195]. Here are key aspects (Figure 7.15) that illustrate the significance of these processes.

7.11 Processing Methods

Solid base catalysts are processed by first being prepared, then activated, and then used in particular processes. Here are a few typical processing techniques.

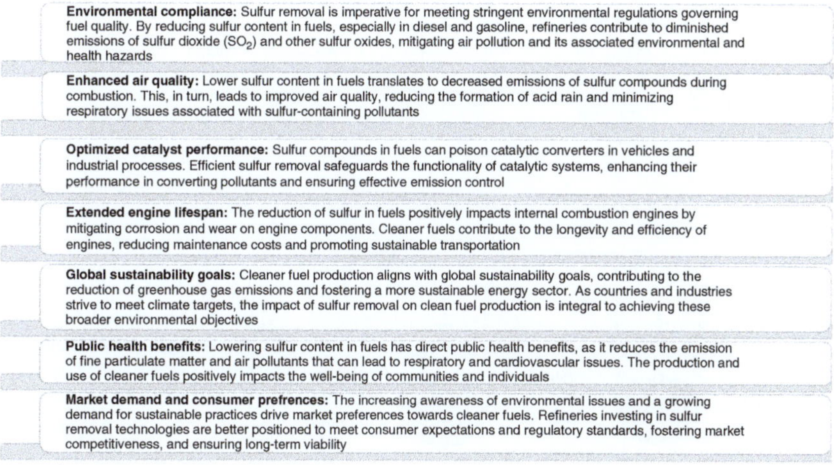

Figure 7.15 Impact of sulfur removal processes on clean fuel production [195–197].

7.11.1 Impregnation Method

Impregnation and drying [198] are widely utilized among other preparation techniques because of their ease of use and minimal waste output. The initial stage involves exposing a porous support to a solution containing a metal precursor [198]. Typical precursors encompass inorganic metal salts like sulfates (SO_4^{2-}), carbonates (CO_3^{2-}), chlorides (Cl^-), nitrates (NO_3^-), or acetates ($CH_3CO_2^-$), as well as organic metal complexes including acetylacetonates [198] ($C_5H_7O_2^-$). Due to the great solubility of many precursors, water is the most frequently utilized solvent for inorganic salts, while organic solvents [198] are primarily used for organometallic precursors. Concentrations below (super) saturation are needed to stop the metal precursor from depositing too soon in bulk solution.

There are two primary techniques for impregnation: pore volume impregnation (PVI) [198, 199], which uses exactly the amount required to fill the pore volume of the support, and wet impregnation (WI) (Figure 7.16), which uses an excess solution. The latter technique maintains a dry nature at the macroscopic level, so it is also referred to as dry impregnation (DI) (Figure 7.16) or incipient wetness impregnation (IWI) [198, 199].

7.11.2 Precipitation and Coprecipitation Method

In this technique, the substance of interest is caused to form a solid precipitate from a uniform mixture. Temperature changes and the addition of basic and acidic liquids cause precipitation. The precipitate is influenced by several factors, including solvent, temperature, pH, and raw material composition. There are four [39] phases involved in the production of a solid from a liquid solution (Figure 7.17).

The fundamental goal of this approach is to keep the pH at a steady level while precipitation occurs. Precipitates with low solubility, easy breakdown, and minimal

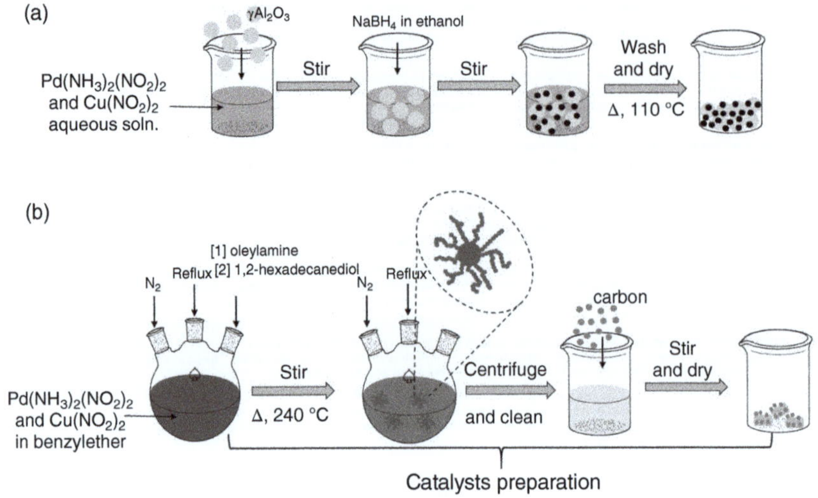

Figure 7.16 Diagram showing (a) dry impregnation; (b) wet impregnation. Source: Adapted from [39] techniques.

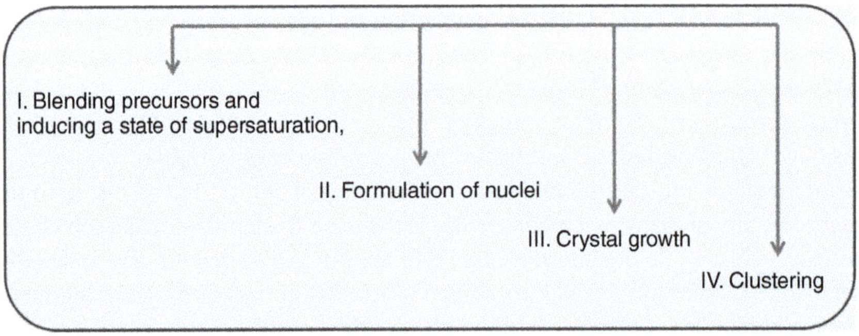

Figure 7.17 Different stages during the formulation of solid from liquid solution [39, 200].

toxicity, such as hydroxides and carbonates, are typically employed. Figure 7.18 illustrates the many steps in the precipitation process used to manufacture solid catalyst [201].

Various examples of solid base catalyst prepared by using coprecipitation method by different researchers are shown in Table 7.7:

7.11.3 Sol–Gel Method

The main step in the sol–gel process is the creation of a stable colloidal mixture called sol, which contains highly pure fine solid particles. The earlier goes through partial hydrolysis along with condensation to create a sol, which is then subjected to polycondensation [206] to create gel. Evaporative drying is used to extract the retained liquid from a gel; the dried product, called xerogel, is then heated to improve its quality [206]. The sol–gel approach is preferable, since it has a positive impact

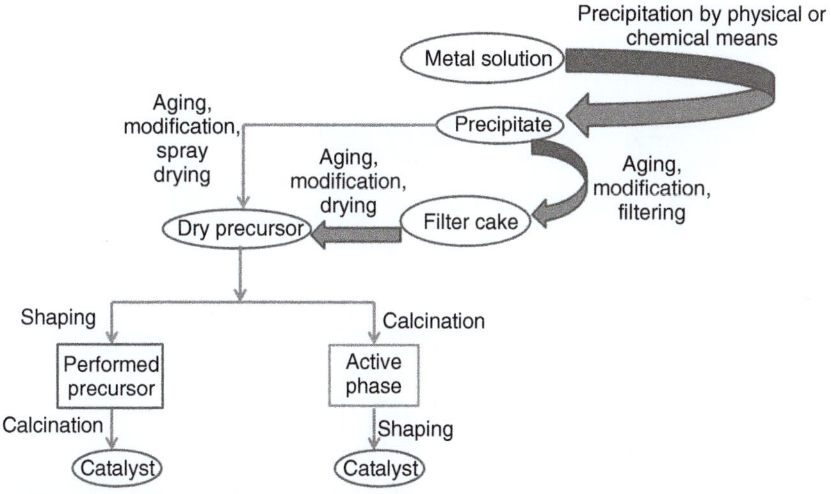

Figure 7.18 Flowchart illustrating the many stages in the precipitation method catalyst preparation. Source: Shahinuzzaman et al. [201], 2017/with permission of Elsevier.

Table 7.7 Examples of solid catalysts prepared by coprecipitation method.

S. no.	Prepared catalyst	Reference
1.	By means of magnesium nitrate hexahydrate [$Mg(NO_3)_2 \cdot 6H_2O$] and aluminum nitrate nonahydrate [$Al(NO_3)_3 \cdot 9H_2O$] as precursors, magnesium–Al oxides were produced via the coprecipitation method at both reduced as well as extremely saturated conditions	[39, 202]
2.	Basic solutions were used for preparing MgZr catalysts. Solid Mg–Zr catalysts were prepared using KOH, K_2CO_3, $H_2N_2O_7Zr$, and magnesium nitrate	[39, 203]
3.	Using the cautious coprecipitation approach, a mixed oxide catalyst of Zn and Ca was created	[39, 200]
4.	The traditional coprecipitation approach was utilized to create the Sr–MgO solid base catalyst. $Mg(NO_3)_2$, $Sr(NO_3)_2$, and $(NH_4)_2CO_3$ were utilized as precipitate-forming medium	[39, 204]
5.	Coprecipitation approach was used to generate MgAlFe-based solid base catalyst. In addition to NaOH and Na_2CO_3, precursors such as $Mg(NO_3)_2 \cdot 6H_2O$, $Fe(NO_3)_3 \cdot 9H_2O$, and $Al(NO_3)_3 \cdot 9H_2O$ were also utilized	[39, 205]

on the uniformity of catalysts as well as their surface and structural characteristics. The concentration of precursors, the kind of solvent used, pH, the concentration of additives [39], and the type and duration of aging process, all have an impact on the finished product. Figure 7.19 is an illustration of the sol–gel process [39, 207].

Various examples of solid base catalyst prepared by using sol–gel method by different researchers are shown in Table 7.8:

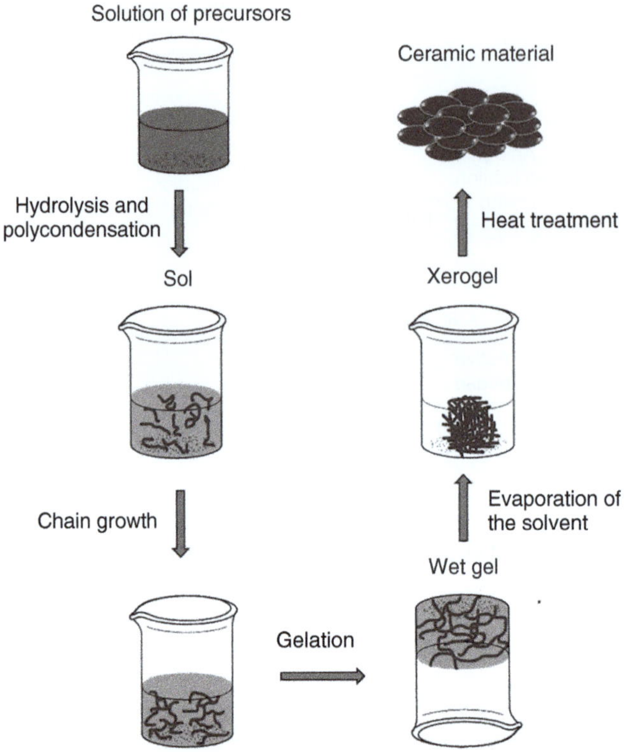

Figure 7.19 A picture that illustrates the sol–gel process [39, 207].

Table 7.8 Examples of solid catalysts prepared by coprecipitation method.

S. no.	Prepared catalyst	Reference
1.	$C_{12}H_{32}O_4Zr$, $C_5H_8O_2$, $Si(OC_2H_5)_4$, and C_3H_8O were utilized in the sol–gel process to generate a ZrO_2–SiO_2-based solid catalyst	[39, 208]
2.	The sol–gel process was employed to manufacture nanocrystalline (Al_2O_3) alumina	[39, 209]
3.	Titanium oxide (TiO_2) and multiwalled carbon nanotubes composite catalysts were created by dissolving 0.1 mol of $C_{12}H_{28}O_4Ti$ in 200 ml of C_2H_5OH	[39, 210]
4.	A new nanoscale magnetic solid base catalyst has been created, consisting of a coated double hydroxide placed on top of a ferrite core	[39, 211]

7.11.4 Hydrothermal Process

In a hydrothermal process, an autoclave is used to treat a homogenous aqueous solution of precursors and alkali to high pressure and temperature [39] for a predetermined amount of time. The high temperatures and pressures [39] caused things to dissolve and recrystallize. Filtration was used to collect the precipitate,

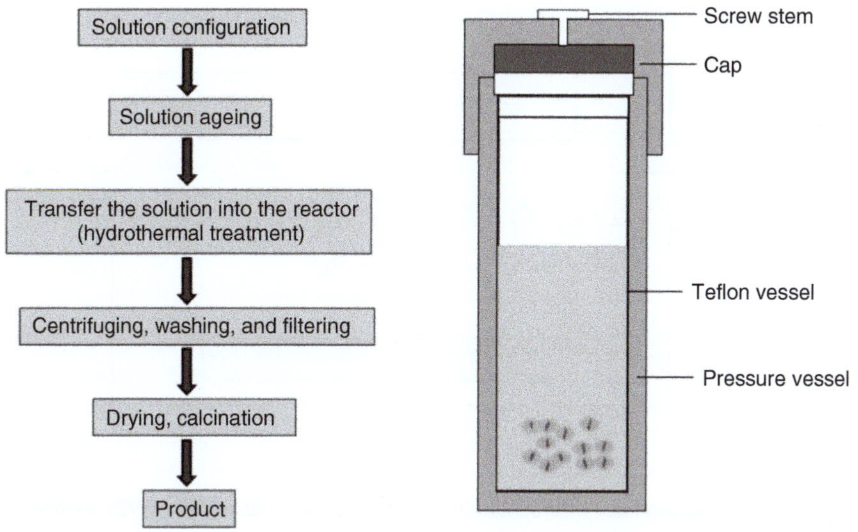

Figure 7.20 A diagram showing the stages in the hydrothermal operation along with the usual instrument required.

which was then cleaned and dried further [206]. Due to the temperature differential between the autoclave's top and bottom sections, the ions or molecules split as soon as the reactants were placed inside. The dissolved materials solidified, and the ions or molecules migrated to areas with lower temperatures. Figure 7.20 depicts several stages in the hydrothermal process.

Here are two examples of solid base catalyst prepared by hydrothermal method:

1) Several magnesium sources were used in the hydrothermal process to create nanocrystalline MgO catalysts [39, 212].
2) Hydrates of magnesium carbonate [39] were produced under hydrothermal conditions, and magnesium oxide catalyst was produced after additional processing [213].

7.11.5 Vapor Phase Deposition Method

In order to coat a substrate with a thin layer, atoms or molecules must typically be in the vapor phase. The item to be coated is put inside the chamber holding the vapors that will be applied to the film's surface. Chemical and physical vapor [39] depositions are the two categories of vapor phase deposition techniques [214]. It is dependent on the coating process, which may be triggered by chemical reactions or by physical impacts. Physical vapor deposition [39] involves subjecting a sample containing the target elements to an electric discharge, which causes the sample to emit atoms and produce a vapor that condenses on the substrate's surface. In contrast, target elements from the sample are supplied in the form of volatile molecules for chemical vapor deposition [39], where the film is created by a sequence of chemical reactions involving precursors, precursor residues, and

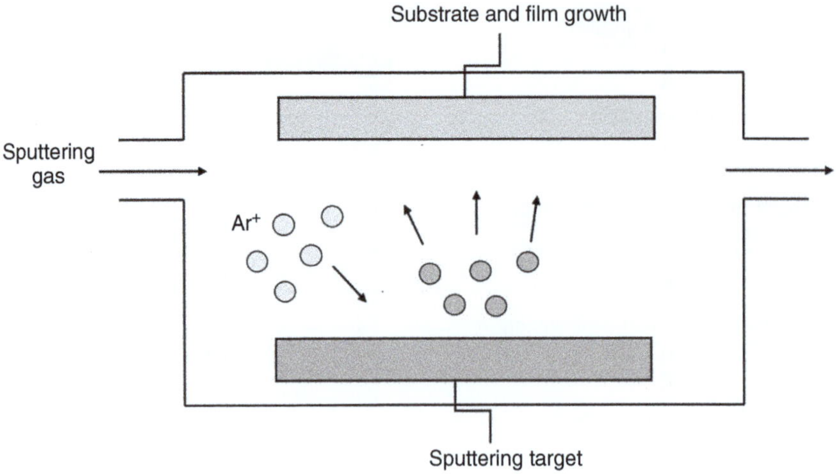

Figure 7.21 Diagrammatic representation of a physical vapor deposition (PVD) method [39, 215].

the substrate [39]. Toxic gases are generated during the chemical vapor deposition process, which can contaminate films and result in poor substrate–film adhesion. Figure 7.21 displays a schematic illustration of the physical vapor deposition process [215].

Here are three examples of solid base catalyst prepared by vapor phase deposition method are

1) $TiCl_4$ vapors were deposited on silica gel powder to create a solid catalyst consisting of Ti and SiO_2.
2) By using the double deposition technique to layer silicon dioxide on calcium oxide, Matsuhashi et al. [39, 216] created a SiO_2/CaO solid base catalyst.
3) Using nickel oxide as a basis, chemical vapor deposition [39] was used to create barium zirconate [217] ($BaZrO_3$) catalyst.

After the synthesis of solid base catalyst using one of the aforementioned methods, the procedure shown in Figure 7.22 should be followed.

The processing methods can vary depending on the specific type of solid base catalyst and the targeted application. The choice of method often depends on factors such as the nature of the catalytic reaction, the desired catalytic activity, and the properties of the catalyst support.

7.12 Use of Solid Base Catalyst in Various Industries

Solid base catalysts are essential to many industrial processes because they provide benefits such as increased selectivity, faster reaction rates, and simpler separation from reaction products. The following are some important sectors where solid base catalysts are used.

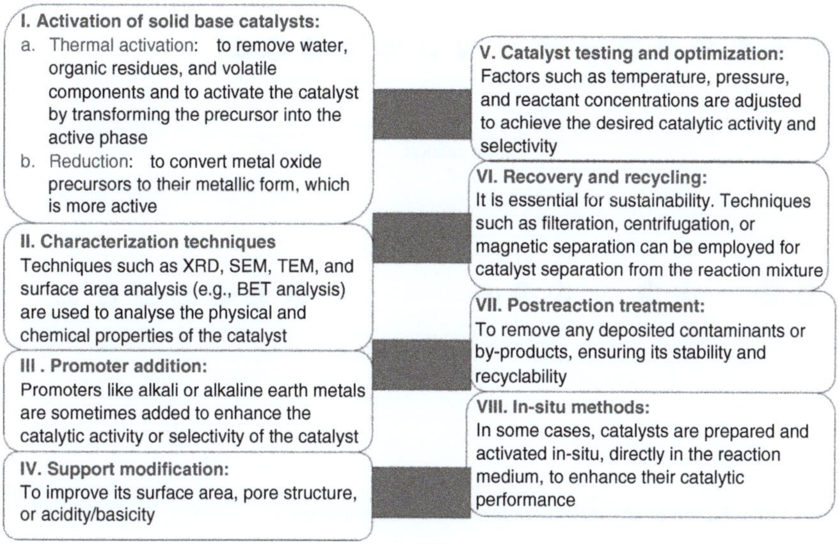

Figure 7.22 Processing procedure of solid base catalysts.

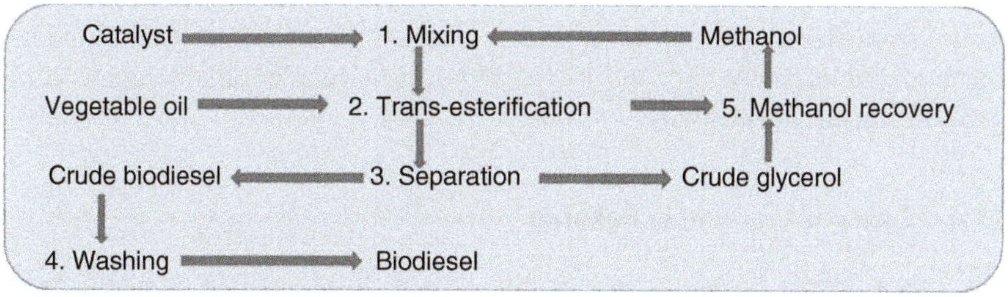

Figure 7.23 General pathway followed for producing biodiesel. Source: Sankar et al. [218], 2017/with permission of Elsevier.

7.12.1 Biodiesel Production (Refer to Section 2)

Triglycerides or vegetable oils are converted into biodiesel through transesterification processes using solid base catalysts such as metal oxides (CaO, MgO). When compared with liquid base catalysts, they offer increased activity, stability, and simpler separation. General steps involved in biodiesel production [218] are displayed in Figure 7.23.

7.12.2 Petrochemical Industries (Refer to Section 6.1)

Solid base catalysts are used in several petrochemical reactions, including reforming, alkylation, and isomerization. For example, in the alkylation of aromatics and the isomerization of hydrocarbons, zeolites and metal oxides function as solid base catalysts. Example of catalysis in petrochemical industry is displayed in Figure 7.24.

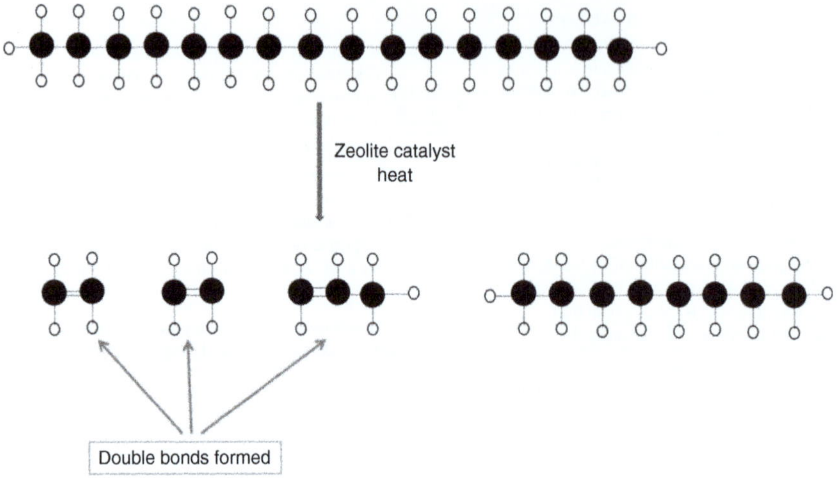

Figure 7.24 Example of catalysis in petrochemical industry.

7.12.3 Environmental Applications (Refer to Section 8)

By taking part in procedures like the catalytic removal of acidic gases (like sulfur dioxide) from flue gas streams, solid base catalysts contribute to environmental protection. Additionally, they aid in the removal of volatile organic compounds (VOCs) during air purification.

7.12.4 Catalytic Cracking in Refining

Solid base catalysts based on zeolite are used in fluid catalytic cracking [219] (Figure 7.25)—a technique in petroleum refining that aids in the breakdown of massive hydrocarbons into useful products like diesel and gasoline.

7.12.5 Biomass Conversion

In biomass conversion processes, such as the synthesis of platform chemicals and biofuels from renewable feedstock, solid base catalysts are used, facilitating reactions like transesterification or hydrolysis. Common catalysts include metal oxides or supported bases, enhancing the efficiency of biomass-to-biofuel or biomass-to-chemicals conversion. Types of solid catalysts used in biomass conversion [7] are displayed in Figure 7.26.

7.12.6 Water Treatment

Solid base materials are applied in water treatment processes to remove contaminants and pollutants [220] through catalytic reactions, improving water quality. Here are some instances of solid base catalyst-mediated reactions in the water treatment industry.

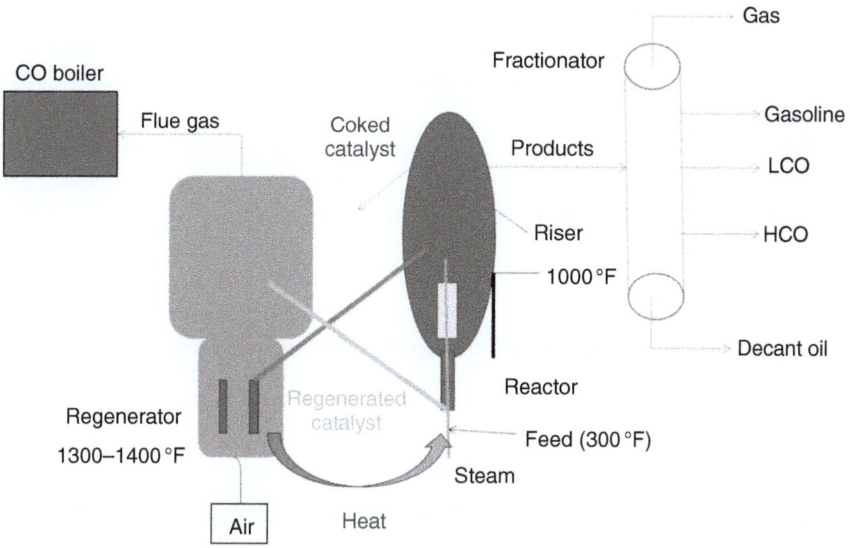

Figure 7.25 Fluid catalytic cracking (FCC).

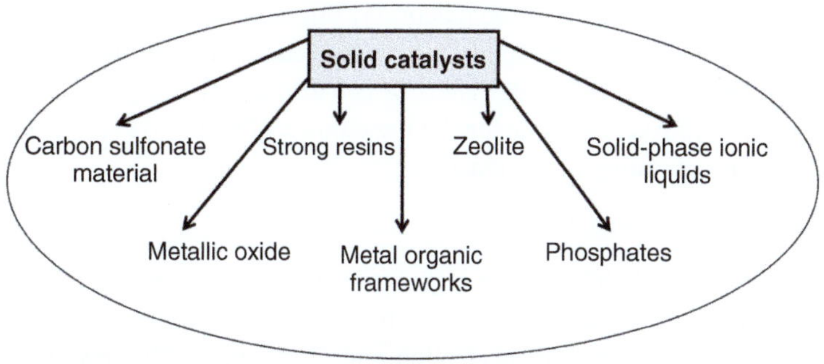

Figure 7.26 Solid catalyst types utilized in the conversion of biomass.

a) Hydrolysis of organic compounds: Catalysts with solid bases, like metal hydroxides or oxides, can be employed for the hydrolysis [26] of organic molecules like esters. In the presence of water and a base catalyst, ester bonds are broken during the reaction

$$RCOOR' + H_2O \rightarrow RCOOH + R'OH$$

b) Transesterification of oils: Solid base catalysts like metal oxides or supported alkali metal catalysts are used in the transesterification of triglycerides to produce biodiesel [221]. The reaction involves the conversion of triglycerides and methanol/ethanol to biodiesel and glycerol

$$\text{Triglyceride} + 3CH_3OH \rightarrow \text{3 fatty acids methyl esters (biodiesel)} + \text{Glycerol}$$

c) Degradation of phenolic compounds: Common contaminants found in industrial effluent, phenolic compounds [222], can be broken down using solid base catalysts. Reactions that convert the phenolic chemicals into less toxic forms can be aided by base catalysis.
d) Aldol condensation for water softening: Solid base catalysts can be used in aldol condensation reactions to remove hardness-causing ions from water. For example, calcium ions can be precipitated as calcium carbonate

$$Ca^{2+} + CO_3^{2-} \rightarrow CaCO_3 \text{ (Precipitate)}$$

These reactions typically involve heterogeneous catalysis, where the solid base catalyst is in a different phase than the reactants. This allows for easier separation of the catalyst from the treated water.

7.12.7 Catalytic Decomposition of Ammonia

The process by which ammonia (NH_3) is broken down into nitrogen (N_2) and hydrogen (H_2) gases is known as catalytic decomposition. Applications for this reaction include the synthesis of hydrogen and the elimination of ammonia from industrial effluents. Solid base catalysts have the potential to aid in this process, as solid base catalysts provide active sites for ammonia [223] adsorption and dissociation. They can enhance the kinetics of the reaction [224], making the process more efficient. The catalysts are often more stable and can be reused in multiple reaction cycles. A metal oxide, like nickel oxide (NiO) or cobalt oxide (Co_3O_4), is one of the most often used solid base catalysts for the breakdown of ammonia [225, 226]. The following is a representation of the reaction

$$2NH_3 \rightarrow N_2 + 3H_2 \text{ (using catalyst)}$$

Metal oxides, especially cobalt [225] and nickel oxide [226], function as solid base catalysts in the ammonia breakdown process. These catalysts' surfaces can serve as active sites for ammonia molecule adsorption and dissociation. The reaction is typically carried out at elevated temperatures to facilitate the decomposition. The temperature range is usually between 400 and 700 °C. The presence of the solid base catalyst allows for the reaction to occur at lower temperatures compared to noncatalytic decomposition.

7.12.8 Aldol Condensation and Knoevenagel Reactions

Solid base catalysts are employed in these reactions in organic synthesis for the production of various chemicals and pharmaceutical intermediates.

7.12.9 Hydrogenation Reactions (Refer to Section 3)

Solid base catalysts are used in hydrogenation reactions, such as the selective hydrogenation of unsaturated compounds in the production of fine chemicals and pharmaceuticals.

7.13 Socioeconomic Impact of Using Solid Base Catalyst

Solid base catalysts are revolutionizing the chemical industry not just through their efficient and selective reactions but also by triggering a ripple effect of positive socioeconomic impacts. The socioeconomic benefits of solid base catalysts are multifaceted and interconnected. Equitable distribution of these benefits is crucial to ensure inclusive and sustainable development. Let us delve deeper into these transformations

Environmental guardian:

- **Green chemistry**: Solid base catalysts often pave the way for cleaner, more sustainable chemical processes. They can reduce energy consumption, minimize waste generation, and enable the use of renewable feedstocks like biomass. Imagine factories with solar panels on their roofs and verdant landscapes surrounding them, instead of smokestacks spewing emissions.
- **Pollution control**: These catalysts can also be employed in emission control technologies, capturing harmful pollutants from industrial processes and improving air quality. Imagine smog-filled skies giving way to clear skies and vibrant cityscapes.

Economic engine:

- **Booming industries**: Imagine factories humming with activity, churning out value-added products like biofuels, pharmaceuticals, and fine chemicals. Solid base catalysts accelerate these processes, leading to increased production, profitability, and job creation.
- **Export potential**: The ability to produce high-quality, sustainable chemicals with solid base catalysts opens doors to new markets and strengthens a nation's economic standing. Imagine "Made with Green Chemistry" becoming a sought-after label, boosting exports and attracting foreign investments.
- **Innovation and research**: The development and use of novel solid base catalysts often drive innovation in the field of catalysis. This can lead to the creation of new industries, technologies, and job opportunities, contributing to economic growth.
- **Competitive advantage**: Industries adopting advanced catalytic processes can gain a competitive advantage by improving efficiency and reducing production costs. This may lead to increased market share and economic success.

Job creation and skill development:

- **Catalyst production and maintenance**: The demand for solid base catalysts creates opportunities for employment in the production, maintenance, and optimization of these catalysts.
- **Technical skills development**: The adoption of advanced catalytic processes requires skilled technicians and engineers, fostering the development of technical expertise and contributing to a skilled workforce.

Health and wellbeing:

- **Life-saving drugs**: Imagine affordable, readily available essential medicines manufactured using efficient solid base catalysts. This can significantly improve access to healthcare, particularly in developing nations.
- **Advanced materials**: These catalysts contribute to the development of innovative materials for medical devices, prosthetics, and implants, enhancing the quality of life for individuals with disabilities or chronic illnesses.

Energy efficiency:

- **Lower energy consumption**: Many reactions catalyzed by solid base catalysts operate under milder conditions compared to traditional processes. This results in lower energy consumption, which is economically beneficial and contributes to overall energy efficiency.

In conclusion, the usage of solid base catalysts has a variety of positive socioeconomic effects, including improved energy and environmental conditions, increased industrial productivity, economic expansion, and the creation of jobs. Positive socioeconomic transformation is expected to be driven by solid base catalysts as long as industries maintain their emphasis on efficiency and sustainability.

7.14 Challenges and Future Prospects

7.14.1 Current Challenges in Solid Base Catalysis

Solid base catalysts have shown promise in several applications, but they are not without difficulties. Currently, solid base catalysis faces the following difficulties:

1) **Limited active sites**: The number of active sites that can be used for catalysis is limited in many solid base catalysts. To improve catalytic efficiency, the trick is to make the most of these active sites. One of the continuing research goals is to increase the density of active sites while maintaining selectivity and stability.
2) **Deactivation and stability**: Several mechanisms, including surface fouling, leaching of active species, and poisoning by reactants or products, can cause solid base catalysts to become inactive over time. One of the main challenges is enhancing the stability of solid base catalysts throughout reactions and avoiding deactivation.
3) **Diffusion limitations:** In solid catalysts, the diffusion of reactants and products to and from active sites may be restricted, particularly in porous materials. Solid base catalysis presents a difficulty in improving mass transport qualities and optimizing the pore shape to minimize diffusion restrictions.
4) **Selectivity problems:** It is difficult for solid base catalysts to achieve good selectivity in complex reactions. An ongoing problem in catalyst design is to selectively activate particular functional groups while preventing unwanted side reactions.

5) **Synthesis and reproducibility**: It can be difficult to synthesize precise and repeatable solid base catalysts. For constant catalytic performance and repeatability, it is essential to manage the structure of the catalyst during synthesis and achieve a homogeneous distribution of active sites.
6) **Poisoning of the catalyst**: Acidic impurities or other contaminants in the reaction mixture have the potential to poison solid base catalysts. One of the current challenges is to improve solid base catalysts' tolerance to contaminants and to develop techniques to prevent catalytic poisoning.
7) **Comprehending reaction mechanisms**: The heterogeneous nature of solid base catalysts makes it difficult to elucidate the specific mechanisms of catalysis. To give an extensive understanding of chemical pathways and surface interactions, advances in computational and experimental methodologies are required.
8) **Scale-up challenges:** Scalability, cost-effectiveness, and preserving catalytic performance under larger-scale circumstances are some of the issues that arise when moving from laboratory-scale reactions to industrial-scale processes. Practical applications depend on the development of scalable processes for the synthesis of solid base catalysts.
9) **Application to various substrates**: It can be difficult to apply solid base catalysis to a variety of substrates and reaction circumstances. Research is now being conducted on tailoring solid base catalysts to suit different functional groups and reaction conditions.

7.14.2 Emerging Technologies and Materials

Several emerging technologies and materials are transforming the field of solid base catalysis, paving the way for more efficient, sustainable, and selective catalytic processes. Here are some notable advancements

1) **Hierarchical porous materials:** Hierarchical porous materials, characterized by their ordered pore structures at multiple length scales, offer enhanced surface area and mass transport properties, leading to improved catalytic performance. These materials can be tailored to specific reaction requirements, optimizing the accessibility of reactants to active catalytic sites.
2) **Single-atom catalysts (SACs)**: Single-atom catalysts (SACs) featuring isolated metal atoms dispersed on a support material exhibit exceptional catalytic activity and selectivity. The precise dispersion of metal atoms on the support provides well-defined active sites, minimizing unwanted side reactions and enhancing catalyst efficiency.
3) **Carbon-based catalysts**: Carbon-based catalysts, particularly graphene and carbon nanotubes, are attracting significant attention due to their unique properties such as high surface area, chemical stability, and electrical conductivity. These materials can be functionalized with various active sites to catalyze a wide range of reactions.

4) **Photocatalysis**: Photocatalysis, utilizing light energy to drive chemical transformations, is gaining traction in solid base catalysis. Photoactive catalysts, such as semiconductors and metal–organic frameworks, can absorb light and generate reactive species that promote catalytic reactions under mild conditions.
5) **In situ characterization techniques**: Advanced in situ characterization techniques, such as operando spectroscopy and microscopy, provide real-time insights into the structure and behavior of catalysts during reaction conditions. This information is crucial for understanding catalytic mechanisms and optimizing catalyst design.
6) **Computational modeling**: Density functional theory (DFT) and other computational methods are playing an increasingly important role in solid base catalysis. These tools allow researchers to predict catalytic activity, selectivity, and stability at the atomic level, guiding rational catalyst design and development.
7) **Machine learning**: Machine learning algorithms are being applied to analyze large datasets of catalytic data, identifying patterns and relationships that can inform catalyst design and reaction optimization. This data-driven approach is accelerating the discovery of new catalysts and reaction conditions.

The emergence of these technologies and materials is revolutionizing solid base catalysis, opening up new avenues for sustainable and efficient chemical synthesis. As research continues, we can expect further advancements in catalyst design, reaction optimization, and process integration, paving the way for a greener and more sustainable future for the chemical industry.

7.14.3 Prospects for Sustainable Industrial Catalysis

Sustainable industrial solid base catalysis holds immense promise for the future of chemical synthesis. The development of novel materials and technologies, such as hierarchical porous materials, SAC, carbon-based catalysts, photocatalysis, in situ characterization techniques, computational modeling, and machine learning, is revolutionizing the field, leading to more efficient, selective, and environmentally friendly catalytic processes. These advancements will enable sustainable production of various chemicals, pharmaceuticals, and materials, reducing environmental impact and contributing to a greener future.

7.15 Conclusion

7.15.1 Summary of Key Points

Solid base catalysts have found widespread applications in various industries, including biodiesel production, hydrogenation and dehydrogenation reactions, bimolecular reactions, methanol and DME synthesis, transesterification of esters, alkylation and isomerization reactions, environmental applications, dehydration reactions, sulfur removal in fuel refining, and organic synthesis. These applications span a diverse range of industries, from pharmaceuticals to petrochemicals, demonstrating the versatility and impact of solid base catalysis.

Solid base catalysts play a crucial role in promoting sustainability and environmental protection. Their use in biodiesel production reduces reliance on fossil fuels and minimizes environmental impact. In environmental applications, solid base catalysts enable efficient sulfur removal from flue gas, reduce nitrogen oxide emissions in catalytic converters, and facilitate waste remediation processes, contributing to cleaner industrial practices.

A significant advantage of solid base catalysts is their reusability, which minimizes waste generation and reduces the need for frequent catalyst replacement. This reusability contributes to cost-effectiveness and environmental sustainability, making solid base catalysis a more environmentally friendly alternative to traditional homogeneous catalysis.

Solid base catalysis continues to drive innovation in various industrial processes. The development of novel catalysts with enhanced activity, selectivity, and stability is crucial for achieving even higher efficiency, sustainability, and product quality. Emerging technologies, such as hierarchical porous materials and SAC, hold promise for overcoming current challenges and further advancing solid base catalysis.

7.15.2 Outlook for Continued Research and Development

The future of solid base catalysis is bright, with several research and development trends expected to drive innovation in the field. These trends include:

- The development of new materials with enhanced catalytic properties, such as hierarchical porous materials and SAC.
- The use of new technologies, such as photocatalysis and in situ characterization techniques, to improve catalyst performance and understanding.
- The application of computational modeling and machine learning to accelerate catalyst design and reaction optimization.
- The integration of solid base catalysis with other emerging technologies, such as nanotechnology and biotechnology, to create new and innovative catalytic processes.
- The development of sustainable and environmentally friendly catalytic processes that minimize the use of hazardous chemicals and solvents.
- The development of cost-effective and scalable catalytic processes for the production of valuable chemicals and materials.
- The development of catalysts that can operate under harsh conditions, such as high temperatures and pressures.

These research and development trends are expected to lead to the development of more efficient, selective, and environmentally friendly catalytic processes. As a result, solid base catalysis will continue to play an important role in the advancement of industrial processes and the development of new and innovative products and materials.

References

1 Ono, Y. and Hattori, H. (2012). *Solid Base Catalysis*, vol. 101. Springer Science & Business Media.

2 Bagreev, A. and Bandosz, T.J. (2002). A role of sodium hydroxide in the process of hydrogen sulfide adsorption/oxidation on caustic-impregnated activated carbons. *Industrial & Engineering Chemistry Research* 41 (4): 672–679.

3 Ottosson, N., Faubel, M., Bradforth, S.E. et al. (2010). Photoelectron spectroscopy of liquid water and aqueous solution: electron effective attenuation lengths and emission-angle anisotropy. *Journal of Electron Spectroscopy and Related Phenomena* 177 (2–3): 60–70.

4 Laszlo, P. (1986). Catalysis of organic reactions by inorganic solids. *Accounts of Chemical Research* 19 (4): 121–127.

5 Climent, M.J., Corma, A., Iborra, S. et al. (2004). Increasing the basicity and catalytic activity of hydrotalcites by different synthesis procedures. *Journal of Catalysis* 225 (2): 316–326.

6 Wang, X., Tseng, Y.H., Chan, J.C., and Cheng, S. (2005). Catalytic applications of aminopropylated mesoporous silica prepared by a template-free route in flavanones synthesis. *Journal of Catalysis* 233 (2): 266–275.

7 Rinaldi, R. and Schüth, F. (2009). Design of solid catalysts for the conversion of biomass. *Energy & Environmental Science* 2 (6): 610–626.

8 Tanabe, K. and Hölderich, W.F. (1999). Industrial application of solid acid–base catalysts. *Applied Catalysis A: General* 181 (2): 399–434.

9 Armor, J.N. (1991). New catalytic technology commercialized in the USA during the 1980's. *Applied Catalysis* 78 (2): 141–173.

10 Wood, A. (1994). Mobil cultivates new zeolites to grow process licensing; finding new applications for ZSM-5 catalysts. *Chemical Week* 155 (6): 34–36.

11 Morris, G. (1995). Solid-acid alkylate catalysts move toward commercial reality. *Chemical Week* 156 (9): 12–12.

12 Bentley, R.W. (2002). Global oil & gas depletion: an overview. *Energy Policy* 30 (3): 189–205.

13 Hocking, M.B. (2012). *Modern Chemical Technology and Emission Control.* Springer Science & Business Media.

14 Micelotta, E.R., Jones, A.P., and Tielens, A.G.G.M. (2010). Polycyclic aromatic hydrocarbon processing in a hot gas. *Astronomy & Astrophysics* 510: A37.

15 Baylis, J. (2019). *Wales and the Bomb: The Role of Welsh Scientists and Engineers in the UK Nuclear Programme.* University of Wales Press.

16 Misono, M. and Nojiri, N. (1990). Recent progress in catalytic technology in Japan. *Applied Catalysis* 64: 1–30.

17 Hölderich, W.F. (1993). New reactions in various fields and production of specialty chemicals. In: *Studies in Surface Science and Catalysis* (ed. H-P. Steinrück), vol. 75, 127–163. Elsevier.

18 Kita, Y., Kishino, K., and Nakagawa, K. (1996). Supported catalyst for synthesis of N-phenylmaleimide. *Nippon Kagaku Kaishi* 23: 269–274.

19 Enze, M. and Peiling, Z. (1993). Progress in catalytic technology in the People's Republic of China during the 1980s. *Applied Catalysis A: General* 95 (1): 1–20.

20 Sinnott, R.A.Y. (2014). *Chemical Engineering Design*, vol. 6. Elsevier.

21 Martín, A.J., Mitchell, S., Mondelli, C. et al. (2022). Unifying views on catalyst deactivation. *Nature Catalysis* 5 (10): 854–866.

22 Frontera, P., Macario, A., Ferraro, M., and Antonucci, P. (2017). Supported catalysts for CO_2 methanation: a review. *Catalysts* 7 (2): 59.

23 Chauvel, A., Delmon, B., and Hölderich, W.F. (1994). New catalytic processes developed in Europe during the 1980's. *Applied Catalysis A: General* 115 (2): 173–217.

24 Maroa, S. and Inambao, F. (2021). A review of sustainable biodiesel production using biomass derived heterogeneous catalysts. *Engineering in Life Sciences* 21 (12): 790–824.

25 Avhad, M.R. and Marchetti, J.M. (2016). Innovation in solid heterogeneous catalysis for the generation of economically viable and ecofriendly biodiesel: a review. *Catalysis Reviews* 58 (2): 157–208.

26 Su, F. and Guo, Y. (2014). Advancements in solid acid catalysts for biodiesel production. *Green Chemistry* 16 (6): 2934–2957.

27 Tomar, R., Kundra, P., Sharma, J. et al. (2024). *Letters in Organic Chemistry* 21 (1): 16–28.

28 Kumar, L., Verma, N., Tomar, R. et al. (2023). *Dalton Transactions* 52 (10): 3006–3015.

29 Tomar, R., Jain, S., Yadav, P. et al. (2022). *Current Organic Synthesis* 19 (3): 414–425.

30 Kumar, L., Verma, N., Sehrawat, H. et al. (2022). *New Journal of Chemistry* 46 (7): 3472–3481.

31 Tomar, R., Singh, N., Kumar, N. et al. (2019). *Catalysis Letters* 149 (6): 1589–1594.

32 Tomar, R., Singh, N., Rathee, G. et al. (2017). *Asian Journal of Organic Chemistry* 6 (12): 1728–1732.

33 Tomar, R., Sharma, J., Nishimura, S., and Ebitani, K. (2016). *Chemistry Letters* 45 (7): 843–845.

34 Arya, K. and Tomar, R. (2013). *Research on Chemical Intermediates* 41 (6): 3389–3400.

35 Tomar, R., Ebitani, K., and Chandra, R. (2019). *ChemistrySelect* 4 (12): 3577–3581.

36 Tomar, R., Rathee, G., Chandra, I. et al. (2018). *ChemistrySelect* 3 (6): 1645–1649.

37 Vignesh, P., Kumar, A.P., Ganesh, N.S. et al. (2021). A review of conventional and renewable biodiesel production. *Chinese Journal of Chemical Engineering* 40: 1–17.

38 Salaheldeen, M., Mariod, A.A., Aroua, M.K. et al. (2021). Current state and perspectives on transesterification of triglycerides for biodiesel production. *Catalysts* 11 (9): 1121.

39 Jambhulkar, D.K., Ugwekar, R.P., Bhanvase, B.A., and Barai, D.P. (2022). A review on solid base heterogeneous catalysts: preparation, characterization and applications. *Chemical Engineering Communications* 209 (4): 433–484.

40 Lee, D.W., Park, Y.M., and Lee, K.Y. (2009). Heterogeneous base catalysts for transesterification in biodiesel synthesis. *Catalysis Surveys from Asia* 13: 63–77.

41 Xie, W., Peng, H., and Chen, L. (2006). Transesterification of soybean oil catalyzed by potassium loaded on alumina as a solid-base catalyst. *Applied Catalysis A: General* 300 (1): 67–74.

42 Suppes, G.J., Dasari, M.A., Doskocil, E.J. et al. (2004). Transesterification of soybean oil with zeolite and metal catalysts. *Applied Catalysis A: General* 257 (2): 213–223.

43 Tsuji, H., Yagi, F., Hattori, H., and Kita, H. (1994). Self-condensation of n-butyraldehyde over solid base catalysts. *Journal of Catalysis* 148 (2): 759–770.

44 Liu, X., He, H., Wang, Y. et al. (2008). Transesterification of soybean oil to biodiesel using CaO as a solid base catalyst. *Fuel* 87 (2): 216–221.

45 Peterson, G.R. and Scarrah, W.P. (1984). Rapeseed oil transesterification by heterogeneous catalysis. *Journal of the American Oil Chemists Society* 61 (10): 1593–1597.

46 Ono, Y. (2003). Solid base catalysts for the synthesis of fine chemicals. *Journal of Catalysis* 216 (1–2): 406–415.

47 Kabashima, H. and Hattori, H. (1998). Cyanoethylation of alcohols over solid base catalysts. *Catalysis Today* 44 (1–4): 277–283.

48 Dossin, T.F., Reyniers, M.F., and Marin, G.B. (2006). Kinetics of heterogeneously MgO-catalyzed transesterification. *Applied Catalysis B: Environmental* 62 (1–2): 35–45.

49 Patil, P.D. and Deng, S. (2009). Transesterification of Camelina sativa oil using heterogeneous metal oxide catalysts. *Energy & Fuels* 23 (9): 4619–4624.

50 Sulaiman, S. and Talha, N.S. (2018). Technique to produce catalyst from egg shell and coconut waste for biodiesel production. *Multifaceted Protocol in Biotechnology* 83–91.

51 Ma, F. and Hanna, M.A. (1999). Biodiesel production: a review. *Bioresource Technology* 70 (1): 1–15.

52 Liu, X., He, H., Wang, Y., and Zhu, S. (2007). Transesterification of soybean oil to biodiesel using SrO as a solid base catalyst. *Catalysis Communications* 8 (7): 1107–1111.

53 Farm Energy (2019). Commercial and Large Scale Biodiesel Production Systems.

54 Rawat, I., Kumar, R.R., Mutanda, T., and Bux, F. (2013). Biodiesel from microalgae: a critical evaluation from laboratory to large scale production. *Applied Energy* 103: 444–467.

55 Leung, D.Y., Wu, X., and Leung, M.K.H. (2010). A review on biodiesel production using catalyzed transesterification. *Applied Energy* 87 (4): 1083–1095.

56 Tafesh, A. and Basheer, S. (2013). Pretreatment methods in biodiesel production processes. *Pretreatment Techniques for Biofuels and Biorefineries* 417–434.

57 Amini, Z., Ilham, Z., Ong, H.C. et al. (2017). State of the art and prospective of lipase-catalyzed transesterification reaction for biodiesel production. *Energy Conversion and Management* 141: 339–353.

58 Atadashi, I.M., Aroua, M.K., and Aziz, A.A. (2011). Biodiesel separation and purification: a review. *Renewable Energy* 36 (2): 437–443.

59 Van Gerpen, J. (2004). *Biodiesel Production Technology*. USDA/NCAUR.

60 Veljković, V.B., Banković-Ilić, I.B., and Stamenković, O.S. (2015). Purification of crude biodiesel obtained by heterogeneously-catalyzed transesterification. *Renewable and Sustainable Energy Reviews* 49: 500–516.

61 Hattori, H. (2015). Solid base catalysts: fundamentals and their applications in organic reactions. *Applied Catalysis A: General* 504: 103–109.

62 Bartholomew, C.H. (2001). Mechanisms of catalyst deactivation. *Applied Catalysis A: General* 212 (1–2): 17–60.

63 Sun, L.B., Liu, X.Q., and Zhou, H.C. (2015). Design and fabrication of mesoporous heterogeneous basic catalysts. *Chemical Society Reviews* 44 (15): 5092–5147.

64 Wang, Z., Chen, G., and Ding, K. (2009). Self-supported catalysts. *Chemical Reviews* 109 (2): 322–359.

65 Izák, P., Bobbink, F.D., Hulla, M. et al. (2018). Catalytic ionic-liquid membranes: the convergence of ionic-liquid catalysis and ionic-liquid membrane separation technologies. *ChemPlusChem* 83 (1): 7–18.

66 Munyentwali, A., Li, H., and Yang, Q. (2022). Review of advances in bifunctional solid acid/base catalysts for sustainable biodiesel production. *Applied Catalysis A: General* 633: 118525.

67 Sabri, M.A., Al Jitan, S., Bahamon, D. et al. (2021). Current and future perspectives on catalytic-based integrated carbon capture and utilization. *Science of the Total Environment* 790: 148081.

68 Jiang, X., Nie, X., Guo, X. et al. (2020). Recent advances in carbon dioxide hydrogenation to methanol via heterogeneous catalysis. *Chemical Reviews* 120 (15): 7984–8034.

69 Veldsink, J.W., Bouma, M.J., Schöön, N.H., and Beenackers, A.A. (1997). Heterogeneous hydrogenation of vegetable oils: a literature review. *Catalysis Reviews* 39 (3): 253–318.

70 List, G.R., Mounts, T.L., Orthoefer, F., and Neff, W.E. (1995). Margarine and shortening oils by interesterification of liquid and trisaturated triglycerides. *Journal of the American Oil Chemists' Society* 72: 379–382.

71 Julkapli, N.M. and Bagheri, S. (2016). Magnesium oxide as a heterogeneous catalyst support. *Reviews in Inorganic Chemistry* 36 (1): 1–41.

72 Stanković, M., Krstić, J., Gabrovska, M. et al. (2017). Supported nickel-based catalysts for partial hydrogenation of edible oils. In: *New Advances in Hydrogenation Processes-Fundamentals and Applications* (ed. M. T. Ravanchi), 131–174.

73 Prasad, R. and Singh, P. (2011). Applications and preparation methods of copper chromite catalysts: a review. *Bulletin of Chemical Reaction Engineering & Catalysis* 6 (2): 63–113.

74 Olah, G.A. and Molnár, Á. (2003). *Hydrocarbon Chemistry*. John Wiley & Sons.

75 Vo, D.V.N. and Adesina, A.A. (2012). A potassium-promoted Mo carbide catalyst system for hydrocarbon synthesis. *Catalysis Science & Technology* 2 (10): 2066–2076.

76 Chen, I. and Chen, F.L. (1990). Effect of alkali and alkaline-earth metals on the resistivity to coke formation and sintering of nickel-alumina catalysts. *Industrial & Engineering Chemistry Research* 29 (4): 534–539.

77 Maddah, H.A. (2018). A comparative study between propane dehydrogenation (PDH) technologies and plants in Saudi Arabia. *American Scientific Research Journal for Engineering, Technology, and Sciences* 45 (1): 49.

78 Polshettiwar, V. and Varma, R.S. (2010). Green chemistry by nano-catalysis. *Green Chemistry* 12 (5): 743–754.

79 Huang, S., Yan, B., Wang, S., and Ma, X. (2015). Recent advances in dialkyl carbonates synthesis and applications. *Chemical Society Reviews* 44 (10): 3079–3116.

80 Lang, X.D. and He, L.N. (2016). Green catalytic process for cyclic carbonate synthesis from carbon dioxide under mild conditions. *The Chemical Record* 16 (3): 1337–1352.

81 Chaban, V.V., Andreeva, N.A., Bernard, F.L. et al. (2023). Chemical similarity of dialkyl carbonates and carbon dioxide opens an avenue for novel greenhouse gas scavengers: cheap recycling and low volatility via experiments and simulations. *Physical Chemistry Chemical Physics* 25 (13): 9320–9335.

82 Hattori, H. and Ono, Y. (2018). Catalysts and catalysis for acid–base reactions. In: *Metal Oxides in Heterogeneous Catalysis*, 133–209. Elsevier.

83 Shaikh, A.A.G. and Sivaram, S. (1996). Organic carbonates. *Chemical Reviews* 96 (3): 951–976.

84 Tamboli, A.H., Chaugule, A.A., and Kim, H. (2017). Catalytic developments in the direct dimethyl carbonate synthesis from carbon dioxide and methanol. *Chemical Engineering Journal* 323: 530–544.

85 Fujita, S.I., Bhanage, B.M., Ikushima, Y., and Arai, M. (2001). Synthesis of dimethyl carbonate from carbon dioxide and methanol in the presence of methyl iodide and base catalysts under mild conditions: effect of reaction conditions and reaction mechanism. *Green Chemistry* 3 (002): 87–91.

86 Hattori, H. (2001). Solid base catalysts: generation of basic sites and application to organic synthesis. *Applied Catalysis A: General* 222 (1–2): 247–259.

87 Caille, S., Cui, S., Faul, M.M. et al. (2019). Molecular complexity as a driver for chemical process innovation in the pharmaceutical industry. *The Journal of Organic Chemistry* 84 (8): 4583–4603.

88 Kontturi, E., Laaksonen, P., Linder, M.B. et al. (2018). Advanced materials through assembly of nanocelluloses. *Advanced Materials* 30 (24): 1703779.

89 Huang, W.C., Huang, M.S., Huang, C.F. et al. (2010). Thermochemical conversion of polymer wastes into hydrocarbon fuels over various fluidizing cracking catalysts. *Fuel* 89 (9): 2305–2316.

90 Azizi, Z., Rezaeimanesh, M., Tohidian, T., and Rahimpour, M.R. (2014). Dimethyl ether: a review of technologies and production challenges. *Chemical Engineering and Processing: Process Intensification* 82: 150–172.

91 Semelsberger, T.A., Borup, R.L., and Greene, H.L. (2006). Dimethyl ether (DME) as an alternative fuel. *Journal of Power Sources* 156 (2): 497–511.

92 Olah, G.A., Goeppert, A., and Prakash, G.S. (2009). Chemical recycling of carbon dioxide to methanol and dimethyl ether: from greenhouse gas to renewable, environmentally carbon neutral fuels and synthetic hydrocarbons. *The Journal of Organic Chemistry* 74 (2): 487–498.

93 Joghee, P., Malik, J.N., Pylypenko, S., and O'Hayre, R. (2015). A review on direct methanol fuel cells – In the perspective of energy and sustainability. *MRS Energy & Sustainability* 2: E3.

94 Mohammadi, R. and Hadizadeh Harandi, M. (2023). Methanol to olefin (MTO) value chain management. *New Applied Studies in Management, Economics & Accounting* 6 (1): 7–17.

95 Ott, J., Gronemann, V., Pontzen, F. et al. (2000). Methanol. In: *Ullmann's Encyclopedia of Industrial Chemistry* (ed. F. Ullmann). Wiley.

96 Erdener, H., Arinan, A., and Orman, S. (2012). Future fossil fuel alternative; DME (A review). *International Journal of Renewable Energy Research* 1 (4): 252–258.

97 Yeoman, A.M. and Lewis, A.C. (2021). Global emissions of VOCs from compressed aerosol products. *Elementa Science of the Anthropocene* 9 (1): 00177.

98 Terasaki, H., Ogura, Y., Kitano, S. et al. (2018). Management of diabetic macular edema in Japan: a review and expert opinion. *Japanese Journal of Ophthalmology* 62: 1–23.

99 Etim, U.J., Song, Y., and Zhong, Z. (2020). Improving the Cu/ZnO-based catalysts for carbon dioxide hydrogenation to methanol, and the use of methanol as a renewable energy storage media. *Frontiers in Energy Research* 8: 545431.

100 Villa, K. and Galán-Mascarós, J.R. (2021). Nanostructured photocatalysts for the production of methanol from methane and water. *ChemSusChem* 14 (9): 2023–2033.

101 Luu, M.T., Milani, D., Bahadori, A., and Abbas, A. (2015). A comparative study of CO_2 utilization in methanol synthesis with various syngas production technologies. *Journal of CO2 Utilization* 12: 62–76.

102 Wei, Y., de Jongh, P.E., Bonati, M.L. et al. (2015). Enhanced catalytic performance of zeolite ZSM-5 for conversion of methanol to dimethyl ether by combining alkaline treatment and partial activation. *Applied Catalysis A: General* 504: 211–219.

103 Wolday, A.K., Gujarathi, A.M., and Ramteke, M. (2023). Multi-objective optimization of methanol production for energy efficiency and environmental sustainability. *Computers & Chemical Engineering* 179: 108426.

104 Jothiramalingam, R. and Wang, M.K. (2009). Review of recent developments in solid acid, base, and enzyme catalysts (heterogeneous) for biodiesel production via transesterification. *Industrial & Engineering Chemistry Research* 48 (13): 6162–6172.

105 Karmakar, G., Ghosh, P., and Sharma, B.K. (2017). Chemically modifying vegetable oils to prepare green lubricants. *Lubricants* 5 (4): 44.

106 Demirbas, A. (2008). Comparison of transesterification methods for production of biodiesel from vegetable oils and fats. *Energy Conversion and Management* 49 (1): 125–130.

107 Andrikopoulos, N.K. (2002). Triglyceride species compositions of common edible vegetable oils and methods used for their identification and quantification. *Food Reviews International* 18 (1): 71–102.

108 Van Gerpen, J. and Knothe, G. (2005). Basics of the transesterification reaction. *The Biodiesel Handbook* 26–41.

109 Gotor, V. (2000). Pharmaceuticals through enzymatic transesterification and enzymatic aminolysis reactions. *Biocatalysis and Biotransformation* 18 (2): 87–103.

110 Shamshuddin, S.M. and Nagaraju, N. (2007). Liquid phase transesterification of methyl salicylate and phenol over solid acids: kinetic studies. *Journal of Molecular Catalysis A: Chemical* 273 (1–2): 55–63.

111 Gok, H.Y.F., Emami, S., Shen, J., and Reaney, M.J. (2013). Polyol-derived alkoxide/hydroxide base catalysts II: transesterification reactions. *Journal of the American Oil Chemists' Society* 90 (2): 299–305.

112 Nisticò, R. (2020). Polyethylene terephthalate (PET) in the packaging industry. *Polymer Testing* 90: 106707.

113 Choi, G. (2005). Polybutylene terephthalate (PBT). In: *Engineering Plastics Handbook*, 2e, 131–154. Blacklick, OH, USA: McGraw-Hill.

114 Jegan Roy, J. and Emilia Abraham, T. (2004). Strategies in making cross-linked enzyme crystals. *Chemical Reviews* 104 (9): 3705–3722.

115 Roli, N.F.M. (2014). Kinetic study of esterification of fatty acids over calcium ferrite catalyst. Doctoral dissertation. UMP.

116 Villegas, J.I., Kumar, N., Heikkilä, T. et al. (2006). Isomerization of n-butane to isobutane over Pt-modified beta and ZSM-5 zeolite catalysts: catalyst deactivation and regeneration. *Chemical Engineering Journal* 120 (1–2): 83–89.

117 Sahu, R., Song, B.J., Im, J.S. et al. (2015). A review of recent advances in catalytic hydrocracking of heavy residues. *Journal of Industrial and Engineering Chemistry* 27: 12–24.

118 Di Cosimo, J.I., Diez, V.K., and Apesteguía, C.R. (1996). Base catalysis for the synthesis of α,β-unsaturated ketones from the vapor-phase aldol condensation of acetone. *Applied Catalysis A: General* 137 (1): 149–166.

119 Raju, S., Moret, M.E., and Klein Gebbink, R.J. (2015). Rhenium-catalyzed dehydration and deoxydehydration of alcohols and polyols: opportunities for the formation of olefins from biomass. *ACS Catalysis* 5 (1): 281–300.

120 Moriggi, J.D. (2001). Solid-phase synthesis of cyclic sulfonamides and sulfamides employing a ring-closing metathesis cleavage strategy. Doctoral dissertation. University of Southampton.

121 Gholami, Z., Tišler, Z., and Rubáš, V. (2021). Recent advances in Fischer-Tropsch synthesis using cobalt-based catalysts: a review on supports, promoters, and reactors. *Catalysis Reviews* 63 (3): 512–595.

122 Fujimoto, K., Kudo, Y., and Tominaga, H.O. (1984). Synthesis gas conversion utilizing mixed catalyst composed of CO reducing catalyst and solid acid: II. Direct synthesis of aromatic hydrocarbons from synthesis gas. *Journal of Catalysis* 87 (1): 136–143.

123 Ito, E. and Van Veen, J.R. (2006). On novel processes for removing sulphur from refinery streams. *Catalysis Today* 116 (4): 446–460.

124 Sánchez-Delgado, R.A. (2002). *Hydrodesulfurization and Hydrodenitrogenation* (ed. M. Poort), 1–34. Netherlands: Springer.

125 Xu, M., Lunsford, J.H., Goodman, D.W., and Bhattacharyya, A. (1997). Synthesis of dimethyl ether (DME) from methanol over solid-acid catalysts. *Applied Catalysis A: General* 149 (2): 289–301.

126 Tu, M. and Davis, R.J. (2001). Cycloaddition of CO_2 to epoxides over solid base catalysts. *Journal of Catalysis* 199 (1): 85–91.

127 Gadakari, V.D. and Gondhali, A.S. (2018). Environmentally benign protocol of Knoevenagel condensation reaction. In: *One Day National Conference on 'Recent trends in chemistry'* (February 2018) held in G.I. Bagewadi Arts, Science and Commerce College, Nipani-591237; Distt. Belagavi, Karnataka State (India).

128 Bagheri, S., Julkapli, N.M., Dabdawb, W.A.Y., and Mansouri, N. (2017). Biodiesel-derived raw glycerol to value-added products: catalytic conversion approach. In: *Handbook of Composites from Renewable Materials, Physico-Chemical and Mechanical Characterization*, 309–366. Wiley.

129 Su, F., Antonietti, M., and Wang, X. (2012). mpg-C 3 N 4 as a solid base catalyst for Knoevenagel condensations and transesterification reactions. *Catalysis Science & Technology* 2 (5): 1005–1009.

130 Bing, W., Zheng, L., He, S. et al. (2018). Insights on active sites of CaAl-hydrotalcite as a high-performance solid base catalyst toward aldol condensation. *ACS Catalysis* 8 (1): 656–664.

131 Boz, N. and Kara, M. (2008). Solid base catalyzed transesterification of canola oil. *Chemical Engineering Communications* 196 (1–2): 80–92.

132 Shanmugam, S., Viswanathan, B., and Varadarajan, T.K. (2004). Esterification by solid acid catalysts—a comparison. *Journal of Molecular Catalysis A: Chemical* 223 (1–2): 143–147.

133 Izumi, Y. (1997). Hydration/hydrolysis by solid acids. *Catalysis Today* 33 (4): 371–409.

134 Sengar, A., Van Santen, R.A., Steur, E. et al. (2018). Deactivation kinetics of solid acid catalyst with laterally interacting protons. *ACS Catalysis* 8 (10): 9016–9033.

135 Iglesia, E., Barton, D.G., Biscardi, J.A. et al. (1997). Bifunctional pathways in catalysis by solid acids and bases. *Catalysis Today* 38 (3): 339–360.

136 Corma, A. (1997). Solid acid catalysts. *Current Opinion in Solid State and Materials Science* 2 (1): 63–75.

137 Mukhtar, A., Saqib, S., Lin, H. et al. (2022). Current status and challenges in the heterogeneous catalysis for biodiesel production. *Renewable and Sustainable Energy Reviews* 157: 112012.

138 Galadima, A., Anderson, J.A., and Wells, R.P.K. (2009). Solid acid catalysts in heterogeneous n-alkanes hydroisomerisation for increasing octane number of gasoline. *Science World Journal* 4 (3): 121.

139 Thomas, J.M. (1992). Solid acid catalysts. *Scientific American* 266 (4): 112–119.

140 Sakthivel, B. and Dhakshinamoorthy, A. (2017). Chitosan as a reusable solid base catalyst for Knoevenagel condensation reaction. *Journal of Colloid and Interface Science* 485: 75–80.

141 Jackson, P.A., Widen, J.C., Harki, D.A., and Brummond, K.M. (2017). Covalent modifiers: a chemical perspective on the reactivity of α, β-unsaturated carbonyls with thiols via hetero-Michael addition reactions. *Journal of Medicinal Chemistry* 60 (3): 839–885.

142 Kabashima, H., Tsuji, H., and Hattori, H. (1997). Michael addition of methyl crotonate over solid base catalysts. *Applied Catalysis A: General* 165 (1–2): 319–325.

143 Baba, T., Kato, A., Yuasa, H. et al. (1998). New Si–C bond forming reactions over solid-base catalysts. *Catalysis Today* 44 (1–4): 271–276.

144 Furgal, J.C. and Lenora, C.U. (2019). Green routes to silicon-based materials and their environmental implications. *Physical Sciences Reviews* 5 (1): 20190024.

145 Weidlich, T. (2021). The influence of copper on halogenation/dehalogenation reactions of aromatic compounds and its role in the destruction of polyhalogenated aromatic contaminants. *Catalysts* 11 (3): 378.

146 Bäckvall, J.E. (ed.) (2011). *Modern Oxidation Methods*. John Wiley & Sons.

147 Clark, J.H. and Rhodes, C.N. (2000). *Clean Synthesis Using Porous Inorganic Solid Catalysts and Supported Reagents*, vol. 4. Royal Society of Chemistry.

148 Montcho Papin, S., Christian, K.T., Pascal, A.D. et al. (2018). Comparative study of transesterification processes for biodiesel production (A review). *Elixir Applied Chemistry* 120: 51235–51242.

149 Faruque, M.O., Razzak, S.A., and Hossain, M.M. (2020). Application of heterogeneous catalysts for biodiesel production from microalgal oil—a review. *Catalysts* 10 (9): 1025.

150 Miceli, M., Frontera, P., Macario, A., and Malara, A. (2021). Recovery/reuse of heterogeneous supported spent catalysts. *Catalysts* 11 (5): 591.

151 Gunatilake, H., Roland-Holst, D., and Sugiyarto, G. (2014). Energy security for India: Biofuels, energy efficiency and food productivity. *Energy Policy* 65: 761–767.

152 Demirbas, A. (2009). Political, economic and environmental impacts of biofuels: a review. *Applied Energy* 86: S108–S117.

153 de Jong, S., Hoefnagels, R., Wetterlund, E. et al. (2017). Cost optimization of biofuel production–The impact of scale, integration, transport and supply chain configurations. *Applied Energy* 195: 1055–1070.

154 Upadhyayula, V.K., Meyer, D.E., Curran, M.A., and Gonzalez, M.A. (2012). Life cycle assessment as a tool to enhance the environmental performance of carbon nanotube products: a review. *Journal of Cleaner Production* 26: 37–47.

155 Howes, R., Skea, J., and Whelan, B. (2013). *Clean and Competitive: Motivating Environmental Performance in Industry*. Routledge.

156 Shi, Q. and Wu, J. (2021). Review on sulfur compounds in petroleum and its products: state-of-the-art and perspectives. *Energy & Fuels* 35 (18): 14445–14461.

157 Mota, G.F., de Sousa, I.G., de Oliveira, A.L.B. et al. (2022). Biodiesel production from microalgae using lipase-based catalysts: current challenges and prospects. *Algal Research* 62: 102616.

158 Sperling, D. and Gordon, D. (2009). *Two Billion Cars: Driving Toward Sustainability*. Oxford University Press.

159 Butterbach-Bahl, K., Baggs, E.M., Dannenmann, M. et al. (2013). Nitrous oxide emissions from soils: how well do we understand the processes and their controls? *Philosophical Transactions of the Royal Society B: Biological Sciences* 368 (1621): 20130122.

160 Gómez-García, M.A., Pitchon, V., and Kiennemann, A. (2005). Pollution by nitrogen oxides: an approach to NOx abatement by using sorbing catalytic materials. *Environment International* 31 (3): 445–467.

161 Han, L., Cai, S., Gao, M. et al. (2019). Selective catalytic reduction of NO_x with NH_3 by using novel catalysts: state of the art and future prospects. *Chemical Reviews* 119 (19): 10916–10976.

162 Board, O.S. and National Academies of Sciences, Engineering, and Medicine (2019). *Environmental Engineering for the 21st Century: Addressing Grand Challenges*. National Academies Press.

163 Clarke, C.J., Tu, W.C., Levers, O. et al. (2018). Green and sustainable solvents in chemical processes. *Chemical Reviews* 118 (2): 747–800.

164 Inchaurrondo, N.S. and Font, J. (2022). Clay, zeolite and oxide minerals: natural catalytic materials for the ozonation of organic pollutants. *Molecules* 27 (7): 2151.

165 Sun, J. and Wang, Y. (2014). Recent advances in catalytic conversion of ethanol to chemicals. *ACS Catalysis* 4 (4): 1078–1090.

166 Kang, M., DeWilde, J.F., and Bhan, A. (2015). Kinetics and mechanism of alcohol dehydration on γ-Al_2O_3: effects of carbon chain length and substitution. *ACS Catalysis* 5 (2): 602–612.

167 Yeung, C.W., Teo, J.Y., Loh, X.J., and Lim, J.Y. (2021). Polyolefins and polystyrene as chemical resources for a sustainable future: challenges, advances, and prospects. *ACS Materials Letters* 3 (12): 1660–1676.

168 Morales, G., Iglesias, J., and Melero, J.A. (2020). Sustainable catalytic conversion of biomass for the production of biofuels and bioproducts. *Catalysts* 10 (5): 581.

169 van Leeuwen, B.N., van der Wulp, A.M., Duijnstee, I. et al. (2012). Fermentative production of isobutene. *Applied Microbiology and Biotechnology* 93: 1377–1387.

170 Guisnet, M. and Pinard, L. (2018). Characterization of acid-base catalysts through model reactions. *Catalysis Reviews* 60 (3): 337–436.

171 Xu, M., Gines, M.J., Hilmen, A.M. et al. (1997). Isobutanol and methanol synthesis on copper catalysts supported on modified magnesium oxide. *Journal of Catalysis* 171 (1): 130–147.

172 Zhang, M. and Yu, Y. (2013). Dehydration of ethanol to ethylene. *Industrial & Engineering Chemistry Research* 52 (28): 9505–9514.

173 Kannan, S. (2006). Catalytic applications of hydrotalcite-like materials and their derived forms. *Catalysis Surveys from Asia* 10: 117–137.

174 Katryniok, B., Paul, S., Bellière-Baca, V. et al. (2010). Glycerol dehydration to acrolein in the context of new uses of glycerol. *Green Chemistry* 12 (12): 2079–2098.

175 Li, C. and Wang, G. (2021). Dehydrogenation of light alkanes to mono-olefins. *Chemical Society Reviews* 50 (7): 4359–4381.

176 Yang, X., Liu, Z., Wei, G. et al. (2022). A critical assessment of the roles of water molecules and solvated ions in acid-base-catalyzed reactions at solid-water interfaces. *Chinese Journal of Catalysis* 43 (8): 1964–1990.

177 Kella, T., Vennathan, A.A., Dutta, S. et al. (2021). Selective dehydration of 1-butanol to butenes over silica supported heteropolyacid catalysts: mechanistic aspect. *Molecular Catalysis* 516: 111975.

178 Gaines, L.L. and Shen, S.Y. (1980). *Energy and Materials Flows in the Production of Olefins and Their Derivatives* (No. ANL/CNSV-9). Argonne, IL, USA: Argonne National Lab.(ANL).

179 Taylor, J.D., Jenni, M.M., and Peters, M.W. (2010). Dehydration of fermented isobutanol for the production of renewable chemicals and fuels. *Topics in Catalysis* 53: 1224–1230.

180 Warkentin, J. and Hine, K.E. (1970). Isobutane from acid-catalyzed dehydration of butanols. *Canadian Journal of Chemistry* 48 (22): 3545–3548.

181 Masoudian, S.K., Sadighi, S., Abbasi, A. et al. (2013). Regeneration of a commercial catalyst for the dehydrogenation of isobutane to isobutene. *Chemical Engineering & Technology* 36 (9): 1593–1598.

182 Akiya, N. and Savage, P.E. (2001). Kinetics and mechanism of cyclohexanol dehydration in high-temperature water. *Industrial & Engineering Chemistry Research* 40 (8): 1822–1831.

183 Sinfelt, J.H. (2002). Role of surface science in catalysis. *Surface Science* 500 (1–3): 923–946.

184 McFarland, B.L., Boron, D.J., Deever, W. et al. (1998). Biocatalytic sulfur removal from fuels: applicability for producing low sulfur gasoline. *Critical Reviews in Microbiology* 24 (2): 99–147.

185 Haruna, A., Merican, Z.M.A., Musa, S.G., and Abubakar, S. (2022). Sulfur Removal Technologies from Fuel Oil for Safe and Sustainable Environment. *Fuel* 329: 125370.

186 Iruretagoyena, D. and Montesano, R. (2018). Selective sulfur removal from liquid fuels using nanostructured adsorbents. *Nanotechnology in Oil and Gas Industries: Principles and Applications* 34: 133–150.

187 Song, C. (2003). An overview of new approaches to deep desulfurization for ultra-clean gasoline, diesel fuel and jet fuel. *Catalysis Today* 86 (1–4): 211–263.

188 Olaremu, A.G., Adedoyin, W.R., Ore, O.T., and Adeola, A.O. (2021). Sustainable development and enhancement of cracking processes using metallic composites. *Applied Petrochemical Research* 11 (1): 1–18.

189 Saleh, T.A. (2022). Global trends in technologies and nanomaterials for removal of sulfur organic compounds: Clean energy and green environment. *Journal of Molecular Liquids* 359: 119340.

190 Aireddy, D.R. and Ding, K. (2022). Heterolytic dissociation of H_2 in heterogeneous catalysis. *ACS Catalysis* 12 (8): 4707–4723.

191 McCue, A.J. and Anderson, J.A. (2014). Sulfur as a catalyst promoter or selectivity modifier in heterogeneous catalysis. *Catalysis Science & Technology* 4 (2): 272–294.

192 Kogan, V.M. and Isaguliants, G.V. (2008). The HDS mechanism: Which "auxiliary" process takes place–sulfur isotopic exchange or replacement–and why is it important to know it? *Catalysis Today* 130 (1): 243–248.

193 Whitehurst, D.D., Isoda, T., and Mochida, I. (1998). Present state of the art and future challenges in the hydrodesulfurization of polyaromatic sulfur compounds. *Advances in Catalysis* 42: 345–471.

194 Cui, T.Y., Rajendran, A., Fan, H.X. et al. (2021). Review on hydrodesulfurization over zeolite-based catalysts. *Industrial & Engineering Chemistry Research* 60 (8): 3295–3323.

195 Yang, J.B., Pan, J.H., Zhu, Y.H. et al. (2022). Two 1D Anderson-type polyoxometalate-based metal–organic complexes as bifunctional heterogeneous catalysts for CO_2 photoreduction and sulfur oxidation. *Inorganic Chemistry* 61 (30): 11775–11786.

196 Li, W., Zhao, S., Yan, Y. et al. (2023). High-efficiency desulfurization adsorbents loaded with uniformly dispersed nano-metal particles prepared from phytoremediation. *Journal of Environmental Chemical Engineering* 11 (5): 110409.

197 Wei, J., Huang, H., Luo, Q. et al. (2022). Synthesis of few layer amorphous 1T/2H MoS_2 by a one-step ethanol/water solvothermal method and its hydrodesulfurization performance. *Catalysis Letters* 152: 263–275.

198 Munnik, P., De Jongh, P.E., and De Jong, K.P. (2015). Recent developments in the synthesis of supported catalysts. *Chemical Reviews* 115 (14): 6687–6718.

199 Van Honschoten, J.W., Brunets, N., and Tas, N.R. (2010). Capillarity at the nanoscale. *Chemical Society Reviews* 39 (3): 1096–1114.

200 Ngamcharussrivichai, C., Totarat, P., and Bunyakiat, K. (2008). Ca and Zn mixed oxide as a heterogeneous base catalyst for transesterification of palm kernel oil. *Applied Catalysis A: General* 341 (1–2): 77–85.

201 Shahinuzzaman, M., Yaakob, Z., and Ahmed, Y. (2017). Non-sulphide zeolite catalyst for bio-jet-fuel conversion. *Renewable and Sustainable Energy Reviews* 77: 1375–1384.

202 León, M., Díaz, E., and Ordóñez, S. (2011). Ethanol catalytic condensation over Mg–Al mixed oxides derived from hydrotalcites. *Catalysis Today* 164 (1): 436–442.

203 Sree, R., Babu, N.S., Prasad, P.S., and Lingaiah, N. (2009). Transesterification of edible and non-edible oils over basic solid Mg/Zr catalysts. *Fuel Processing Technology* 90 (1): 152–157.

204 Tantirungrotechai, J., Thepwatee, S., and Yoosuk, B. (2013). Biodiesel synthesis over Sr/MgO solid base catalyst. *Fuel* 106: 279–284.

205 Wang, S.H., Wang, Y.B., Dai, Y.M., and Jehng, J.M. (2012). Preparation and characterization of hydrotalcite-like compounds containing transition metal as a solid base catalyst for the transesterification. *Applied Catalysis A: General* 439: 135–141.

206 Lee, D.W. and Yoo, B.R. (2014). Advanced metal oxide (supported) catalysts: synthesis and applications. *Journal of Industrial and Engineering Chemistry* 20 (6): 3947–3959.

207 Esposito, S. (2019). "Traditional" sol-gel chemistry as a powerful tool for the preparation of supported metal and metal oxide catalysts. *Materials* 12 (4): 668.

208 Zhan, Z. and Zeng, H.C. (1999). A catalyst-free approach for sol–gel synthesis of highly mixed ZrO_2–SiO_2 oxides. *Journal of Non-Crystalline Solids* 243 (1): 26–38.

209 Wang, J.A., Bokhimi, X., Morales, A. et al. (1999). Aluminum local environment and defects in the crystalline structure of Sol–Gel alumina catalyst. *The Journal of Physical Chemistry B* 103 (2): 299–303.

210 Wang, W., Serp, P., Kalck, P., and Faria, J.L. (2005). Photocatalytic degradation of phenol on MWNT and titania composite catalysts prepared by a modified sol–gel method. *Applied Catalysis B: Environmental* 56 (4): 305–312.

211 Zhang, H., Qi, R., Evans, D.G., and Duan, X. (2004). Synthesis and characterization of a novel nano-scale magnetic solid base catalyst involving a layered double hydroxide supported on a ferrite core. *Journal of Solid State Chemistry* 177 (3): 772–780.

212 Ding, Y., Zhang, G., Wu, H. et al. (2001). Nanoscale magnesium hydroxide and magnesium oxide powders: control over size, shape, and structure via hydrothermal synthesis. *Chemistry of Materials* 13 (2): 435–440.

213 Sutradhar, N., Sinhamahapatra, A., Pahari, S.K. et al. (2011). Controlled synthesis of different morphologies of MgO and their use as solid base catalysts. *The Journal of Physical Chemistry C* 115 (25): 12308–12316.

214 Pedersen, H. and Elliott, S.D. (2014). Studying chemical vapor deposition processes with theoretical chemistry. *Theoretical Chemistry Accounts* 133: 1–10.

215 Koley, G. and Cai, Z. (2006). Growth of gallium nitride nanowires and nanospirals. In: *MRS Online Proceedings Library (OPL)*, 963 (conference conducted by Kansas State University in 2006).

216 Matsuhashi, H. and Kawamura, A. (2012). Preparation of a novel solid base catalyst of CaO covered with SiO_2. *Catalysis Today* 185 (1): 236–240.

217 Seo, M., Kim, S.Y., Kim, Y.D. et al. (2018). Highly stable barium zirconate supported nickel oxide catalyst for dry reforming of methane: From powders toward shaped catalysts. *International Journal of Hydrogen Energy* 43 (24): 11355–11362.

218 Sankar, S., Manivarma, V.K., and Thampi, A. (2017). Performance and emission analysis of a CI engine fuelled with JOME-JOEE-diesel blends. *Applied Mechanics and Materials* 867: 261–271.

219 Bai, P., Etim, U.J., Yan, Z. et al. (2018). Fluid catalytic cracking technology: current status and recent discoveries on catalyst contamination. *Catalysis Reviews* 12: 496.

220 Centi, G. and Perathoner, S. (2005). *Use of Solid Catalysts in Promoting Water Treatment and Remediation Technologies*, 46–71. Cambridge, UK: Royal Society of Chemistry Publishing.

221 Chouhan, A.S. and Sarma, A.K. (2011). Modern heterogeneous catalysts for biodiesel production: a comprehensive review. *Renewable and Sustainable Energy Reviews* 15 (9): 4378–4399.

222 Anku, W.W., Mamo, M.A., and Govender, P.P. (2017). Phenolic compounds in water: sources, reactivity, toxicity and treatment methods. *Phenolic Compounds-Natural Sources, Importance and Applications* 12: 419–443.

223 Knözinger, H. and Kochloefl, K. (2000). Heterogeneous catalysis and solid catalysts. In: *Ullmann's Encyclopedia of Industrial Chemistry* (ed. C. Ley), Wiley Online Library: 1–117.

224 Islam, A., Taufiq-Yap, Y.H., Chan, E.S. et al. (2014). Advances in solid-catalytic and non-catalytic technologies for biodiesel production. *Energy Conversion and Management* 88: 1200–1218.

225 Xu, J.M. and Cheng, J.P. (2016). The advances of Co_3O_4 as gas sensing materials: a review. *Journal of Alloys and Compounds* 686: 753–768.

226 Dey, S. and Mehta, N.S. (2020). Oxidation of carbon monoxide over various nickel oxide catalysts in different conditions: a review. *Chemical Engineering Journal Advances* 1: 100008.

8

Silica-Supported Heterogenous Catalysts: Application in the Synthesis of Tetrazoles

Suman Swami[1], *Neelam Sharma*[2], *and Rahul Shrivastava*[2]

[1] *Department of Chemistry, Chandigarh University, NH-05, Ludhiana-Chandigarh State Highway, Mohali, Punjab, 140413, India*
[2] *Department of Chemistry, Manipal University Jaipur, VPO-Dehmi Kalan, Off Jaipur Ajmer Express Way, Jaipur, Rajasthan, 303007, India*

8.1 Introduction

The chemical industry is striving to pursue greener protocols to produce various products such as pharmaceuticals, food additives, energy materials, fertilizers, and pesticides. This urge is because of environmental issues, legal obligations, sustainability objectives, energy efficiency, resource conservation, innovation potential, risk mitigation, public perception, competitive advantages, and the need to secure long-term economic viability. In the current global economy, adopting greener practices is considered as both an ethical and strategic need. Common synthetic techniques involve harsh reaction conditions such as high temperature, pressure, hazardous chemicals, catalysts, and the production of enormous amounts of chemical waste that are detrimental to the environment. The scientific community is encouraged to develop new tactics that reduce the consumption of auxiliary compounds to achieve purification and recycle the catalyst as environmental rules have been more stringent in recent years. Additionally, there is a strong demand for the design of eco-compatible protocols in line with sustainability due to the health issues brought on by harmful chemicals. The use of heterogenous catalyst-based protocols is imperative in attaining low temperature and pressure, a minimum number of toxic by-products, better efficiency, selectivity, catalyst recyclability, simple product purification, and environmental sustainability in the aspect to achieve sustainable protocols. Due to these advantages, heterogeneous catalysis is a desirable solution for numerous industrial applications [1, 2].

The two main groups of catalysts, **(I)** homogenous catalysts and **(II)** heterogenous catalysts, are majorly used in organic transformation. A homogeneous catalyst is in the same phase as the reactants. It is typically a dissolved substance in the reaction mixture, meaning it is usually in the form of a gas or a liquid. Homogeneous catalysts interact directly with the reactants in the same phase [3, 4]. They can form intermediate complexes with reactants during the reaction. In contrast,

Solid Base Catalysts: Synthesis, Characterization, and Applications, First Edition.
Edited by Ravi Tomar, K.K. Pant, and Ramesh Chandra.

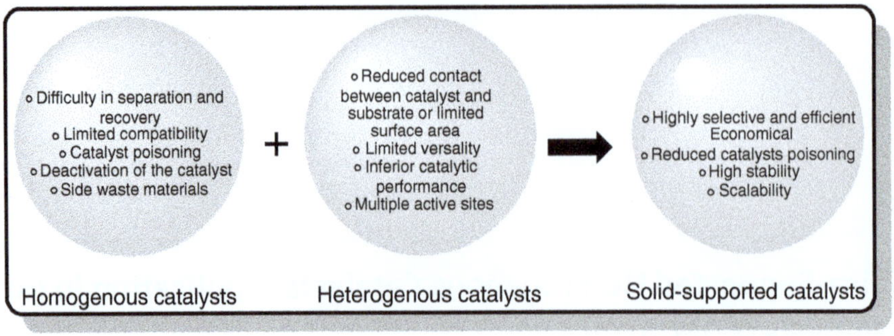

Figure 8.1 Advantages of solid-supported catalyst over homogenous and heterogenous catalysts.

heterogeneous catalysts are in a different phase from the reactants. They are usually solid or immobilized substances, which can be in the form of a powder, a supported metal catalyst on a solid surface, or a structured catalyst such as zeolites. Heterogeneous catalysts do not mix with the reactants at the molecular level. Instead, they interact with the reactants at the catalyst's surface, where the reactants adsorb and react [5, 6]. Although homogeneous catalysts provide better activity, selectivity, ease of spectroscopic monitoring, and regulated and tunable reaction sites, separating the catalyst from the product is a difficult operation that necessitates further technical efforts prerequisite once the reaction is completed [7]. Conversely, heterogenous catalysts have several benefits and adhere to the green chemistry principles that assure environmental sustainability by facilitating easy separation, which frequently reduces operating costs and environmental load [8, 9]. Organic synthesis, which encounters the benefits of both homogeneous and heterogenous catalysts, is relevant in the further incorporation of acid or basic functionality into solid surfaces (Figure 8.1).

8.1.1 General Synthetic Protocol for Tetrazoles

Heterocycles are enormously important in the field of medicinal chemistry, as they are a common structural motif found in many pharmaceuticals and bioactive compounds [10], with more than 75% of drugs available in the market being based on heterocyclic structures. Many heterocyclic compounds have demonstrated their effectiveness as bioactive agents. Among these bioactive compounds, tetrazoles and their derivatives have gained significant attention in the realm of drug design. Tetrazole (CH_2N_4) exhibits electron-rich, flat structural characteristics owing to its five-membered ring, which comprises four nitrogen atoms, one carbon atom, and two hydrogen atoms. It should be noted that the characteristics of tetrazoles can vary significantly depending on the number, kind, and position of substituents inside the heterocycle. 1*H*-tetrazoles, including 1- and 1,5-disubstituted tetrazoles, are typically more polar and thermally stable compared to 2*H*-tetrazoles, which include 2- and 2,5-disubstituted tetrazoles. Tetrazole serves as a bio-isostere for the carboxylic acid group, making it an appropriate replacement in therapeutic compounds to improve bioavailability, lipophilicity, and minimize side effects [11, 12]. Tetrazoles can interact via non-covalent interactions with a range of enzymes and receptors in living organisms [13], exhibiting a broad spectrum of biological effects, including antibacterial [14], anticancer [15], fungicidal [16], anti-inflammatory [17], and antiviral activities [18]. Furthermore, the tetrazole scaffold can serve as an appealing linker for combining or stabilizing different pharmacophore fragments, enabling the creation of specialized functionalized molecules. In addition to medicinal utility, some tetrazole derivatives are used in the development of energetic materials, including explosives and propellants. The nitrogen-rich nature of tetrazoles contributes to their high energy content, making them useful in pyrotechnics and the defense industry. Hence, owing to wide range of applications, versatility, reactivity, and unique properties, in the past years, numerous tetrazole-based derivatives have been successfully progressed and widely used as medicinal drugs for treating various diseases, energetic materials, corrosion inhibitors, and in ligand design. Examples of such drugs include antihypertensives such as Losartan and Valsartan [19], antibiotics such as Flomoxef and Cefonicid, and the anti-nociceptive drug Alfentanil [20].

Generally, 1-substituted 1*H*-tetrazoles are produced through several methods. A widely used method entails an acid-catalyzed cycloaddition reaction among isocyanides and trimethylsilyl azide, or a cycloaddition concerning hydrazoic acid and isocyanides. Another option for their synthesis involves a cyclization reaction between primary amines or their salts, ortho-carboxylic acid esters, and sodium azide in the presence of an acidic catalyst [12]. Nevertheless, an alternative and more feasible approach relies on the interaction of amines, triethyl orthoformate, and sodium azide with the application of various catalysts, as illustrated in Scheme 8.1. However, the simplest way to synthesize 5-substituted 1*H*-tetrazoles involves a [3 + 2] cycloaddition, where an azide ion reacts with the analogous organic nitriles, as illustrated in Scheme 8.2. Traditional methods have some drawbacks, including the need for expensive and toxic metal-organic azides as a source of azide in the reaction and the risk of forming explosive hydrazoic acid during the reaction process. Given the significance of the tetrazole scaffold, classical methods have been adapted to incorporate more sustainable multicomponent reactions.

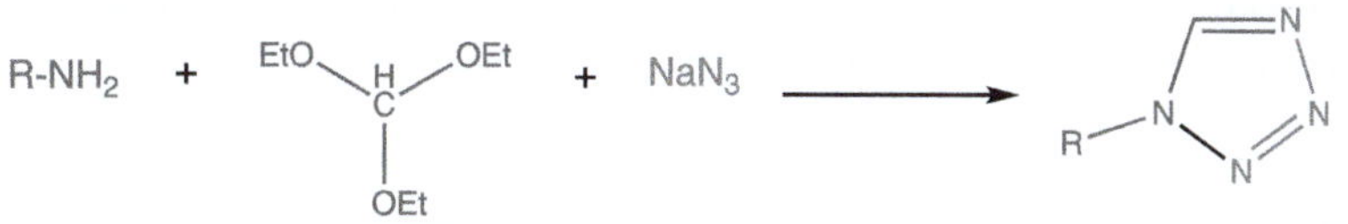

Scheme 8.1 Synthetic protocol of tetrazole synthesis via means of a one-pot multi-component reaction involving distinct amines, sodium azide, and triethyl orthoformate.

Furthermore, owing to limitations associated with homogenous catalysts, homogenous catalysts were replaced or modified with heterogenous catalysts to encounter the benefits of both homogeneous and heterogenous catalysts, via further incorporation of acid or basic functionality into solid surfaces. Materials that are incorporated or functionalized with acid or base functionalities exhibit remarkable

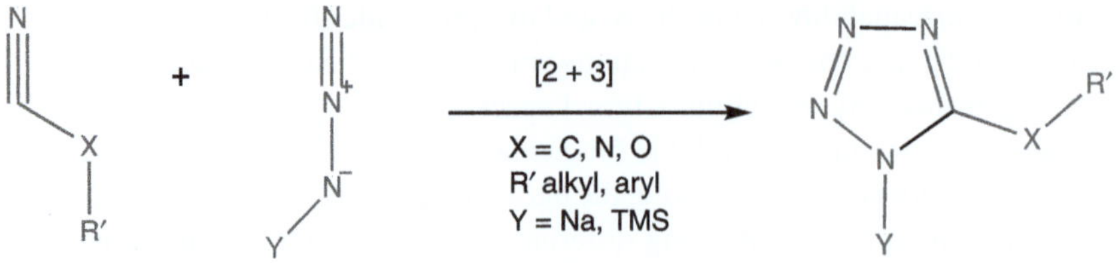

Scheme 8.2 Synthetic protocol of tetrazole synthesis via the conventional [3 + 2] cycloaddition reaction, which involves the combination of isocyanides with either trimethyl azide or hydrazoic acid.

potential in replacing liquid catalysts, offering an eco-friendly heterogeneous alternative. The primary consideration when it comes to solid-supported materials is their ability to maintain stability during high temperature reactions. Owing to their impressive stability at elevated temperatures, a broad array of catalysts supported on solid surfaces has been documented in the literature regarding the production of tetrazoles. Among the titled solid-supported catalysts, silica-supported heterogenous catalysts were effectively explored by researchers as promising catalysts for tetrazole synthesis. Numerous investigations have demonstrated that the properties of supporting materials, including their acid/base characteristics, morphological features, and pore structures, exert a substantial influence on catalytic performance. Silica is frequently considered one of the most straightforward supporting materials and is widely utilized in a variety of applications. Conventional silica supports, such as silica gel, colloidal silica, and ordered mesoporous silica, are typically non-crystalline and exhibit a variety of silanol groups and siloxane bridges on their surfaces. In contrast to numerous other supporting materials, silica is commonly seen as chemically inert and offers the flexibility for additional modifications to introduce various functionalities, including acid and base sites, which demonstrate remarkable characteristics. These include a large specific surface area, adjustable acidity and basicity, stability, a high specific pore volume, ease of preparation, environmental friendliness, and wide-ranging. This chapter provides an overview of documented catalysts based on silica or supported by silica, serving as an effective and eco-friendly catalytic system for tetrazole synthesis. This chapter briefly covers the preparation of silica-supported catalysts, their characterization, and outlines their comparative advantages in relation to other catalysts.

8.2 Silica-Supported Heterogenous Catalysts for Tetrazole Synthesis

Silica-supported catalyst, also known as a silica-supported heterogeneous catalyst, is a type of catalyst in which the catalytic active species is anchored or immobilized on a silica or silica-supported material [21]. Silica, also known as silicon dioxide (SiO_2), is a common supporting material for heterogeneous catalysts due to its high surface area, thermal stability, and chemical inertness [22, 23]. Silica-supported catalysts are typically employed to facilitate the formation of tetrazoles through a variety of reactions, with one of the common routes being the reaction of nitriles with azides. The catalyst, often composed of a metal or metal compound, helps activate the reactants and promote the formation of tetrazoles. In silica-supported catalysts, silica as supporting material could be utilized in the form of silica gel, silica nanoparticles, or other silica-based materials. The selection of the supporting material can influence the performance of the catalyst. The use of silica support provides a high surface area, which increases the active sites available for the reaction [24]. It also enhances the stability and reusability of the catalyst, reducing the need for frequent replacement. Owing to the effectiveness of silica-supported catalysts, widely utilized such catalysts in tetrazole synthesis. For instance, Sajadi established a silica-supported heterogeneous catalyst, MoO_3-SiO_2, for catalytic application in 5-substituted 1*H*-tetrazole synthesis. In typical experimental procedure, reaction initiated via combining nitriles with sodium azide in the presence of MoO_3-SiO_2 as an efficient and reusable heterogeneous catalyst. This approach offers several notable benefits, including high yield, elimination of hazardous hydrazoic acid, simple work-up, industrial scalability, and sustainability [25].

A silica coating on magnetic nanoparticles could enhance further catalytic activity with easiness of catalyst separation from reaction mixture via an external magnet. Since Fe_3O_4 surfaces are easily alterable, it allows the attachment of various desirable functional groups. For example, a novel silica-supported heterogeneous catalyst for the production of 5-substituted-1*H*-tetrazoles in PEG by immobilizing Cu(II) onto Fe_3O_4@SiO_2@L-arginine have developed as illustrated in Scheme 8.3. The catalyst was prepared by first coating silica-coated iron oxide nanoparticles with L-arginine and then treating the Fe_3O_4@SiO_2@L-arginine with $Cu(NO_3)_2{\cdot}3H_2O$. The titled step resulted in the immobilization of Cu(II) on Fe_3O_4@SiO_2@L-arginine. The Cu(II) anchored to Fe_3O_4@SiO_2@L-arginine, which was prepared, has shown exceptional efficiency and reusability as a nanocatalyst

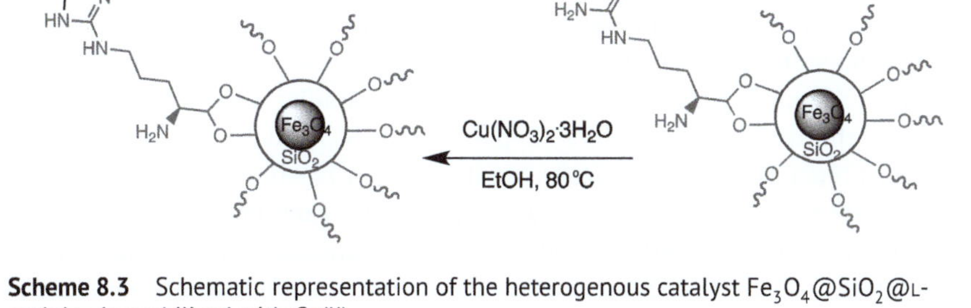

Scheme 8.3 Schematic representation of the heterogenous catalyst Fe_3O_4@SiO_2@L-arginine immobilized with Cu(II).

in the production of 5-substituted-1*H*-tetrazoles. Another noteworthy aspect of the approach was its straightforward operation and providing diverse heterocyclic compounds. Moreover, the nanoparticle-based catalyst can be readily retrieved using an external magnetic field and can be employed for a minimum of four cycles without a substantial decrease in its catalytic performance [26].

Furthermore, Habibi and Nasrollahzadeh introduced a silica-supported heterogeneous catalyst, $ZnCl_2/AlCl_3$/silica (abbreviated as ZAS) for the cyclization reaction of different cyanamides with sodium azide to create arylaminotetrazoles under thermal conditions. The typical process for catalyst preparation involves combining anhydrous $AlCl_3$ and $ZnCl_2$ with silica gel in a CCl_4 solvent under a nitrogen atmosphere and in the absence of light. The resultant mixture is then filtered and dried under vacuum. The method offers several notable advantages, including high yields, short reaction durations, synthesis of regiospecific compounds, removal of hazardous hydrazoic acid, a straightforward and uncomplicated work-up route, cost-effectiveness, and ease of catalyst preparation and application [27]. In a similar approach, Habibi et al. investigated the use of a silica-based heterogeneous catalyst, known as silica sulfuric acid (SSA), to synthesize 1*H*-1,2,3,4-tetrazoles with a 1-substitute. The synthetic protocol includes reaction of primary amines with triethyl orthoformate and sodium azide underneath both thermal and solvent-free conditions. SSA possesses additional benefits, including its affordability, ease of preparation, recyclability, and insolubility in most types of organic solvents [28].

Dehghani et al. investigated the utilization of functionalized magnetic–silica nanoparticles containing a Cu(II) salen complex ($Fe_3O_4@SiO_2$/salen) to produce 1- and 5-substituted 1*H*-tetrazoles from amines and nitriles [29]. The $Fe_3O_4@SiO_2$/salen complex of Cu(II) preparation involved a sequential process. The structural characteristics of prepared magnetic silica particles were determined using scanning electron microscopy (SEM), transmission electron microscopy (TEM), and magnetic properties via VSM instruments. To optimize the synthesis of tetrazoles, under the catalytic activity of Cu(II) $Fe_3O_4@SiO_2$/salen complex, researchers used *p*-methoxy aniline, triethyl orthoformate, and sodium azide as model reactants. The recyclability of the catalyst was also assessed by measuring the amount of copper leaching into the solution during the reaction. Using inductively coupled plasma analysis, the copper loading was measured before and after each cycle of the process. The results showed that the amount of Cu leaching after the first cycle was only 0.2%, but it grew to 5.4% after seven repeated recycling cycles. Therefore, the $Fe_3O_4@SiO_2$/salen complex of Cu(II) demonstrated its effectiveness as a heterogeneous catalyst in this protocol, as it could be easily retrieved using an external magnet and reprocessed for at least seven consecutive reactions with minimal deterioration in catalytic activity [29]. Furthermore, in a similar approach, Habibi et al. examined the application of a silica-supported P_2O_5-SiO_2 catalyst in the production of 1-substituted 1*H*-1,2,3,4-tetrazoles via the one-pot reaction of primary amines, triethyl orthoformate, and sodium azide [30]. This experiment was conducted using both traditional methods and ultrasound irradiation, all in a solvent-free environment. The catalyst's preparation involved a straightforward procedure in which phosphorous pentoxide and silica gel (230–400 mesh) were combined in a container and stirred for four hours, resulting in the formation of a white powder. This powder was stored in a desiccator due to its sensitivity to moisture and used as needed. This catalyst offered several benefits, including yielding relatively high product yields, requiring short-reaction times, facilitating the isolation of products, being cost-effective, and allowing for the recovery and reuse of the catalyst. The adoption of sustained catalysts and solvent-free conditions was a strategic choice since these reagents not only simplified the purification procedure but also aided in preventing the release of toxic by-products into the environment. Additionally, the research team led by Habibi et al. explored a similar silica-supported catalyst "$FeCl_3$-SiO_2" to produce 1-aryl-1*H*-1,2,3,4-tetrazoles over solvent-free processes that entail aryl amines, sodium azide, and triethyl orthoformate [31]. The reaction was conducted under both conventional heating and ultrasound irradiation conditions. The typical procedure for preparing the catalyst involved introducing silica gel into a solution of ferric chloride hexahydrate at room temperature under reduced pressure. This resulted in the formation of a yellow catalyst powder, which was stored under nitrogen and utilized as required in subsequent reactions. The approach offered several notable advantages, including achieving high product yields, removal of hazardous and noxious hydrazoic acid, a simple post-reaction procedure, and the ease of catalyst preparation and management. The catalyst can be retrieved by filtration and employed again for subsequent reactions [31].

Esmaeilpour et al. conducted a research on a silica-based heterogeneous catalyst with superparamagnetic properties, denoted as (Fe_3O_4@SiO_2/ligand/Cu(II)), and utilized for the production of 1- and 5-substituted 1*H*-tetrazoles from nitriles and amines [32]. They initiated this research by synthesizing magnetite particles, as illustrated in Scheme 8.4, via a straightforward co-precipitation method in an aqueous

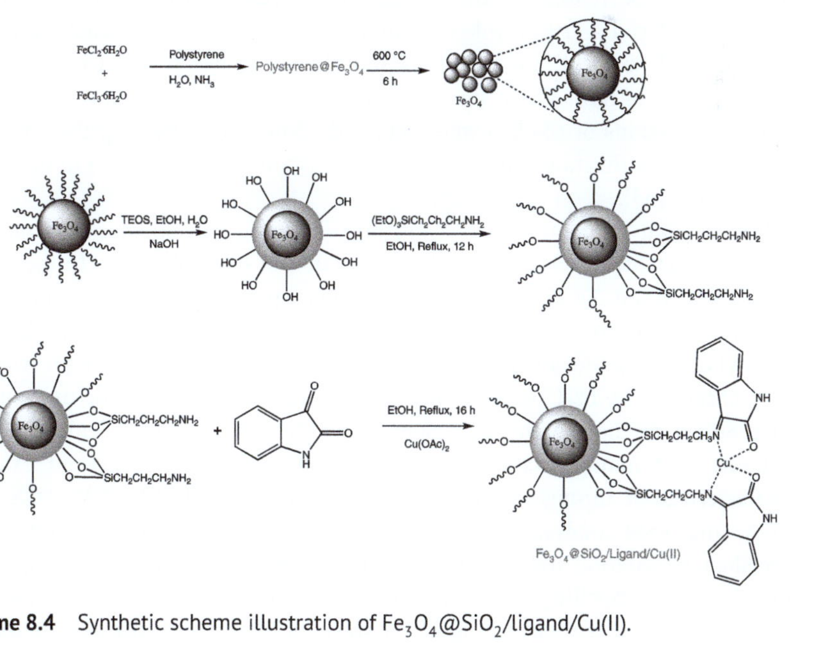

Scheme 8.4 Synthetic scheme illustration of Fe_3O_4@SiO_2/ligand/Cu(II).

solution. Following this, Fe_3O_4@SiO_2 nanospheres were crafted, using nano-Fe_3O_4 as the central core, TEOS used as a silica source, and a surfactant (polyethylene glycol). These silica-coated Fe_3O_4 nanoparticles were subsequently enveloped with a complex ligand containing Cu(II). The prepared magnetic core–shell nanoparticles (MNPs) produced in this manner were subject to a range of characterization techniques, including Fourier transform infrared (FTIR), scanning electron microscope (SEM), transmission electron microscopy (TEM), X-ray diffraction (XRD), thermogravimetric analysis (TGA), dynamic light scattering (DLS), and N_2 adsorption/desorption. The copper concentration was determined using an inductively coupled plasma analyzer. After characterization, the immobilized copper complex was utilized as an efficient catalyst for production of 1- and 5-substituted 1*H*-tetrazoles from nitriles and amines. The outcomes of the study demonstrated that this catalyst, supported on a solid matrix, could be conveniently retrieved using an external magnetic field and reused for subsequent reactions. Remarkably, it maintained its catalytic activity for at least six cycles with minimal degradation.

Similarly, a silica-supported lanthanum triflate catalyst ($Ln(OTf)_3$-SiO_2) for the production of 5-substituted 1*H*-tetrazoles through a [3 + 2] cycloaddition reaction between aromatic/heteroaromatic nitriles and sodium azide were investigated by Meshram et al. Various silica-supported Lewis acids, including $Ln(OTf)_3$-SiO_2, $BiCl_3$-SiO_2, and $SbCl_3$-SiO_2, also explored to compare their efficiency with $Ln(OTf)_3$-SiO_2. Research outcomes revealed that when $Ln(OTf)_3$-SiO_2 was used in a solvent mixture of dimethylformamide (DMF) and methanol (MeOH) in a 4 : 1 ratio at a reaction temperature of 100 °C, it was found the most sustainable and efficient catalytic system, providing the anticipated product in significantly higher yields. Additionally, the catalytic system demonstrated nontoxicity, recoverability, and the ability to be reused for three consecutive cycles. Furthermore, the amount of catalyst played a crucial role in determining the yield of the reaction. In their experimental protocol, the use of 200 mg of the catalyst resulted in the highest product yield of 88%. When the amount of catalyst was reduced to 100–150 mg, lower conversions of 65–72% were obtained. However, increasing the amount of catalyst beyond this range did not lead to a substantial improvement in the yield of the reaction [33].

Naeimi and Mohamadabadi contributed to the development of magnetic nanoparticles based on silica for the production of tetrazoles [34]. In their study, they focused on Fe_3O_4@SiO_2 composite nanoparticles for the production of 1-substituted 1*H*-tetrazoles through a 3-C-multicomponent reaction, involving triethyl orthoformate, an amine, and sodium azide, all conducted in a solvent-free environment. The Fe_3O_4@SiO_2 composite nanoparticles were made using a modified Stöber method, with Fe_3O_4 spheres as the core and a shell made of silica sulfonic acid nanoparticles. These Fe_3O_4@SiO_2 catalysts underwent extensive characterization through various techniques, including FTIR, SEM, DLS, XRD, TGA, and VSM analysis. The catalysts displayed remarkable catalytic activity, excellent reusability, and a straightforward magnetic separation process, rendering them a compelling choice for the cost-effective synthesis of 1-substituted 1*H*-tetrazoles, all without the need for solvents. It is important to note that the

(a) (b)

Figure 8.2 (a) Scanning electron microscopy (SEM) image of silica. (b) Scanning electron microscopy (SEM) image showing silica-incorporated molybdic acid. Source: Reproduced from Ahmed and Siddiqui [35], with permission of Royal Society of Chemistry.

Fe_3O_4@SiO_2 catalyst could be reclaimed for up to six successive reactions with no substantial deterioration in its catalytic performance. Ahmed and Siddiqui reported a silica-supported catalyst, known as silica molybdic acid (SMA), which was designed for synthesizing tetrazoles through a three-component reaction involving aldehydes, malononitrile, and sodium azide in a water medium facilitated by microwaves [35]. The preparatory protocol for the catalyst includes incorporation of molybdic acid on silica. Effectiveness of incorporating molybdate moieties onto the silica surface was assessed by examining the spectra of sodium molybdate and SMA. In the FTIR spectra of SMA (sulfuric acid-modified aluminosilicate), distinct peaks were observed that originate from both the silica and molybdate components. Specifically, the absorption peaks at 873 and 975 cm^{-1} corresponded to the stretching frequencies of the Mo—O bonds in $[MoO_4]^{2-}$ species. Additionally, there was a broad peak spanning from 1000 to 1200 cm^{-1}, which could be attributed to the antisymmetric stretching of Si—O—Si bonds. The symmetric stretching vibration of Si—O— Si bonds appeared at 803 cm [1]. To examine the surface morphology of the catalyst, SEM imaging was analyzed, as illustrated in Figure 8.2. These images clearly revealed the alterations in the surface structure of silica resulting from the introduction of molybdate groups. Noteworthy, characteristics of this catalyst include its straightforward synthesis and the potential for reusing it in up to seven cycles with minimal deprivation in catalytic performance.

A silica-supported heterogeneous catalyst called perlite was developed by Jahanshahi and Akhlaghinia for synthesizing 5-substituted 1*H*-tetrazoles from nitriles and [bmim]N_3 under solvent-free conditions [36]. "Perlite" is a naturally occurring glassy volcanic rock with a typical content of silica and 2–6% water. This approach offers numerous benefits, including excellent yields, quick reaction times, straightforward procedures, and convenient product recovery. The catalyst could be easily retrieved through filtration and reused effectively, resulting in consistently good yields. In a similar manner, as depicted in Scheme 8.5, Cu(II) was immobilized onto aminated epichlorohydrin-activated silica, referred to as CAES, to create an effective heterogeneous nanocatalyst for the production of 5-substituted-1*H*-tetrazoles [37].

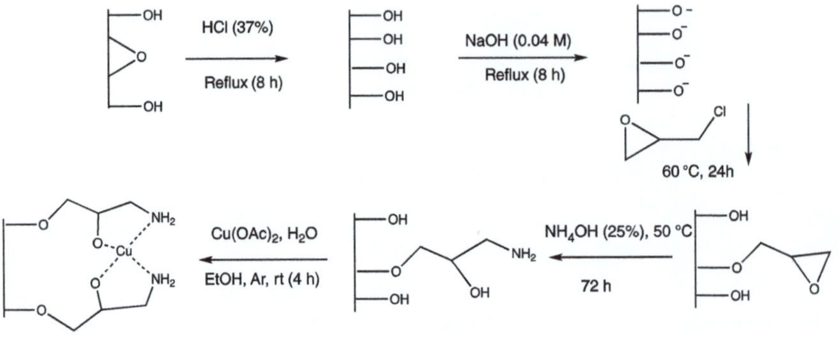

Scheme 8.5 Synthesis of a Cu(II) immobilized catalyst on silica activated by aminated epichlorohydrin (referred to as CAES) involves a specific preparation process.

This involves a five-step synthetic method as depicted in Scheme 8.5. The prepared catalyst undertook characterization through various methods, including SEM analysis as shown in Figure 8.3a,b. The SEM images revealed that the nanocatalyst exhibited uniform particles with a spherical morphology. TEM images further indicated that a significant portion of the nanoparticles prepared were spherical in shape and had sizes ranging from 10 to 20 nm, findings that align well with the results obtained from SEM. When used at a concentration of 0.1 mol%, the CAES catalyst yielded satisfactory tetrazole products via the [3 + 2] cycloaddition reaction between different organic nitriles and sodium azide (Figure 8.4a,b).

In another study, researchers reported a nanocatalyst, Fe_3O_4@SiO_2-DAQ-Cu(II), through the immobilization of 1,4-dihydroxyanthraquinone-copper(II) onto superparamagnetic Fe_3O_4@SiO_2 nanoparticles. This particular catalyst was utilized for the production of 1- and 5-substituted 1*H*-tetrazoles, as depicted in Scheme 8.6 [38]. The research demonstrated that a catalyst loading of 0.9 mol% was adequate for the production of 5-substituted 1*H*-tetrazoles, while 0.8 mol% was suitable for 1-substituted 1*H*-tetrazoles. The study revealed that an inclusive range of aromatic nitriles, including those with electron-donating and electron-withdrawing groups,

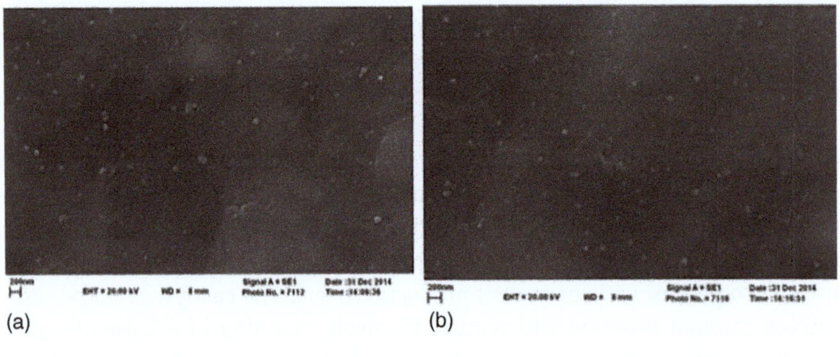

Figure 8.3 (a,b) SEM images depicting "Cu(II) immobilization on aminated epichlorohydrin-activated silica." Source: Reproduced from Razavi and Akhlaghinia [37], with permission of Royal Society of Chemistry.

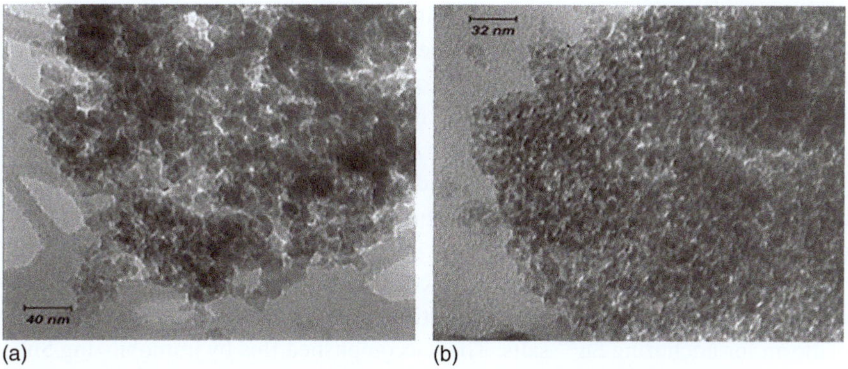

Figure 8.4 (a,b) TEM images illustrating "Cu(II) immobilization on aminated epichlorohydrin-activated silica." Source: Reproduced from Razavi and Akhlaghinia [37], with permission of Royal Society of Chemistry.

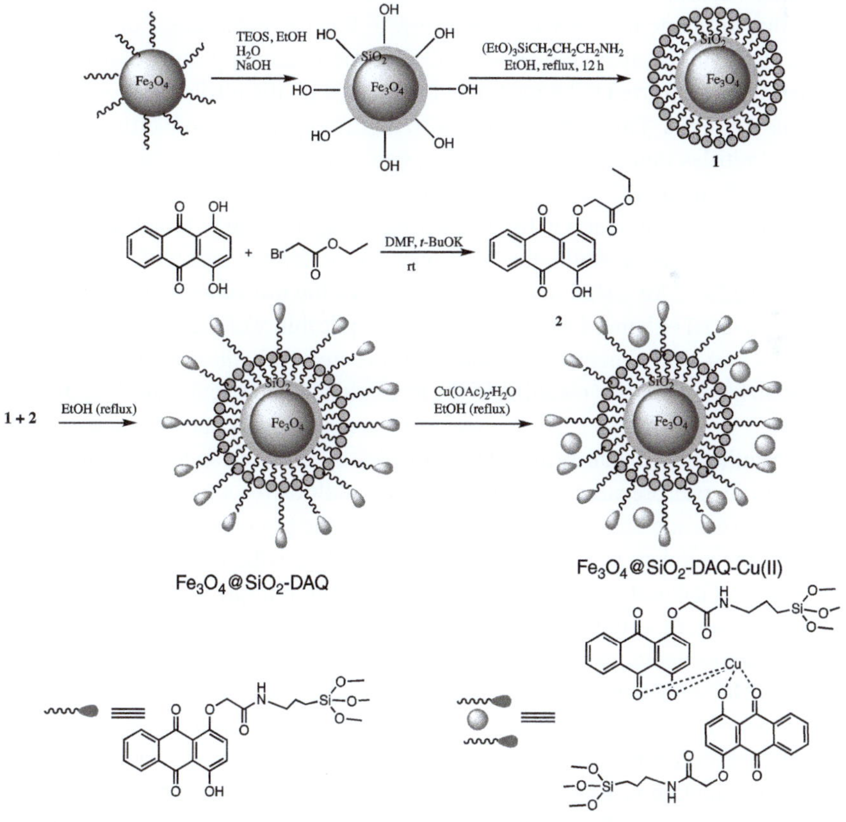

Scheme 8.6 Synthetic protocol illustration of $Fe_3O_4@SiO_2$-DAQ-Cu(II).

were effectively employed in the tetrazole synthesis process, leading to outstanding yields of the final products. The nature of the substituents on benzonitriles significantly influenced the reaction time. As an example, reactions that included electron-deficient aromatic and heteroaromatic nitriles were accomplished within a short time frame, whereas those with electron-rich substituents necessitated an extended reaction duration. The Fe_3O_4@SiO_2-DAQ-Cu(II)-catalyzed synthesis protocol offers several advantages, including its operational simplicity, the ability to be reused for up to six runs with slight copper leaching, and the production of products in excellent yields.

Furthermore, owing to the high surface area, substantial pore volume, and the ability to be recycled, Kumar et al. investigated the use of silica nanoparticles as a platform for anchoring Sn^{2+} salts. They accomplished this by immobilizing $SnCl_2$ through a reaction with hydroxyl groups present on the silica surface. As a result, the initially homogeneous $SnCl_2$ catalyst was transformed into a heterogeneous catalyst with enhanced properties for synthesizing 5-substituted 1*H*-tetrazoles [39]. The silica-supported heterogeneous catalyst, referred to as $SnCl_2$-nano-SiO_2, found extensive utility in the formation of 5-substituted 1*H*-tetrazoles from aliphatic, aromatic, heteroaromatic, and sterically hindered nitriles. Additionally, it was noted that aromatic nitriles with electron-withdrawing substituents demonstrated higher reactivity compared to those with electron-donating groups. Furthermore, aromatic nitriles proved to be more reactive than aliphatic nitriles in the context of tetrazole synthesis. This approach was also highly effective in converting sterically hindered nitriles into their corresponding tetrazoles. Thus, the optimal conditions for the $SnCl_2$-nano-SiO_2 catalyst were found to be a 5 mmol% catalyst loading at 100 °C for a four-hour reaction time, resulting in high efficiency, broad applicability, and recyclability. Meanwhile, the $SnCl_2$ catalyst performed best with a 10 mmol % loading at 120 °C for a six-hour reaction time. So, the utilization of silica as a supporting material enhanced the efficiency and applicability of the catalyst compared to $SnCl_2$. In a study, magnetic nanoparticles coated with silica were modified by immobilizing L-histidine, resulting in the formation of Fe_3O_4@SiO_2@L-histidine (as illustrated in Scheme 8.7). This novel heterogeneous nanocatalyst was employed in the synthesis of 5-substituted 1*H*-tetrazoles. A diverse range of 5-substituted 1*H*-tetrazoles was prepared through [3+2] cycloaddition reactions involving various organic nitriles and sodium azide, with polyethylene glycol serving as the solvent. Notably, the catalyst exhibited exceptional stability, with no evidence of metal leaching, and it could be easily recovered and reused multiple times [40].

Furthermore, in a study conducted by Esmaeilpour et al., they developed a composite heterogeneous catalyst by immobilizing superparamagnetic Fe_3O_4@SiO_2 nanoparticles with dendrimer-encapsulated Cu(II) nanoparticles [41]. This catalyst was designed for the synthesis of 5-substituted 1*H*-tetrazoles. Comprehensive characterizations of the composite catalyst were carried out using various analytical techniques, including FTIR, XRD, DLS, UV-vis, EDX, TGA, and VSM. Detailed examinations through field emission scanning electron microscopy (FE-SEM) and TEM revealed that the particles exhibited predominantly spherical shapes with an average size of 50 nm (as shown in Figure 8.5). Moreover, the copper content in

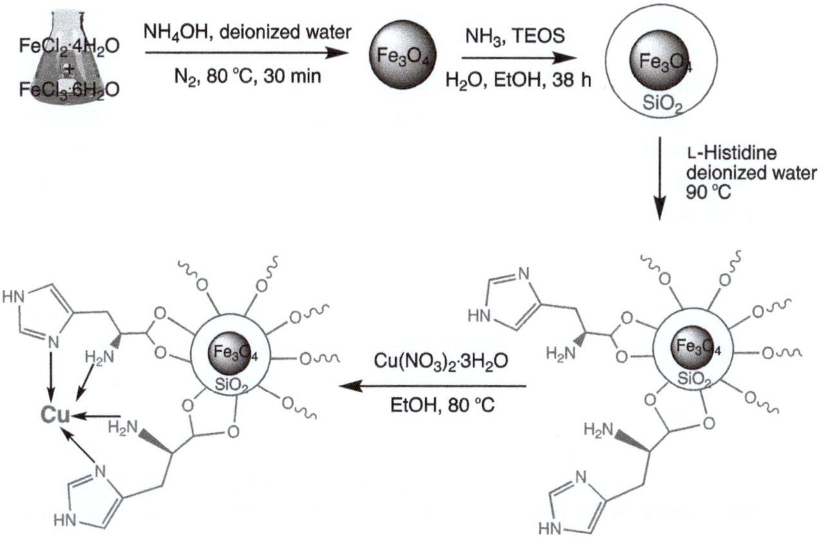

Scheme 8.7 Synthetic protocol illustration of Fe_3O_4@SiO_2@L-histidine.

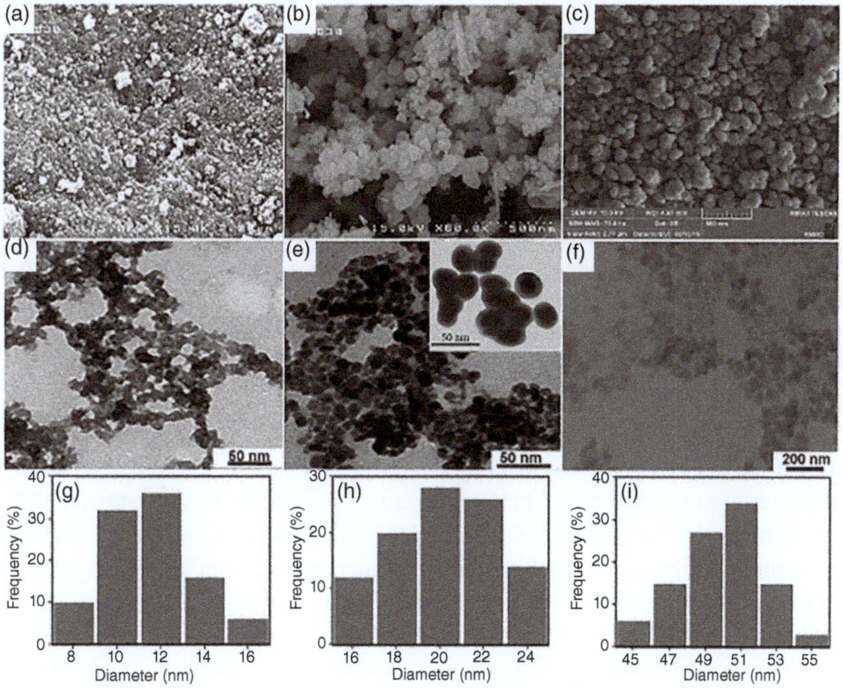

Figure 8.5 (a) Scanning electron microscope (SEM) image of Fe_3O_4, (b) SEM image of Fe_3O_4@SiO_2, (c) SEM image of Fe_3O_4@SiO_2 enveloped by dendrimer-encapsulated Cu(II), (d) transmission electron microscope (TEM) image of Fe_3O_4, (e) TEM image of Fe_3O_4@SiO_2, (f) TEM representation of Fe_3O_4@SiO_2 enclosed with dendrimer-encapsulated Cu(II), and (g) size distributions of Fe_3O_4, (h) Fe_3O_4@SiO_2, and (i) Fe_3O_4@SiO_2 coated with dendrimer-encapsulated Cu(II). Source: Reproduced from Esmaeilpour et al. [41], with permission of John Wiley & Sons.

the catalyst was quantified at 0.51 mmol g^{-1} through inductively coupled plasma analysis. This catalyst was utilized in N-arylation reactions, combining nitrogen heterocycles with aryl halides (iodine and bromine) as well as aryl boronic acids. Furthermore, it played a key role in a one-step, three-component process that facilitated the eco-friendly and efficient production of 5-substituted 1*H*-tetrazoles. This reaction involved diverse aldehydes, hydroxylamine hydrochloride, and sodium azide in a water-based solution. Notably, the catalyst retained its outstanding performance over seven cycles without a substantial decrease in its effectiveness.

A core–shell nanosphere, comprising Fe_3O_4 at its core, SiO_2-SO_3H in the middle, and MS-$NHCOCH_3$ on the outermost layer, was developed to serve as a hybrid composite catalyst with both acidic and basic properties [42]. The catalyst was employed in a three-component cyclization reaction involving aromatic aldehyde, nitroalkane, and sodium azide to produce NH-1,2,3-triazoles. The bifunctional solid catalyst offers several advantages, including high reactivity, excellent selectivity without the release of toxic HN_3, and improved product yield due to the silica shell's ability to restrict by-product formation. Additionally, the incorporation of the silica-encapsulated acidic component enhances safety during operation. Furthermore, its magnetic properties and structural stability ensure that the catalytic activity remains intact even after multiple uses. The catalyst also demonstrates a broad range of applicability, successfully converting various aldehydes into 5-aryl-NH-1,2,3-triazoles with yields as high as 98% (Figure 8.6).

In a research report by Mahajan et al., heterogeneous catalyst called silica-supported melamine trisulfonic acid (SMTSA) was reported for synthesizing 5-substituted 1*H*-tetrazole through a [2 + 3] cycloaddition reaction involving organic nitrile and sodium azide. The use of melamine trisulfonic acid as a catalyst resulted in lower yields of 5-substituted 1*H*-tetrazole. However, when they employed SMTSA, it achieved an 85% yield within a six-hour time frame. After completing the reaction, the catalyst could be retrieved via filtration and utilized again with only a minor decrease in its activity under identical reaction conditions. This method successfully converted various aliphatic and aromatic nitriles into corresponding 5-substituted 1*H*-tetrazoles with excellent yields and minimal contamination [43]. Samanta and his team synthesized a well-structured mesoporous silica of the MCM-41 type, known for its large pore structure, referred to as l-MSN. They then impregnated this l-MSN with samarium nitrate, creating Sm@l-MSN. The synthesis of Sm@l-MSN was achieved using a wet impregnation method, and it involved the use of pore-expanded mesoporous silica of the MCM-41 type, l-MSN. To assess the properties of the synthesized materials, researchers employed various techniques, including N2-sorption, SEM, XRD, EDS, and TEM. The catalytic efficiency of Sm@l-MSN was comprehensively assessed in the production of 5-substituted 1*H*-tetrazoles, a process that proved to be efficient, atom-economical, and eco-friendly. It notably occurred under aerobic conditions, eliminating the need for additional additives. The catalyst exhibited versatility with various nitriles and resulted in relatively short reaction times, producing yields ranging from very good to excellent. The catalyst's ability to be reused was thoroughly tested, and it exhibited consistent performance throughout five consecutive cycles,

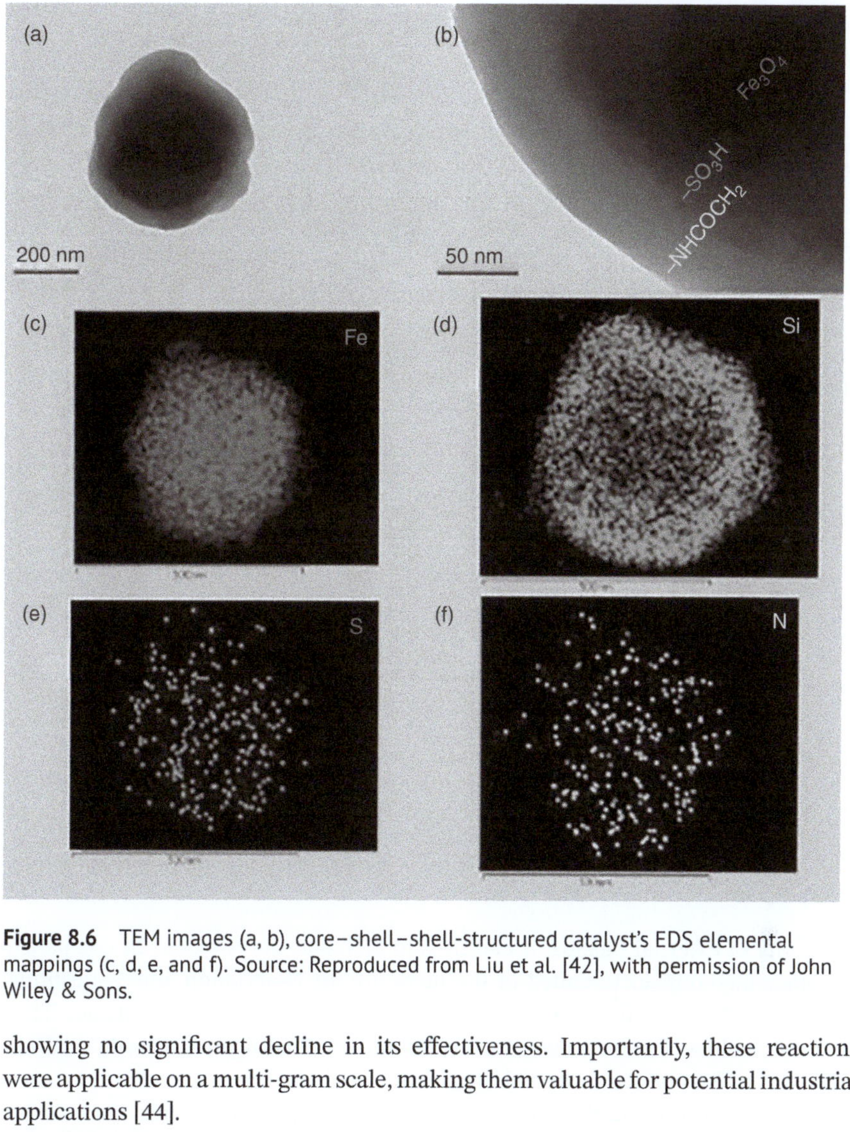

Figure 8.6 TEM images (a, b), core–shell–shell-structured catalyst's EDS elemental mappings (c, d, e, and f). Source: Reproduced from Liu et al. [42], with permission of John Wiley & Sons.

showing no significant decline in its effectiveness. Importantly, these reactions were applicable on a multi-gram scale, making them valuable for potential industrial applications [44].

Sardarian et al. have introduced a novel heterogeneous catalyst based on silica for the eco-friendly one-step production of 5-substituted 1*H*-tetrazoles. This process involves a one-pot reaction of aldehydes, hydroxylamine hydrochloride, and sodium azide in a water-based environment. The preparation of this catalyst involves a sequential process, which includes the surface modification of Fe_3O_4@SiO_2 with polyvinyl alcohol, followed by the immobilization of a copper(II) complex. The catalyst has undergone thorough characterization using an array of techniques, encompassing FTIR, UV-Vis, FE-SEM, XRD, TEM, DLS, EDX, X-ray photoelectron spectroscopy (XPS), TGA, VSM, NMR, ICP, and elemental analysis (as shown in Figure 8.7). This catalyst was efficiently utilized in the N-arylation of nitrogen-containing heterocycles and amines using aryl halides, all without the

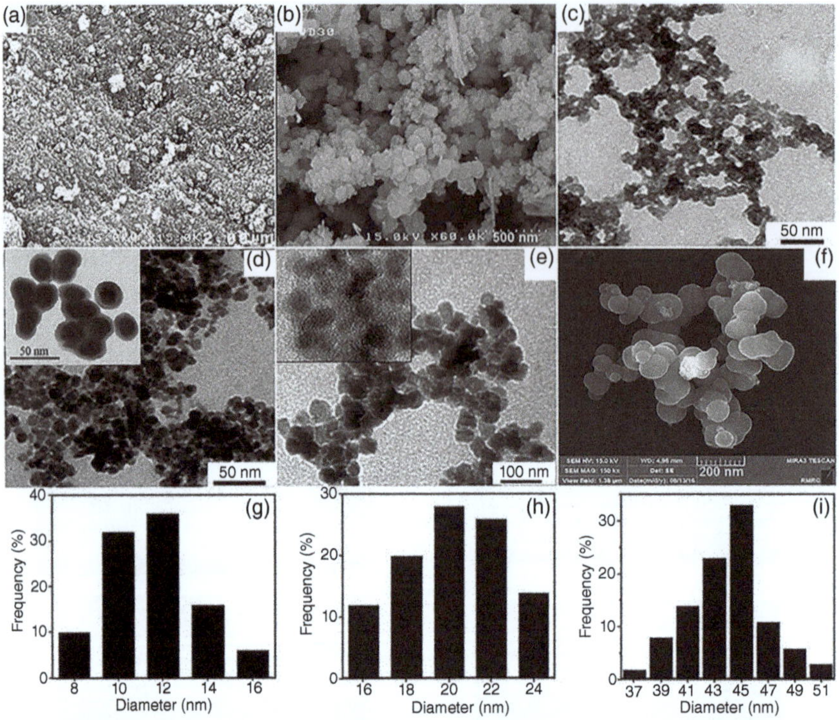

Figure 8.7 "(a) SEM image Fe_3O_4, (b) SEM image of $Fe_3O_4@SiO_2$, (c) SEM image of $Fe_3O_4@SiO_2$ modified with TCT-PVA and encapsulated with Cu(II), (d) TEM image of Fe_3O_4, (e) TEM image of $Fe_3O_4@SiO_2$, (f) TEM image of $Fe_3O_4@SiO_2$ with TCT-PVA and Cu(II), and (g) size distributions of (g) Fe_3O_4, (h) $Fe_3O_4@SiO_2$, and (i) $Fe_3O_4@SiO_2$ with TCT-PVA and Cu(II), respectively." Source: Reproduced from Sardarian et al. [45], with permission of John Wiley & Sons.

necessity of external ligands or additives to facilitate the reaction. In comparison to other methods documented in the literature for N-arylation and 5-substituted tetrazole synthesis, this catalytic approach offers numerous advantages. These benefits include the exclusion of additives and external ligands, readily separated with the assistance of an external magnet, the ability to reuse the catalyst for seven consecutive cycles, and the generation of outstanding product yields of the desired products [45].

Ahmadi et al. have presented a nanocomposite known as Fe_3O_4@MCM-41-SB-Cu. This composite comprises a magnetic core surrounded by a mesoporous silica shell that has been modified with a Schiff base and its copper complex [46]. The development of this nanocomposite followed a stepwise procedure, involving the preparation of Fe_3O_4@MCM, functionalization with *N*-(2-aminoethyl)-3 aminopropyltrimethoxysilane (AEAPS), synthesis of Schiff base-grafted nanoparticles, and subsequent introduction of $Cu(CH_3COOH)_2{\cdot}H_2O$. The described process involved the synthesis of MNPs functionalized with a Cu(II) Schiff base complex, resulting in the creation of an innovative organic–inorganic hybrid nanocomposite (Scheme 8.8). This nanocomposite underwent thorough characterization

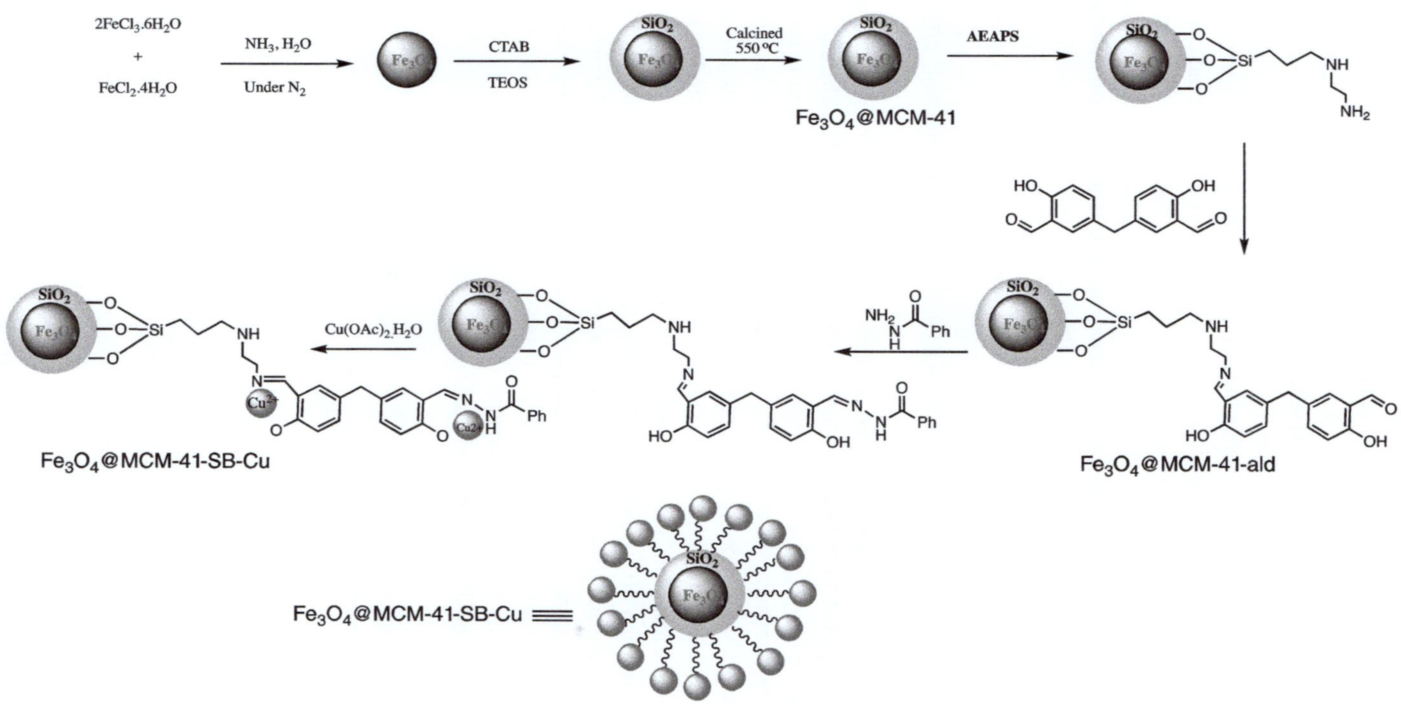

Scheme 8.8 Synthetic protocol illustration of Fe_3O_4@MCM-41-SB-Cu synthesis.

using a variety of analytical techniques, including FTIR, AAS, TGA, XRD, VSM, FE-SEM, HRTEM, Brunauer–Emmett–Teller (BET), and EDX methods. The resulting magnetic nanocatalyst was effectively applied to expedite the synthesis of 5-substituted 1*H*-tetrazole derivatives through one-pot three-component reactions, which included different aldehydes, hydroxylamine hydrochloride, and sodium azide. The advantage of this catalyst is its ability to be easily separated from the reaction mixture by employing an external magnet. Furthermore, it displayed the potential for reuse in at least five reaction cycles without a significant decline in its catalytic activity. Furthermore, the research delved into an examination of the antimicrobial characteristics of magnetic nanoparticles loaded with streptomycin against both Gram-positive bacteria such as *Staphylococcus aureus* and Gram-negative bacteria such as *Escherichia coli*. This evaluation was conducted under conditions with and without the influence of a magnetic field. The findings revealed that when a magnetic field was present, the nanocomposites exhibited an enhanced bacteriostatic effect. Moreover, the nanoparticles' ability to immobilize the amylase enzyme was investigated, demonstrating their effective enzyme immobilization capability.

A magnetic nanostructure composed of silica (Fe_3O_4@SiO_2–APTES–TFA) were prepared as depicted in Scheme 8.9 and characterized via SEM, TEM and XRD analysis (Figure 8.8a–c). This nano-catalyst proved to be highly efficient to produce tetrazoles. This was achieved through a cycloaddition reaction involving nitriles and azides. Remarkably, a wide range of reactants, including those featuring electron-donating and electron-withdrawing groups, could be employed under gentle reaction conditions. This method consistently yielded impressive results, with typically more than 90% product yields, often falling into the excellent or good category. This was achieved with a minimal amount of catalyst (0.05 g) and for a brief reaction duration. The magnetic-supported catalyst enabled straightforward, uncomplicated, and rapid recovery of the catalyst, making it highly suitable for reuse. This catalytic system is anticipated to attract both academic and industrial interest due to its environmental and economic advantages [47].

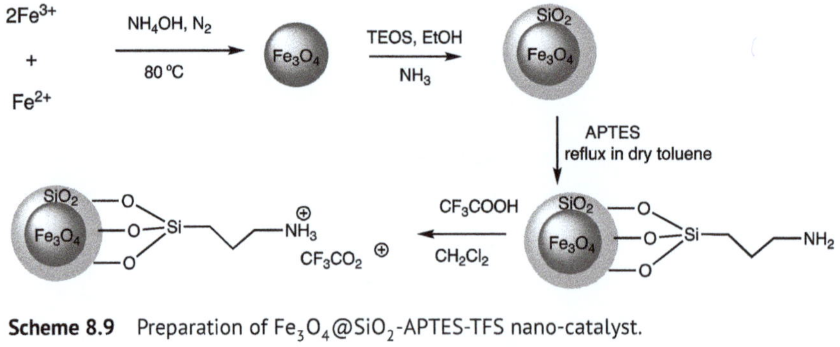

Scheme 8.9 Preparation of Fe_3O_4@SiO_2-APTES-TFS nano-catalyst.

Similarly, Rezaei et al. developed a magnetic nanocatalyst by covalently attaching a Cu(II) complex of aza-crown ether onto a silica-coated iron oxide support, as

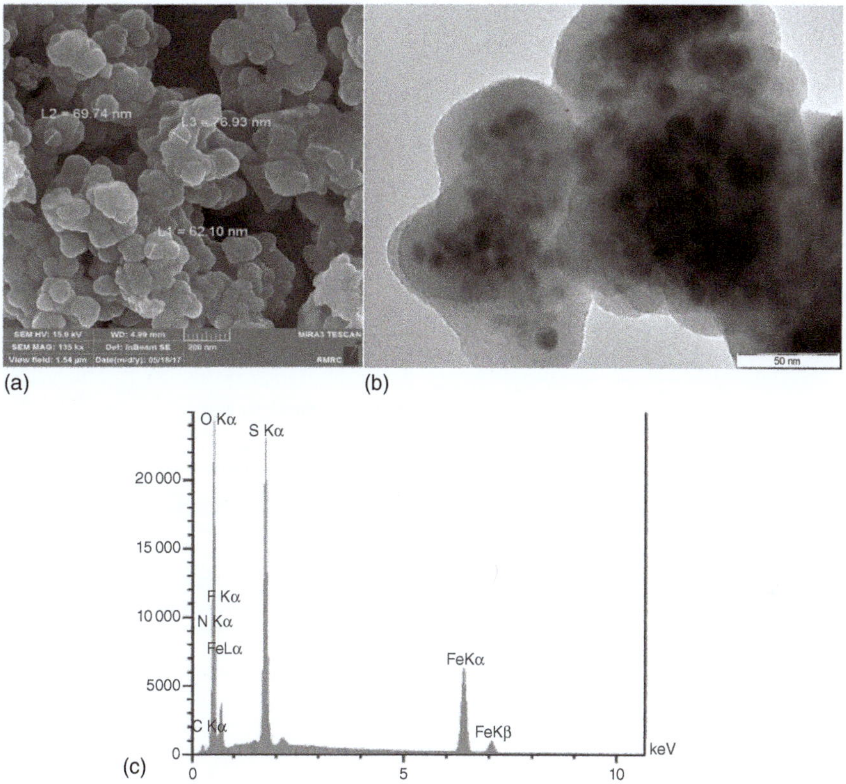

Figure 8.8 (a) SEM and (b) TEM images and (c) XRD spectrum of Fe_3O_4@SiO_2–APTES–TFA nanostructures. Source: Reproduced from Fatahi et al. [47], with permission of John Wiley & Sons.

illustrated in Scheme 8.10. The nanocatalyst underwent thorough characterization through a variety of techniques, including FTIR, SEM, TEM, XRD, TGA, and ICP. This led to the formation of Fe_3O_4@SiO_2/aza-crown ether-Cu(II), an inorganic/organic composite that was evaluated for its catalytic performance in the one-pot synthesis of 1,4-disubstituted-1,2,3-triazoles, 5-substituted 1*H*-tetrazoles, and 1-substituted 1*H*-tetrazoles. The catalyst demonstrated exceptional efficiency, producing the desired products quickly. Additionally, this magnetic nanocatalyst could be effortlessly retrieved via an external magnetic field and can be employed in subsequent reactions over the course of at least five cycles, the catalyst consistently maintained its catalytic effectiveness, exhibiting no notable reduction in performance [48].

A silica-based catalyst (BNPs@SiO_2-TPPTSA) was employed in the production of various 5-substituted 1*H*-tetrazoles. The synthesis of the BNPs@SiO_2-TPPTSA catalyst is outlined in Scheme 8.11, and its characterization is done via different techniques including FTIR, TEM, SEM, XRD, EDX, and thermogravimetric analysis–differential thermal analysis (TGA-DTA). The TEM and SEM images (Figure 8.9) of the catalyst exhibit size and morphology; a spherical distribution

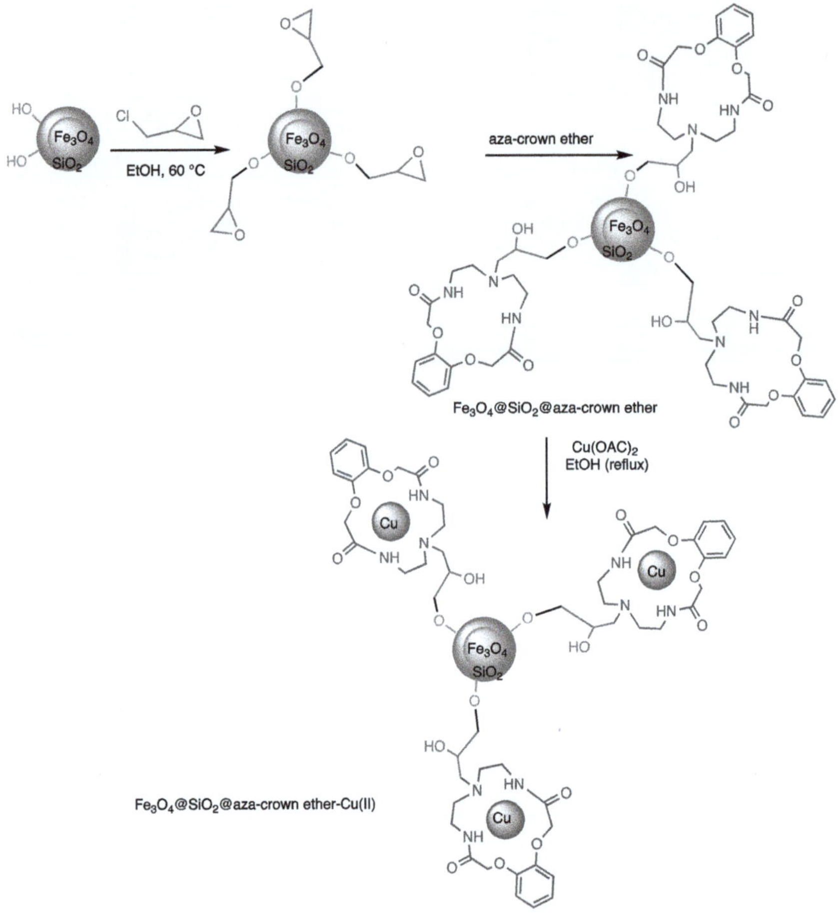

Scheme 8.10 Synthetic scheme of Fe_3O_4@SiO_2/aza-crown ether-Cu(II) heterogenous catalyst.

observed in SEM and TEM images, SEM images exhibit particles size ranges from 10 to 50 nm. Furthermore, the catalytical results exhibited that this catalytic system produced tetrazoles with yields of up to 90%. Moreover, it is possible to employ the catalyst for a minimum of five successive cycles without experiencing a notable decline in its catalytic performance, as depicted in Figure 8.10a, SEM images of the catalyst after fifth run and FTIR spectrum Figure 8.10b, reveal no noticeable changes [49].

8.2.1 Generalized Reaction Mechanism of Silica-Supported Heterogenous-Catalyzed Tetrazole Synthesis

The production of tetrazoles utilizing silica-supported heterogeneous catalysts entails a sequence of chemical reactions. In this context, we succinctly examine two generic reaction mechanisms for the synthesis of tetrazoles: one through a [3 + 2]

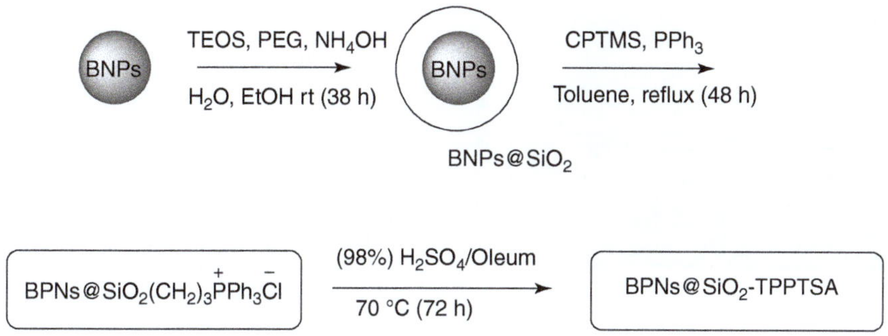

Scheme 8.11 Schematic for synthesis of BPNs@SiO_2-TPPTSA.

Figure 8.9 (a) TEM image of the BNPs@SiO_2-TPPTSA (b) FE-SEM of the BNPs@SiO_2-TPPTSA. Source: Reproduced from Khodamorady et al. [49], with permission of John Wiley & Sons.

cycloaddition and the other via a one-pot multicomponent reaction involving amine, triethyl orthoformate, and azide.

8.2.1.1 Via [3 + 2] Cycloaddition

The [3 + 2] cycloaddition reaction is a key protocol in tetrazole synthesis, as it involves the combination of an azide and a nitrile to form the tetrazole ring. The silica-supported heterogeneous-catalyzed tetrazole synthesis via [3 + 2] cycloaddition reactions typically initiated with activation of catalyst. Silica-supported catalyst, which may contain metal nanoparticles or other active sites, is activated, often by heating or pretreatment. Certainly, after activation of catalyst, it interacts with nitrile group and enhances the nucleophilic attack susceptibility at the C-end of the nitrile group. Subsequently, this accelerates [3 + 2] cycloaddition reaction of nitrile and an azide **(A)**, this addition reaction facilitates formation of nitrilimine intermediate (**B**). Thereafter, the nitrilimine intermediate undergoes an intramolecular cyclization via 1,3-*H* shift with acid work-up and tetrazole is formed. Finally,

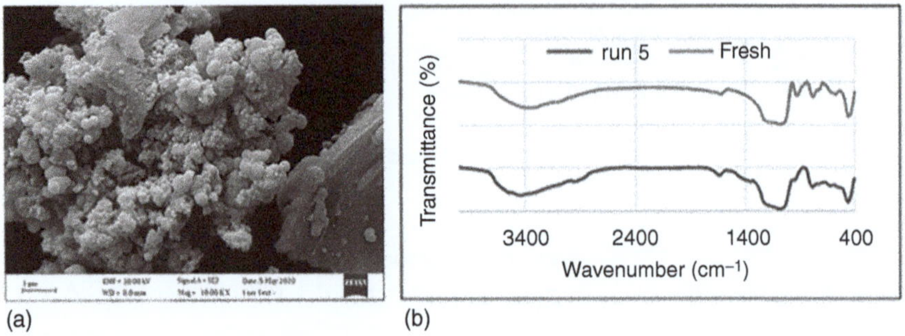

Figure 8.10 (a) Scanning electron microscopy (SEM) depiction of the catalyst after the 5th cycle and (b) FTIR spectra of the fresh catalyst and after the 5th cycle for the BNPs@SiO_2-TPPTSA. Source: Reproduced from Khodamorady et al. [49]/with permission of John Wiley & Sons.

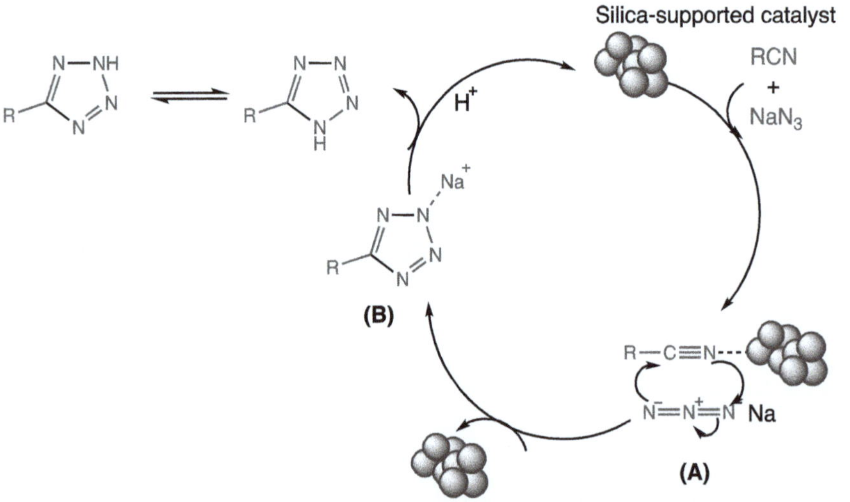

Figure 8.11 Generalized [3 + 2] cycloaddition reaction mechanism for tetrazole synthesis.

catalyst can be regenerated by releasing the product and regenerated catalyst is then ready for another catalytic cycle [50] (Figure 8.11).

8.2.1.2 Via One-Pot Multicomponent Reaction of Amine, Triethyl Orthoformate, and Azide

The synthesis of tetrazoles via a one-pot multicomponent reaction is another alternative route, where amines or amino acids, triethyl orthoformate, and sodium azide react together in a single reaction vessel. The reaction mechanism is initiated with formation of an intermediate **(I)** under the influence of the silica-supported catalyst. As a result, this process enables the nucleophilic attack of the amino group on an intermediate. Subsequently, two molecules of ethanol are successively eliminated, followed by the nucleophilic attack of azide to yield Intermediate **(II)**.

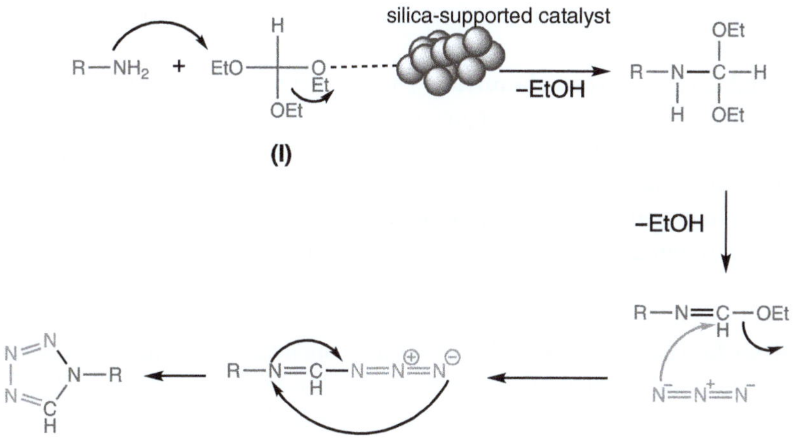

Figure 8.12 Generalized one-pot multicomponent reaction plausible mechanism for tetrazole synthesis.

The azido nitrile Intermediate II undergoes an intramolecular cyclization reaction, resulting in the formation of the tetrazole ring [50] (Figure 8.12).

8.3 Future Perspective of Silica-Supported Catalysts in Tetrazole Synthesis

The use of silica-supported catalysts in tetrazole transformations has been a subject of significant research and development in the field of catalysis. Silica-supported catalysts offer several advantages, including improved stability, reusability, and selectivity in tetrazole synthesis. The future perspective of silica-supported catalysts in tetrazole synthesis holds great promise and potential for advancements in organic chemistry. Some of the promising points are summarized as follows:

- **Catalyst design and tuning**: Researchers are expected to continue designing and fine-tuning silica-supported catalysts for tetrazole synthesis. This may involve optimizing the types of active sites, ligands, and metal complexes immobilized on silica supports to enhance catalytic activity, selectivity, and stability.
- **Mechanistic insights**: A deeper understanding of the mechanistic aspects of tetrazole synthesis with silica-supported catalysts will likely emerge. This knowledge can aid in the rational design of catalysts and the development of more efficient reaction pathways.
- **Catalyst immobilization methods**: Advances in the immobilization techniques of active species on silica supports may lead to catalysts with improved performance. This includes developing novel methods for controlled deposition of catalytic species and precise control over their spatial arrangement.
- **Tandem and multicomponent catalysis**: Future research may focus on exploring tandem and multicomponent catalysis using silica-supported catalysts

for tetrazole synthesis. These strategies can lead to more efficient and selective transformations.

- **Sustainability and green chemistry**: The development of silica-supported catalysts that align with principles of green chemistry will continue to be a priority. This involves minimizing waste, using safer solvents, and reducing energy consumption in tetrazole synthesis.
- **Catalysis in continuous flow systems**: The integration of silica-supported catalysts in continuous flow systems is likely to gain momentum. Flow chemistry allows for better control over reaction parameters, safety, and scalability, which is advantageous for large-scale tetrazole synthesis.
- **Catalysis for drug discovery**: Tetrazoles have applications in pharmaceuticals, and silica-supported catalysts will play a crucial role in the synthesis of drug intermediates and active compounds. The design and development of catalysts for enantioselective tetrazole synthesis can be a particularly valuable area of exploration.
- **Scale-up and industrial applications**: As the demand for tetrazole derivatives grows in various industries, efforts will be made to scale up silica-supported catalyst-based processes for industrial production. This involves addressing issues of catalyst recycling and stability under continuous operation.

8.4 Conclusion

In conclusion, the utilization of silica-supported heterogeneous catalysts for tetrazole synthesis represents a significant advancement in the field of organic chemistry. This chapter has delved into the various aspects of this catalytic approach, highlighting its key attributes and future potential. The synthesis of tetrazoles, important compounds with versatile applications in pharmaceuticals, materials science, and beyond, has been greatly facilitated by the introduction of silica-supported catalysts. These catalysts offer several advantages, including ease of handling, reusability, and efficient catalytic activity. The immobilization of active species on a silica support not only enhances the stability of the catalyst but also allows for fine-tuning of reaction conditions and selectivity. Furthermore, this chapter has provided insights into the diverse reactions and strategies employed in the synthesis of tetrazoles using silica-supported catalysts, with a focus on their mechanisms and the role of the support material. By harnessing the unique properties of silica and the versatility of heterogeneous catalysis, researchers have achieved remarkable progress in developing efficient and sustainable methods for tetrazole synthesis.

Acknowledgment

The authors express their deep gratitude to Chandigarh University, Punjab, for the e-resources made available for this research.

References

1 Giannakakis, G., Mitchell, S., and Perez-Ramirez, J. (2022). *Trends in Chemistry* 4 (4): 264–276.
2 Swami, S. and Shrivastava, R. (2020). *Nanotechnology for Energy and Environmental Engineering* 21: 477–493.
3 Van Leeuwen, P.W. (2006). *Homogeneous Catalysis: Understanding the Art*. Springer Science & Business Media.
4 Bhaduri, S. and Mukesh, D. (2014). *Homogeneous Catalysis: Mechanisms and Industrial Applications*. John Wiley & Sons.
5 George, S.M. (1995). *Chemical Reviews* 95: 475–476.
6 Van Santen, R.A. (2017). *Modern Heterogeneous Catalysis: An Introduction*. John Wiley & Sons.
7 Karan, R., Bhatia, R., and Rawal, R.K. (2021). *Applications of Homogeneous Catalysis in Organic Synthesis*, 159–188. Elsevier.
8 Guo, J. and Jiang, D. (2020). *ACS Central Science* 6: 869–879.
9 Sharma, N., Swami, S., Shrivastava, V. et al. (2021). *Materials today: Proceedings* 43: 3309–3317.
10 Heravi, M.M. and Zadsirjan, V. (2020). *RSC Advances* 10: 44247–44311.
11 Bredael, K., Geurs, S., Clarisse, D. et al. (2022). *Journal of Chemistry* 2022: 1–21.
12 Neochoritis, C.G., Zhao, T., and Dömling, A. (2019). *Chemical Reviews* 119: 1970–2042.
13 Sharma, N., Swami, S., Pathak, S. et al. (2023). *Research on Chemical Intermediates* 1–19.
14 Dai, L.-L., Zhang, H.-Z., Nagarajan, S. et al. (2015). *MedChemComm* 6: 147–154.
15 Zhang, J., Wang, S., Ba, Y., and Xu, Z. (2019). *European Journal of Medicinal Chemistry* 178: 341–351.
16 Wang, S.-Q., Wang, Y.-F., and Xu, Z. (2019). *European Journal of Medicinal Chemistry* 170: 225–234.
17 Bhaskar, V. and Mohite, P. (2011). *Journal of Optoelectronics and Biomedical Material* 3: 7–16.
18 Poonian, M.S., Nowoswiat, E.F., Blount, J.F. et al. (1976). *Journal of Medicinal Chemistry* 19: 286–290.
19 Hedner, T., Oparil, S., Rasmussen, K. et al. (1999). *American Journal of Hypertension* 12: 414–417.
20 Baba, H. and Paterson, D.L. (2010). *Kucers' The Use of Antibiotics Sixth Edition: A Clinical Review of Antibacterial, Antifungal and Antiviral Drugs*, 295.
21 Rostamnia, S. and Doustkhah, E. (2014). *RSC Advances* 4: 28238–28248.
22 Bibby, A. and Mercier, L. (2002). *Chemistry of Materials* 14: 1591–1597.
23 Yu, X. and Williams, C.T. (2022). *Catalysis Science & Technology* 12: 5765–5794.
24 Kaur, M., Sharma, S., and Bedi, P.M. (2015). *Chinese Journal of Catalysis* 36: 520–549.
25 Sajadi, S.M. (2012). *Chemistry and Materials Research* 2: 12–16.
26 Ghorbani-Choghamarani, A., Shiri, L., and Azadi, G. (2016). *RSC Advances* 6: 32653–32660.

27 Habibi, D. and Nasrollahzadeh, M. (2012). *Monatshefte für Chemie-Chemical Monthly* 143: 925–930.

28 Habibi, D., Nabavi, H., and Nasrollahzadeh, M. (2013). *Journal of Chemistry* 2013.

29 Dehghani, F., Sardarian, A.R., and Esmaeilpour, M. (2013). *Journal of Organometallic Chemistry* 743: 87–96.

30 Habibi, D., Nasrollahzadeh, M., Mehrabi, L., and Mostafaee, S. (2013). *Monatshefte für Chemie-Chemical Monthly* 144: 725–728.

31 Habibi, D., Mostafaee, S., and Mehrabi, L. (2013). *Journal of Chemical Research* 37: 464–466.

32 Esmaeilpour, M., Javidi, J., Dodeji, F.N. et al. (2014). *Journal of Molecular Catalysis A: Chemical* 393: 18–29.

33 Meshram, G.A., Deshpande, S.S., Wagh, P.A., and Vala, V.A. (2014). *Tetrahedron Letters* 55: 3557–3560.

34 Naeimi, H. and Mohamadabadi, S. (2014). *Dalton Transactions* 43: 12967–12973.

35 Ahmed, N. and Siddiqui, Z.N. (2015). *RSC Advances* 5: 16707–16717.

36 Jahanshahi, R. and Akhlaghinia, B. (2015). *RSC Advances* 5: 104087–104094.

37 Razavi, N. and Akhlaghinia, B. (2015). *RSC Advances* 5: 12372–12381.

38 Esmaeilpour, M., Javidi, J., and Zahmatkesh, S. (2016). *Applied Organometallic Chemistry* 30: 897–904.

39 Kumar, A., Kumar, S., Khajuria, Y., and Awasthi, S.K. (2016). *RSC Advances* 6: 75227–75233.

40 Azadi, G., Ghorbani-Choghamarani, A., and Shiri, L. (2017). *Transition Metal Chemistry* 42: 131–136.

41 Esmaeilpour, M., Sardarian, A.R., and Firouzabadi, H. (2018). *Applied Organometallic Chemistry* 32: e4300.

42 Liu, L., Ai, Y., Li, D. et al. (2017). *ChemCatChem* 9: 3131–3137.

43 Mahajan, D.H., Mistry, D., and Tailor, H. (2018). *Chemical Science* 7: 676–686.

44 Samanta, P.K., Biswas, R., Das, T. et al. (2019). *Journal of Porous Materials* 26: 145–155.

45 Sardarian, A.R., Eslahi, H., and Esmaeilpour, M. (2018). *ChemistrySelect* 3: 1499–1511.

46 Ahmadi, A., Sedaghat, T., Motamedi, H., and Azadi, R. (2020). *Applied Organometallic Chemistry* 34: e5572.

47 Fatahi, H., Jafarzadeh, M., and Pourmanouchehri, Z. (2019). *Journal of Heterocyclic Chemistry* 56: 2090–2098.

48 Rezaei, F., Amrollahi, M.A., and Khalifeh, R. (2019). *Inorganica Chimica Acta* 489: 8–18.

49 Khodamorady, M., Ghobadi, N., and Bahrami, K. (2021). *Applied Organometallic Chemistry* 35: e6144.

50 Swami, S., Sahu, S.N., and Shrivastava, R. (2021). *RSC Advances* 11: 39058–39086.

9

Theoretical Insights on Reduction of CO_2 Using Functionalized Ionic Liquid at Gold Surface

Shanmugasundaram Kamalakannan[1], *Muthuramalingam Prakash*[1], *and Majdi Hochlaf*[2]

[1] *SRM Institute of Science and Technology, Computational Chemistry Research Laboratory (CCRL), Department of Chemistry, Kattankulathur-603203, Tamil Nadu, India*
[2] *Université Gustave Eiffel, COSYS/IMSE, 5 Bd Descartes, 77454, Champs Sur Marne, France*

9.1 Introduction to Heterogeneous Catalysts for CO_2RR Applications

Carbon dioxide (CO_2) is the main source of global pollution due to its steady growth in the atmosphere, global warming and greenhouse effect [1–3]. Recently, a wide range of technologies for the significant reduction of atmospheric CO_2 and its conversion have attracted attention from the scientific community [4]. It is commonly admitted that the amine scrubbing method is a promising solution for large-scale CO_2 capture [5–8]. Indeed, this flue gas was captured using aqueous monoethanolamine (MEA), which forms a chemical bond with CO_2 resulting in the formation of carbamate. However, this scrubbing process has several drawbacks, in terms of stability, energy, and environmental concerns [9]. These issues can be solved with the use of ionic liquid (IL) co-catalysts or IL electrolytes for the CO_2 reduction reaction [10]. ILs (pairs of a cation and an anion) were invented in 1914 as an alternative solution to MEA technology [11]. They are widely employed as high-performance liquid electrolytes in a wide range of applications including energy technology, catalysis, and carbon capture and sequestration (CCS) [12]. Owing to their distinctive property such as alterability, wide electrochemical window, excellent specific conductance, and ILs have distinguished themselves from polymeric and amine solvents. Also, there are large number of cations such as pyrazolium, imidazolium, quaternary phosphonium, pyridinium, and quaternary ammonium/anionic pairs, which have been reported and utilized for the above-mentioned applications. In particular, imidazolium (Im^+)-based ILs were explored as potential co-catalysts for CO_2 capture and conversion [13]. Consequently, they have generated more interest in carbon capture than traditional aqueous solutions. This was attributed to the lower energy barrier to CO_2 reduction, recyclability, substantial capture, and sustainability [14].

Solid Base Catalysts: Synthesis, Characterization, and Applications, First Edition.
Edited by Ravi Tomar, K.K. Pant, and Ramesh Chandra.

To date, there are two approaches to capture CO_2 molecules: the first one makes use of standard (non-functional) ILs [15] and the second one involves functionalized ILs (e.g. NH_2, OH, and CF_3) [16]. The main drawback with standard ILs is that they offer limited active sites for both CO_2 capture and activation. Whereas cation-functionalized ILs are greatly viewed to overcome such limitations and they are thus effective compounds to activate and convert CO_2. For these purposes, researchers employed a broad variety of functional groups [17]. Among them, NH_2-functionalized imidazole cations $(Im)^+$ offered a high affinity for reducing CO_2 molecules with lower energy consumption [18]. Recently, different series of ILs have been synthesized by mixing Im^+ with various functional groups [19].

The main challenge with CO_2 capture by either bulk or isolated ILs is that self-diffusion, high viscosity, and high regeneration energy all hinder CO_2 reduction. To overcome such challenge, the adsorption-based strategy for CO_2 capture has gained more prominence. These adsorption processes require a small amount of energy to have CO_2 molecules over the catalytic medium [20]. This can be achieved by incorporating functionalized co-catalysts onto solid adsorbents (e.g. metal surface). Hence, these solid adsorbents offer several benefits, including high surface area and selective binding, which reduce the dynamic movement of CO_2@IL. In addition, electrocatalytic reduction of CO_2 is a simple and cost-effective process for the reduction and conversion of CO_2 into valuable products [21, 22]. Moreover, an electrocatalytic material offers several advantages, such as recyclability, reduced corrosion, and significant CO_2 reduction and transformation efficiency [23]. Recently, Luo et al., [24] proposed that NH_2-incorporated ILs can significantly promote the selective capture of CO_2. These authors also showed that effective CO_2 capture occurs via intramolecular H-bonding involving the amino group and via proton transfer mechanisms within the complexes. In addition, the anionic groups have a significant influence on the reached adsorption capacity (of 1.96 mol mol^{-1} ILs) and their good selectivity. Moreover, Cho et al., [25] examined the catalytic transformation of CO_2 onto a nitrogen-functionalized silica surface. They found that silanol structures on the surface of silica productively assist in the capture of CO_2 through direct proton interaction with the amine group. The nudged elastic band (NEB) simulation results illustrate that surface hydrogen donor migrations are the proton transfer route under dry conditions, whereas under humid conditions water acts as a hydrogen transport bridge to convert the CO_2 molecule. This amine-functionalized cation performs quite well in terms of CO_2 reduction. In fact, CO_2 conversion is not limited to anion-NH_2 functionalization; it can also occur via cationic NH_2 functionalization. For instance, Pan and Wang [26] pioneered the use of NH_2-functionalized phenolic anions for CO_2 conversion. These authors showed that the conversion of CO_2 to carbamic acid occurs via intramolecular proton exchange at the oxygen-containing site of the phenolic anion.

The reduction of CO_2 at the ILs@Bi interface was investigated by Medina-Ramos et al. [27]. According to the results of experiments, force field, and density functional theory (DFT) based studies, it was shown that the production of Bi-$[Im]^+$ at the cathodic potential is essential for the CO_2 activation on the interfacial material [27]. Additionally, Chernyshova and Ponnurangam demonstrated using DFT methods that an external electric field and a co-adsorbed alkali cation can activate CO_2 at an electrode–electrolyte interface [28]. Furthermore, Niemann et al. revealed the interfacial arrangement of hydroxyl-modified ILs on a negative mica layer using atomic force tunneling spectroscopy [29]. In addition, Parameswari et al. treated the adsorption and capture of CO_2 over the amino acid-decorated TiO_2 surface, showing that the cooperative dispersion interactions stabilize the gas molecule over this interface. Furthermore, the sorbent-anchored hybrid catalysts are not significantly affected by the aqueous microenvironment, demonstrating the suitability of the catalyst for the activation process [30]. However, the in-depth understanding of the reactivity of functional groups within the ILs at metal surface for the CO_2 activation/conversion is still scarce. Thus, in the present study, we focus on studying the behavior of various functional groups on CO_2 activation at the electrode–catalytic interface structure/metal surface toward the CO_2 reduction reaction (CO_2RR) applications. This is still an open question. First principles theoretical methodologies attempt to investigate the interfacial stability of ILs (i.e. NH_2-functionalized) at the interface (i.e. electrodes). These model interfaces are considered as electrode–electrolyte interfaces for the investigation of CO_2 capture, activation, and conversion approach. In the last decade, we used such methodologies, at the DFT level, to identify at the molecular level the reactive sites, and adsorption mechanisms at model solid–liquid interfaces. We thus use various indicators and analyses based on binding energies, electron and spin densities, Bader charge transfer, and electric field effects.

9.2 Computational Methodology

First, the stabilization of CO_2 at the functionalized ILs was carried out using the Gaussian16 [31] package and the minimum energy structures were obtained after geometry optimizations, followed by evaluation of the vibrational frequencies (all >0). The interaction between the functionalized ILs and CO_2 was treated by applying the PBE + D3 method and the atoms were described by a 6-311++G** basis set. The binding energy (BE) of CO_2 on the functionalized ILs was calculated using the following equation:

$$BE_{(CO_2)} = E_{CO_2@IL} - E_{CO_2@IL} - E_{CO_2} - E_{IL} \tag{9.1}$$

We have considered the basis set superposition error (BSSE) using the counterpoise (CP) method of Boys and Bernardi [32]. Wherein $E_{CO_2@IL}$, E_{CO_2}, and E_{IL} represent the total energies of the CO_2 attached to the IL, CO_2, and bulk phase of the functionalized ILs, respectively. These energies are calculated in the complete basis set of the CO_2@IL system.

The periodic simulations were performed with the Vienna Ab-initio Simulation Program (VASP) using the generalized gradient approximation-Perdew–Burke–Ernzerhof (GGA-PBE) electron exchange-correlation functional [33–36]. The electron–ion interaction was treated using the projected augmented-wave approach [37]. The exchange and correlated parts were treated by the (GGA) [38].

The $3 \times 3 \times 1$ unit Au(111) surface was constructed with three layers and stabilized to find the most stable structure. To avoid mutual coupling between the neighboring periodic images, a 15 Å vacuum region was kept in the Z-direction.

It is known that the top layer is involved in catalysis with the foreign molecule, so the bottom layer was fixed throughout the simulation and all were allowed to move freely along the X, Y, and Z directions. To determine the electronic structure of the system, the Brillouin zone was evaluated using a Monkhorst–Pack *k*-point mesh of $3 \times 3 \times 1$ for structure optimization and a mesh of $6 \times 6 \times 1$ for self-consistent calculations [39]. The geometric optimization threshold was set at 10^{-5} eV in energy and 0.01 eV Å in forces. Then, the adsorption energies (E_{ads}) of the CO_2 species on IL@Au(111) were evaluated using the following equation,

$$E_{ads}(CO_2@IL@Au(111)) = (E_{CO_2@IL\text{-}Au(111)} - E_{CO_2} - E_{IL@Au(111)}) \tag{9.2}$$

where, the $E_{CO_2@IL\text{-}Au(111)}$, $E_{IL@Au(111)}$, and E_{CO_2}represent the total energies of the CO_2 at the IL-Au(111) interface, IL-Au(111) interface, and bulk phase of CO_2 molecule, respectively. Further, the density contour maps were generated from the most stable geometries by using this equation,

$$\Delta\rho = \rho_{CO_2@IL@Au(111)} - \rho_{IL@Au(111)} - \rho_{CO_2} \tag{9.3}$$

where, $\rho_{CO_2@IL@Au(111)}$, $\rho_{IL@Au(111)}$, and ρ_{CO_2}correspond to the electron densities of CO_2 at the IL-Au(111) interface, IL at the Au(111) surface, and CO_2 molecule, respectively. The stable geometries have been used to calculate the band structure, projected density of states (PDOS) and Bader charge analyses [40, 41]. The reaction mechanism for the CO_2 conversion is carried out by the climbing image-nudged elastic band (CI-NEB) method as implemented in the CP2K code [42, 43].

9.3 Characterization of Functionalized Ionic Liquids Interacting with CO_2

9.3.1 Studies of CO_2 Interacting with ILs in Gas Phase

A common cation, 1-ethyl-3-methylimidaoilum ($EMIm^+$), was used for CO_2 absorption within a pair involving different anions [i.e. dicyanamide and thiocyanate]. The role of the substituents attached to the cation has a significant influence on the physicochemical properties of the ILs. In this chapter, we consider compounds where the methyl of the ethyl moiety is substituted with two types of functional groups: (ii) electron-donating —CH_3 and —NH_2; and (ii) electron-accepting —CF_3 groups. The functional group modifications in the ILs influence the CO_2 absorption capacity of these ILs. The PBE+D3/6-311++G(d,p) dispersion incorporated method is used to calculate all functionalized IL properties with and without the presence of CO_2. In particular, we will present the significance of the substituents on the effective capture and conversion of CO_2 [44, 45].

Figure 9.1 shows the minimum energy geometries of these CO_2-IL systems in gas phase. The interaction of $(CO_2)O\cdots H(CH_3)$ of IL forms weak hydrogen bond with

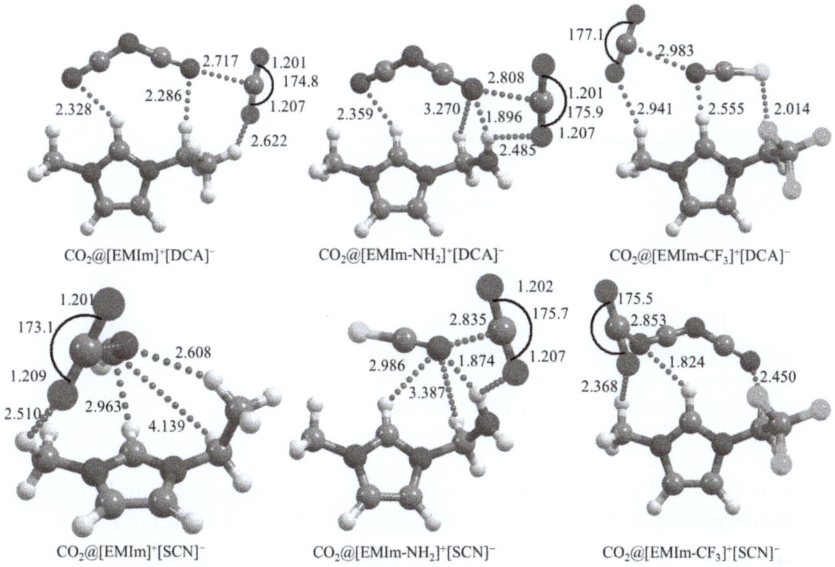

Figure 9.1 The computed geometries of CO_2@NH_2-functionalized ILs at gas phase. Source: Kamalakannan et al. [46]/with permission of IOP Publishing.

distance of 2.622 Å. This length is lowered once the —NH_2 moiety is attached to the 1-ethyl-3-methylimidazolium rather than the methyl group. The —NH_2 group considerably stimulates the negatively charged O atom of the CO_2 molecule, increasing the strength of the (CO_2)O····IL interaction. Furthermore, this aforementioned interaction generates major alteration in the C—O bond lengths (~1.2 Å) and slight bending of the O—C—O in-plane angle (~176°) of CO_2 (cf. Figure 9.1). Generally, several intermolecular interactions contribute toward the stabilization of CO_2 interacting with functionalized ILs.

In addition, we found that the incorporation of X-substituents (X = SCN^- or DCA^-) on the cationic ring impacts the H-bonds occurring between the C of CO_2 and the anions. In particular, these intermolecular bonds involve the outermost electron pair in the N of the SCN^- attracting the highly electropositive C atom of the CO_2 moiety. A similar effect was found in other SCN-containing complexes functionalized ILs [45]. The minimal interatomic separations between the C of CO_2 and the N of anions are 2.717 Å for CO_2@$[EMIm]^+[DCA]^-$, 2.808 Å for CO_2@$[NH_2\text{-EMIm}]^+[DCA]^-$, for 2.675 Å CO_2@$[EMIm]^+[SCN]^-$, and 2.835 Å for CO_2@$[NH_2\text{-EMIm}]^+[SCN]^-$.

On the other hand, the in-plane CO_2 angle reduces to 173° (~7°) from 180° at the abovementioned ILs [47]. Table 9.1 shows that the binding energies (BEs) of CO_2 decrease by ~8.0 kJ mol^{-1} upon incorporating the NH_2 group into the cationic ring. Thus, the presence of amino group has a significant effect on the adsorption capacity of CO_2 to ILs. It is well known that CO_2 adsorbs, at bulk-ILs, in random manner as a result of intermolecular couplings between the substituent of the Im^+ part and CO_2. This coupling process leads to the transfer of positive charges from ILs to CO_2.

Table 9.1 Calculated BEs (kJ mol^{-1}) of the CO_2 interacting with ILs in gas phase.

CO_2@IL	BE (kJ mol^{-1})
$[EMIm]^+[SCN]^-$	−28.66
$[EMIm\text{-}NH_2]^+[SCN]^-$	−19.70
$[EMIm\text{-}CF_3]^+[SCN]^-$	−15.02
$[EMIm]^+[DCA]^-$	−23.84
$[EMIm\text{-}NH_2]^+[DCA]^-$	−19.03
$[EMIm\text{-}CF_3]^+[DCA]^-$	−12.01

Source: Kamalakannan et al. [46]/with permission of IOP Publishing.

This process limits the activation capacity of CO_2 at the bulk medium. In contrast, these issues related with capture and conversation of CO_2 at bulk medium can be solved by attaching the IL at the electrode surface i.e. Au(111) surface.

9.3.2 Geometries and Energetics of CO_2 Interacting with Solid–Liquid Interface

The DFT approach is used to study the activation mechanisms of CO_2 at the functionalized and non-functionalized ILs adsorbed on Au(111) surface. The cationic and anionic composition of ILs have a profound effect on the reactivity of CO_2 molecules. Therefore, we have introduced several functional species onto the $EMIm^+$ ring, comprising donor and accepting functional units, as well as two different anions (i.e. DCA^- and SCN^-), each having an N-lone pair. These functionalized IL cation and anion sites are used for CO_2 adsorption and activation. Computations show that the activation of CO_2 has minimal impact on functionalized ILs since these ILs have close equilibrium structures in the isolated or liquid-surface interface, whereas this is not the case with CO_2. Indeed, the solid–liquid interface could considerably stimulate and activate CO_2 molecules via several types of intermolecular interactions [45]. Periodic DFT calculations using the dispersion incorporated PBE+D3 approach are used to compute all our interfacial systems. Figures 9.2 and 9.3 show the optimized geometries of all systems of interest in the present study. These figures demonstrates that Im^+ cation ring remains parallel to the Au solid surface. It interacts with this surface via π-stacking, which can be used to stabilize incoming CO_2 molecules at such interfacial medium [46].

In addition, the sites for adsorption of anions via either N—Au or S—Au bonds on the Au(111) surface are influenced by the substituents (i.e. electron donating or withdrawing) bound to the cationic part of the ILs. For instance, the lone pair nitrogen of DCA^- significantly couples to the top sites of the gold surface via strong electrostatic contacts, while S atom in the other anion binds to the top/bridge sites of the gold surface, which results in the tuning/improving of the reactivity of the catalyst. These interfacial anion/surface bindings control the electrostatic behavior

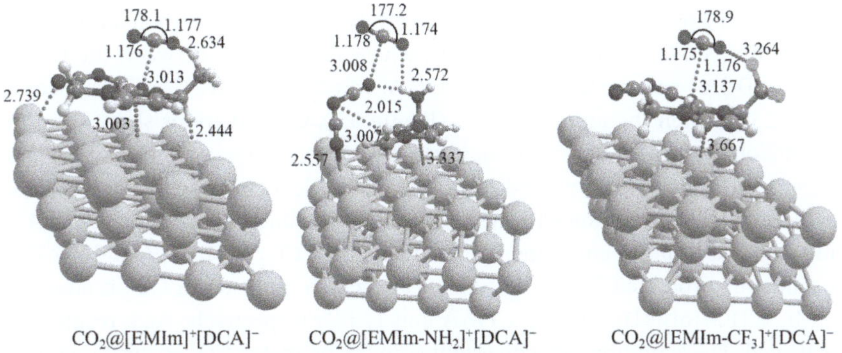

Figure 9.2 Minimum energy structures of CO_2@DCA^- based ILs decorated at the Au(111) surface, as optimized at the DFT-PBE+D3 method. Source: Kamalakannan et al. [46]/with permission of IOP Publishing.

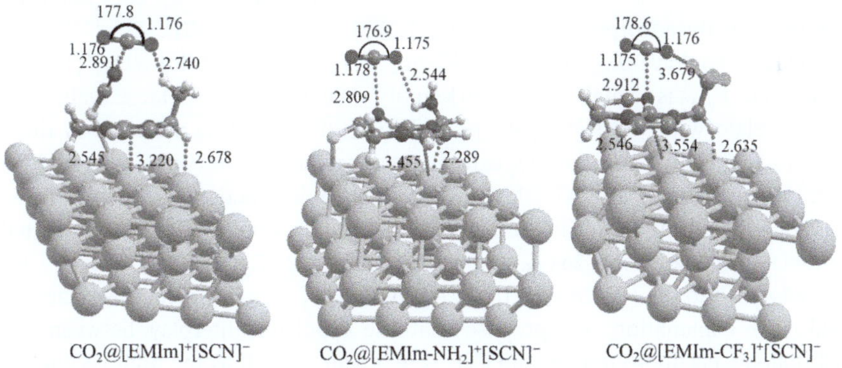

Figure 9.3 PBE+D3/TZVP minimum energy structures of CO_2@SCN^- based ILs adsorbed to Au(111). Source: Kamalakannan et al. [46]/with permission of IOP Publishing.

of the gold surface, which has significant effects on the activation of CO_2 molecules (vide infra) [45, 46, 48].

9.4 CO_2 Catalytic Activation at IL@Au(111) Liquid–Solid Interface Model

9.4.1 CO_2 interacting with [EMIm-Z]$^+$[DCA]$^-$@Au(111) Interface

First, we treated the CO_2 activation and adsorption close to [EMIm-Z]$^+$[DCA]$^-$@Au(111). The incoming CO_2 molecule interacts with this electrode/electrolyte medium via several long-range interactions. For understanding the CO_2 adsorption/activation at this interface, we performed geometry optimizations for the CO_2@[EMIm-Z]$^+$[DCA]$^-$@Au(111) system.

Figure 9.2 shows that one O of CO_2 exhibits a strong H-bond with the incorporated functional moieties at the cation of ILs where the O ····CH_3 and O ····NH_2 distances are computed 2.634 Å and 2.572 Å, respectively. In addition, the carbon atom of CO_2 links to nitrogen-lone pair of the anions, where the distance between the O=CN moiety of DCA^- and the C of CO_2 is ~3.01 Å, whereas we have C—N distances of ~2.8 to 2.9 Å when the anion is SCN^- (Figure 9.2). Computations show also that the functionalization of the cation significantly activates CO_2, via the weakening of the C=O bond and the slight bending of O=C=O angle. For instance, the CO_2 bond lengths and angle are reduced in CO_2 at the amine-functionalized ILs@Au(111). The calculated in-plane angle and CO bond lengths of CO_2 are 1.178 Å and 177.2° at the amine-incorporated DCA^- ILs at Au(111), which differ when CO_2 is adsorbed with IL = $[EMIm]^+[DCA]^-$ are 1.176 Å and 178.1°, respectively. As a result, CO_2 is activated. These findings are consistent with the predicted CO_2 adsorption on the other Au surfaces [49]. For instance, Zhu et al. [49] showed that the bent form of CO_2 on the gold surface reduces the energetic barrier of the activation of gas molecules.

To get more insights on the electron donor and acceptor character of substituents (such as $—CH_3$, $—CF_3$, and $—NH_2$) in the activation of CO_2, we computed the energetics i.e. the adsorption energies (E_{ads}) (Table 9.2). For CO_2@$[EMIm]^+[DCA]^-$@Au(111), we have E_{ads} of $-15.46\,kJ\,mol^{-1}$. For CO_2@$[EMIm\text{-}NH_2]^+[DCA]^-$@Au(111), E_{ads} is equal to $-12.26\,kJ\,mol^{-1}$. Such relatively large values of E_{ads} correlate with the significant perturbation of CO_2 while attached to these ILs@Au(111). The C atom of CO_2 exhibits weak van der Waals (vdWs) interactions with the nitrogen of $[DCA]^-$. Meanwhile, dipolar interactions involving the O atom of CO_2 with the fluorine of the perfluoro moiety are induced. The estimated O=C=O····F distance is 3.264 Å, and we determined an E_{ads} of $-4.60\,kJ\,mol^{-1}$ for CO_2 adsorbed to the $[EMIm\text{-}CF_3]^+[DCA]^-$@Au(111) interface, which is relatively weak. For explanation, we propose that the repulsive interplay between the hydrophobic perfluoro moiety and CO_2 is in action, as well as the nonexistence of any donating moiety within the cationic group, which may lower such E_{ads}. This is further confirmed by the close-to-linear CO_2 angle (=178.9°) and C=O bond lengths (=1.175 Å) of CO_2, which are close to those of free CO_2. These are signatures of negligible effects on this molecule when approaching such interfacial system.

The present findings allow us to state that the CO_2 reduction initial phase at these IL@ Au(111) interfaces is the gain of electrical charge inducing geometrical modifications in CO_2, resulting thus in its activation, followed eventually by its conversion. Indeed, when an electron is transferred from ILs@Au(111) into CO_2,

Table 9.2 Calculated E_{ads} ($kJ\,mol^{-1}$) of CO_2@$[EMIm\text{-}Z]^+[X]^-$@Au(111) solid–liquid interface.

	E_{ads} ($kJ\,mol^{-1}$)	
$[EMIm\text{-}Z]^+$	$[X]^-$ = $[DCA]^-$	$[X]^-$ = $[SCN]^-$
$[EMIm]^+$	−15.46	−15.40
$[EMIm\text{-}NH_2]^+$	−12.26	−14.94
$[EMIm\text{-}CF_3]^+$	−4.60	−6.75

Source: Kamalakannan et al. [46]/with permission of IOP Publishing.

the latter is activated by the interplay between anion–metal bonding. From these observations, we can state that CO_2 can be significantly activated/tuned by electron effects mainly from the donor part of cationic moiety. Our results are in agreement with previous DFT observations on the non-covalent binding of CO_2 on amino acid-functionalized TiO_2 surfaces [30].

9.4.2 CO_2 Interacting with $[EMIm\text{-}Z]^+[SCN]^-$@Au(111) Interface

CO_2 adsorption on the $[EMIm]^+[SCN]^-$ ILs decorated Au(111) is illustrated in Figure 9.3. This figure shows that the CO_2 moiety adsorbs on both the nitrogen atom of the anionic moiety and on the active sites of the cationic ($[EMIm]^+$) functional groups (Z = NH_2 and CF_3). The computed distances between CO_2 and nitrogen-lone pairs of the anionic parts of $[EMIm\text{-}CF_3]^+[SCN]^-$, $[EMIm]^+[SCN]^-$, and $[EMIm\text{-}NH_2]^+[SCN]^-$ confirm that the interactions between both entities are strengthened by changing the electron-donating groups rather than the electron-withdrawing substituents.

For $[X]^-=[SCN]^-$, the computed distances between N of SCN^- and C of CO_2 are 2.912 Å, 2.891 Å, and 2.809 Å for $[EMIm]^+$, $[EMIm\text{-}CF_3]^+$, and $[EMIm\text{-}NH_2]^+$ cations containing ILs, respectively. Furthermore, one of the CO_2 moiety oxygens is involved in considerable non-covalent interactions at the functional sites of the $[EMIm]^+$ cations. The identified interactions are O=C=O····F (for $[EMIm\text{-}CF_3]^+$), O=C=O····H (for $[EMIm]^+$), and O=C=O····H (for $[EMIm\text{-}NH_2]^+$). The corresponding distances are calculated as 3.679, 2.740, and 2.544, in Å. The lowest interaction values show a significant electrostatic impact, and that the donating substituents have a strong influence on the behavior of CO_2 close to these solid–liquid interfaces.

Further, we determined the E_{ads} of CO_2 at the interface (Table 9.2). While the computed E_{ads} for —CF_3 group containing IL is −6.75 kJ mol^{-1} (Table 9.2), those for electron-donating species are −15.40 kJ mol^{-1} and −14.94 kJ mol^{-1} for $[EMIm]^+[SCN]^-$ and $[EMIm\text{-}NH_2]^+[SCN]^-$, respectively. Besides, the predicted CO_2 binding at the interfacial layer is in line with the optimized geometries, and E_{ads} are reduced to half when compared to the electron donor group.

Besides, we examined the lengths of the bonds and in-plane angle of CO_2. We have a slight change in C=O bond lengths (1.175 Å) and O=C=O angle (178.6°) compared to isolated species. These small deviations confirm that the —CF_3 group in the cation reduces the CO_2 reactivity close to the interface. In addition, C=O bond lengths and CO_2 angle are more affected in $[EMIm\text{-}NH_2]^+[SCN]^-$ (of 1.178 Å and 176.9°) and in $[EMIm]^+[SCN]^-$ (of 1.176 Å and 177.8°). These findings evidence that the activation of CO_2 can be effectively tuned by the donating moieties at the cationic ring of the ILs/Au(111) interface. Furthermore, the changes in C=O bond lengths and the O=C=O angle lead to CO_2 activation. As said above, these effects are responsible for the activation/reduction of CO_2 at these interfaces.

It is interesting to note that the activation of CO_2 is not only due to the non-covalent interactions at the interface. For example, the injection of an electronic charge from the interface material to CO_2 through the S—Au bond also contributes to the overall catalytic process. The sulfur atom of $[SCN]^-$ is in fact closely connected to Au(111), where the S—Au distances are ~2.5 to 2.8 Å. The resulting S—Au interaction enhances the density of electrons on the Au(111) support, thereby inducing the donor activity of the ILs and the subsequent activation of CO_2. According to our calculations, the functionally substituted ILs can significantly influence the reactivity of CO_2 compared to the unmodified ILs at the interfacial region. This has arisen from the CO_2 electronic redistribution of charges at the heterogeneous surface and electronic charge transfers at the interface to get more insights into the adsorption mechanism, charge transfer process, and charge density analyses are useful.

9.5 Charge Transfer and Charge Density Analyses at the Interface

9.5.1 Charge Redistribution Between CO_2 and Interfacial Medium

We use electron density difference analysis to examine the charge density and overlapping of CO_2 with the electrode–electrolyte interfacial medium. Figure 9.4 depicts the redistribution of electron density contours between CO_2 and the interface. This picture documents the electron density accumulation/depletion of CO_2 at the interface.

The smallest E_{ads} is for CO_2@[EMIm-CF_3]$^+$. It is primarily due to the dual <0 charge distribution surrounding fluorine atoms of this cation and the O of CO_2 molecule. Thus, the CO_2 adsorption strength at the solid–liquid interface is lowered, in good agreement with the predicted geometrical distances and adsorption energies. Whereas, opposite charges are observed on alkyls and the functional groups of the cation chain when Z = —CH_3 or —NH_2. Meanwhile, negative charges are enriched on nitrogen of both anions, which favor the interaction with the electropositive C atom of CO_2.

Additionally, the [EMIm]$^+$ cation binding mode on the gold surface enhances the fixation of CO_2. It leads to a remarkable CO_2 activation by increasing the C=O bond lengths and reducing the CO_2 in-plane angle. In addition, total density analyses of CO_2 complexed to IL@Au(111), as shown in Figure 9.4, indicate the weak physical binding of CO_2 at interfaces as well as the chemical attraction between the anion and the Au(111) surface via establishing of the Au—S bond. Furthermore, we found that the electron distribution across CO_2 and the interface strongly stimulates the CO_2 molecule, in contrast to previously reported results [28].

9.5.2 Interfacial Charge Transfer Analysis of CO_2@Interface

We performed Bader charge calculations for all CO_2@Interface complex structures to quantify the interface-CO_2 charge transfers. Table 9.3 presents the obtained total and partial charges of CO_2 at the interfacial systems. As expected, the binding of CO_2 at ILs@Au(111) causes the carbon atom of the CO_2 to acquire a larger positive charge. This allows CO_2 to bind more strongly to the solid–liquid interface via the

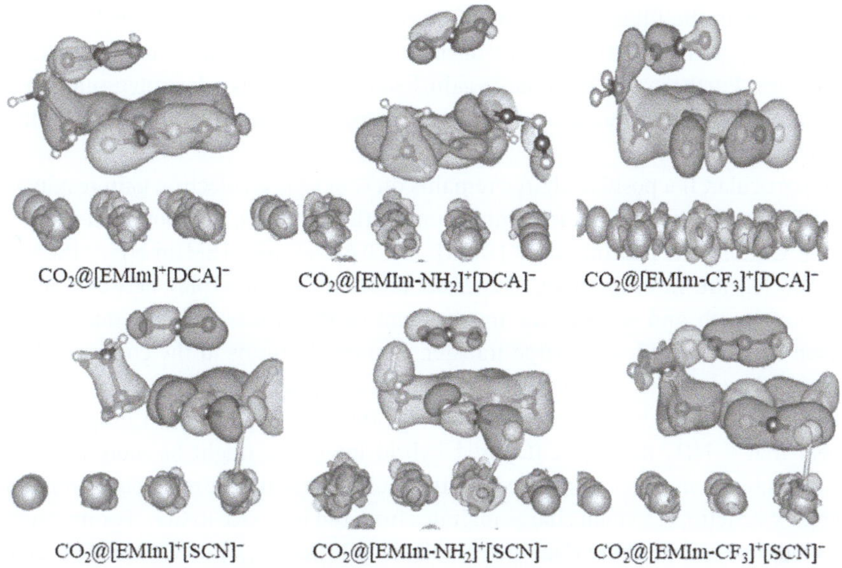

Figure 9.4 Charge density difference contours of CO_2@Interface. Isosurfaces are mapped for all complexes at 0.001 e $Å^{-3}$. Yellow (cyano) corresponds to accumulation (depletion). Source: Kamalakannan et al. [46]/with permission of IOP Publishing.

Table 9.3 CO_2 total and partial charges (in |e|) while adsorbed ILs@Au(111).

Ils	**CO_2**	**C**	**O_1**	**O_2**
$[EMIm]^+[SCN]^-$	−0.055	0.584	−0.324	−0.315
$[EMIm\text{-}NH_2]^+[SCN]^-$	−0.051	0.663	−0.356	−0.358
$[EMIm\text{-}CF_3]^+[SCN]^-$	−0.038	0.570	−0.319	−0.288
$[EMIm]^+[DCA]^-$	−0.043	0.619	−0.333	−0.329
$[EMIm\text{-}NH_2]^+[DCA]^-$	−0.061	0.648	−0.323	−0.385
$[EMIm\text{-}CF_3]^+[DCA]^-$	−0.031	0.618	−0.333	−0.315

Source: Kamalakannan et al. [46]/with permission of IOP Publishing.

establishment of bonds involving the nitrogen-lone pair. This reduces the amount of CO_2 and the amount of energy required to reduce CO_2. The nature of the specific functional moieties of Im^+ cations influence the quantity of positive charges on C atom of CO_2. It is observed that more positive charges are acquired by this atom by donating moieties at the ILs@Au(111) interface i.e. for $[EMIm\text{-}NH_2]^+[DCA]^-$ and $[EMIm\text{-}NH_2]^+[SCN]^-$. Also, more negative charges are acquired by O of CO_2 in the presence of electron donor molecules near the interface. These findings agree with the computed respective binding energies. We have shown that when CO_2 is absorbed at the interface, a substantial quantity of negative electronic charge transfer occurs through the interface due to the anti-bonding orbital of CO_2.

This results in the filling of this anti-bonding orbital, which reduces the bounding within CO_2 and thus the amount of activation energy required to reduce CO_2 molecules. Furthermore, the charge gain/loss on CO_2 at the electrolyte–electrode interface can be controlled by varying the functional moieties in the $[EMIm]^+$ cationic ring.

In particular, if a positive charge remains on C of CO_2 an electron gain results in the effective CO_2 reduction. In comparison to other systems, the computed charges of the C atoms in $[EMIm\text{-}NH_2]^+[DCA]^-$ (of 0.64 |e|) and $[EMIm\text{-}NH_2]^+[SCN]^-$ (of 0.66 |e|) are more positive. The effect is due to the interplay between anion and metal with and without the involvement of the N-lone pairs. It may be the reason for this significant charge transfer. Besides, O atoms in the corresponding interfacial systems become more negative, with charges of −0.38 |e| and of −0.35 |e| for $[EMIm\text{-}NH_2]^+[DCA]^-$ and $[EMIm\text{-}NH_2]^+[SCN]^-$, respectively. Computations reveal that —NH_2 decorated IL-based hybrid interfaces might be more efficient for the activation of CO_2 compared to other substituted ILs. Furthermore, in other complex structures, a small charge migrates from the interface to CO_2. For instance, the electrical charge of the C atom is 0.56 |e| in $[EMIm\text{-}CF_3]^+[SCN]^-$ and 0.58 |e| in $[EMIm]^+[SCN]^-$. Accordingly, there is more electron migration from the interfacial region to CO_2. In summary, it appears that NH_2-moiety incorporated ILs/Au(111) would be an ideal candidate for CO_2 reduction reactions.

9.5.3 Electronic Structure Analysis of CO_2 at the IL@Au(111) Interface

Insights on CO_2 electronic structure at solid–liquid interfacial materials can be obtained after examination of the density of states (DOS). Figure 9.5a–d shows the zero-band gap of all CO_2@Interface complex systems under investigation. This zero gap is due to the existence of Au within these systems and it confirms the metallic character of such catalyst materials. Moreover, after the CO_2 molecule is adsorbed, bands in the valance band maximum (VBM) region are found. Transitions to such higher energy levels cause the VBM to be partially filled. Therefore, non-bonded connections at the interface should stimulate the CO_2 molecule, as stated above. The observation of a less intense peak in close proximity to the Fermi zone validates the presence of vdWs and charge transfer interactions between the solid–liquid interface and the incoming CO_2. Additionally, these peaks attest that the interfacial material improves the catalytic activity of the CO_2 reduction. Indeed, once CO_2 adsorbed, bands in the VBM region change to higher energy levels, causing the VBM to be partially filled. Again, weak interactions activate the CO_2 molecule at the interface.

Additionally, the PDOS shown in Figure 9.6a–d reveals that atomic orbitals such as C(2p), Au(4d), N(2p), and O(2p) make the conduction and valence bands of both groups. The $CO_2@[EMIm\text{-}NH_2]^+[SCN]^-$ and $CO_2@[EMIm]^+[SCN]^-$ valence and conduction bands are made of C(2p), Au(4d), N(2p), O(2p), and S(2p) orbitals in that order. The primary cause of the metallic character introduced for each system is due to the presence of the Au(4d) orbital. Further, CO_2 interacts with the interfacial system when it is immobilized on the interface, as evidenced by the considerable

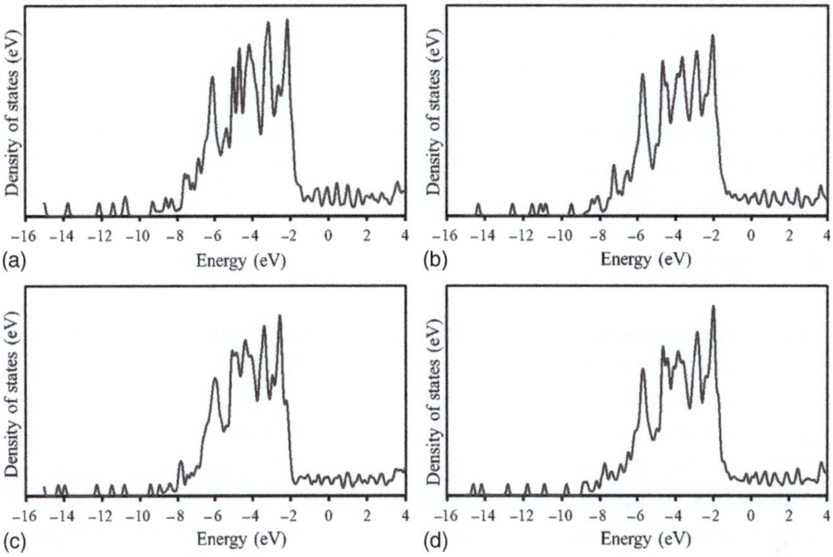

Figure 9.5 Computed DOS for (a): CO_2@[EMIm]$^+$[DCA]$^-$; (b): CO_2@[EMImNH$_2$]$^+$[DCA]$^-$; (c): CO_2@[EMIm]$^+$[SCN]$^-$; and (d) CO_2@[EMIm-NH$_2$]$^+$[SCN]$^-$ under investigation. Source: Adapted from Ref. [45].

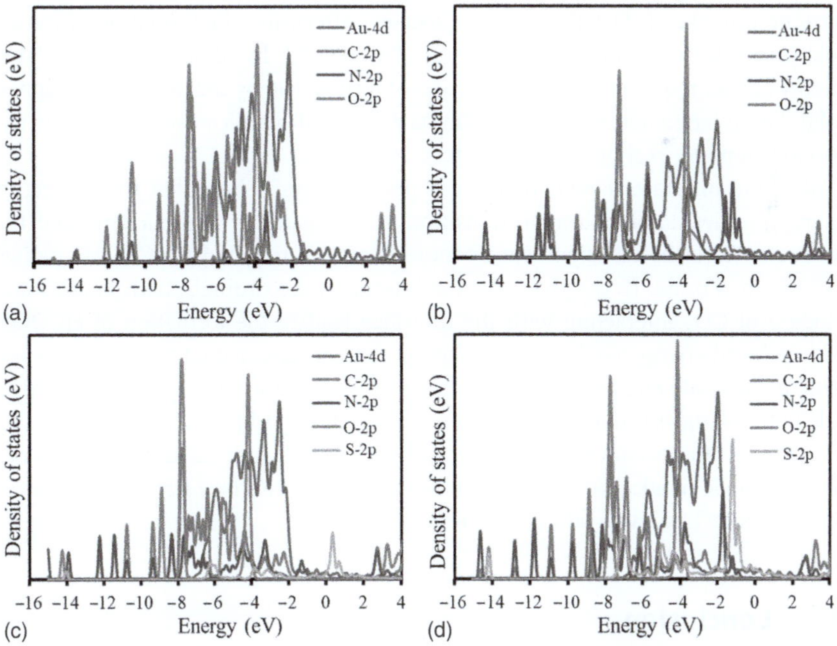

Figure 9.6 Computed PDOS of CO_2 at interfaces: (a): CO_2@[EMIm]$^+$[DCA]$^-$; (b): CO_2@[EMImNH$_2$]$^+$[DCA]$^-$; (c): CO_2@[EMIm]$^+$[SCN]$^-$; and (d) CO_2@[EMIm-NH$_2$]$^+$[SCN]$^-$. Source: Kamalakannan et al. [46]/with permission of IOP Publishing.

shift in the DOS and the movement of CO_2 energy range to a lower level, between −16.0 and −8 eV. The interplay of C(2p) and N(2p) orbitals throughout each of the valence states is the most intriguing result of PDOS investigations. This suggests that a strong interaction takes place between C of CO_2 and N of NH_2 of the ILs, which are electrophilic and nucleophilic, respectively. Our optimized complex system energies and geometries have a strong correlation with the computed PDOS plots.

9.6 Application: Conversion of CO_2 into HCOOH

In some of our studies, we investigated the CO_2 conversion into value-added chemicals (e.g. HCOOH) at the IL@Au(111) interface models [45]. For the conversion process, we therefore used screened $[C_6MIm]^+[Cl]^-$ IL and $[C_6MIm]^+[Cl]^-$@Au(111) interfaces. For illustration, Figure 9.7 shows the case of the $H_2 + CO_2 \rightarrow$ HCOOH reaction at the $[C_6MIm]^+[Cl]^-$@Au(111) interface. The interaction between H_2 and CO_2 molecules at the Im^+ ring initiates the reaction. The $CO_2 + H_2 \rightarrow$ HCOOH reaction occurs then via the addition of an H_2 molecule into a CO bond. The associated chemical pathway consists of various steps, where CO_2 and H_2 are complexed to the Im^+ cation. The activated O atoms of CO_2 and H, which are in an ionized state, interact at the intermediate stage, producing the COOH* radical, which has the highest energy in the reaction profiles (Figure 9.7).

After the formation of —OH, a second H atom interacts with the C atom of CO_2, forming thus the HCOOH* molecular species. Afterward, the formation of formic acid proceeds through the isomerization of HO–COH. In addition, the calculated energy barrier required for the production of formic acid by the use of the designed solid–liquid-based heterogeneous catalysts is lowered significantly compared to the previous reports, as shown in Table 9.4.

Also, the interface models or the gas phase analysis show that the respective energy profiles admit a unique transition state separating the reactants and the products. Nevertheless, significantly more energy is required for COOH* radical to be formed in the IL gas phase than at interfaces. Accordingly, both Mode 1 and Mode 2 of CO_2 interacting with the interface lead to the formation of HCOOH with reduced energy barriers. This suggests that our engineered interfacial system with an adequate energy barrier can stimulate or convert the CO_2 molecules. Also, calculations suggest that the adsorption of CO_2 through Mode 2 is more favorable than Mode 1 for its conversion into formic acid. In sum, our calculations point out that the mode of adsorption of CO_2 molecules at the interface strongly influences its conversion into value-added products.

9.7 Conclusion

Our findings demonstrate that IL-impregnated Au(111) surface might efficiently stimulate/activate CO_2. The existence of non-covalent interactions directly affects the CO_2 binding and activation processes at the interface. Indeed, this activation is

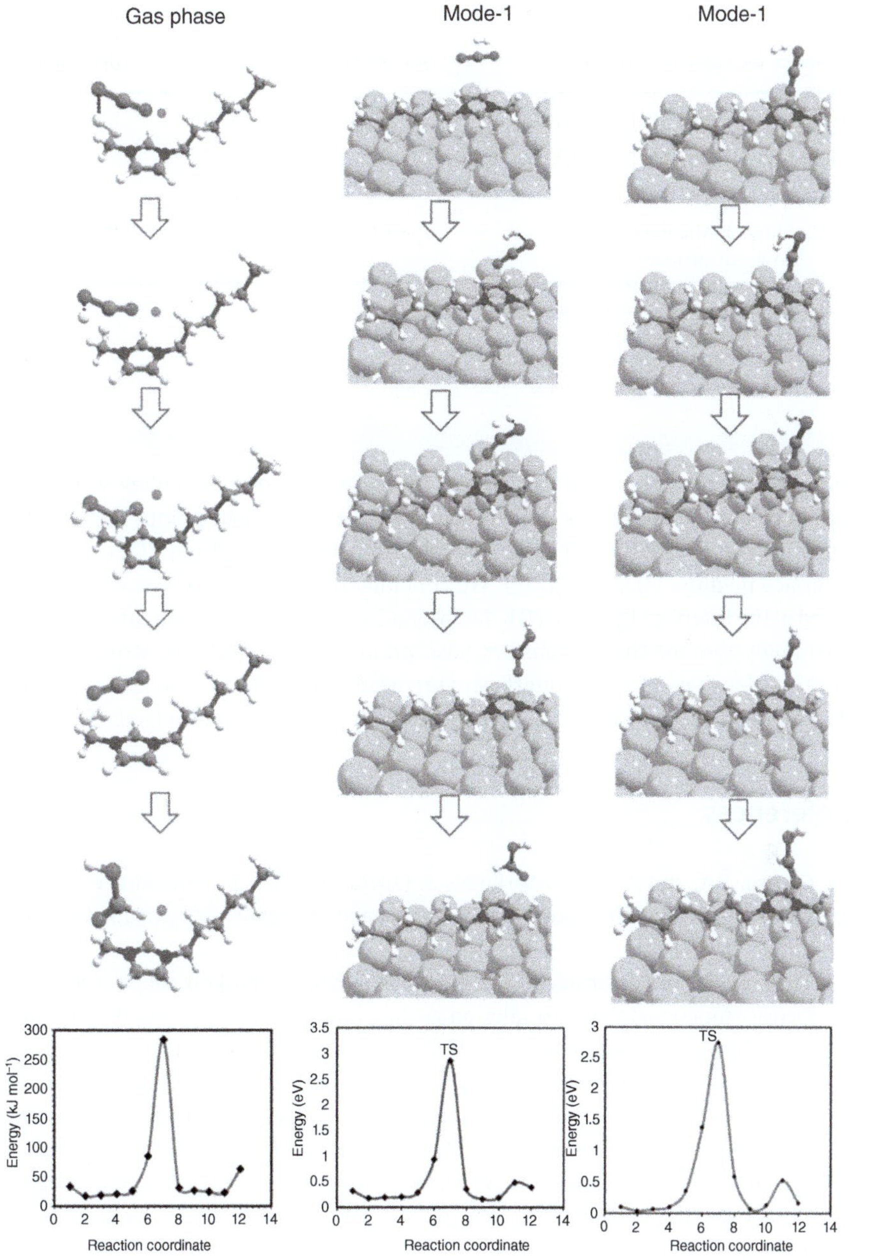

Figure 9.7 Hydrogenation of CO_2 to form formic acid at gas phase and solid–liquid interfaces in two different modes ([i] Model-1 and [ii] Mode-2). Source: Kamalakannan et al. [45]/with permission of Elsevier.

Table 9.4 Comparison of the energetics for the $H_2 + CO_2 + e^- \rightarrow HCOOH^*$ reaction at the present solid–liquid interface and earlier studies.

Name of the catalytic systems	Type of reaction process	Energy barrier (in kJ mol^{-1})
Mode-1: $[C_6MIm]^+[Cl]$@Au(111)	$H_2 + CO_2 + e^- \rightarrow HCOOH^*$	21.8[a)]
Mode-2: $[C_6MIm]^+[Cl]^-$@Au(111)	$H_2 + CO_2 + e^- \rightarrow HCOOH^*$	9.0[a)]
Bare-Au(111) surface	$H_2 + CO_2 + e^- \rightarrow HCOOH^*$	58.85[b)]
4-Pyridinylethane mercaptan (4-PEM)@Au surface	$H_2 + CO_2 + e^- \rightarrow HCOOH^*$	40.52[b)]

The energy barrier is given in kJ mol^{-1}.
a) This work. Reference [45].
b) Previous report. Reference [50].
Source: Reported values are taken from Refs. [45, 50].

enhanced by the donor character of the —NH_2 group in ILs. Nonetheless, the activation of CO_2 is reduced by the electron-acceptor substituent groups (e.g. —CF_3). Furthermore, we found that there is an electron charge transfer to CO_2 close to the interface medium (ILs@Au(111)). We conclude that CO_2 can be successfully activated at the interface by the —NH_2 functionalized ILs. These findings are confirmed after analysis of the charge transfers, adsorption energies, electronic structures, and electron densities of CO_2@Interface. Our results will offer a new avenue for the development of functional materials that effectively capture and activate CO_2.

References

1 Owusu, P.A. and Asumadu-Sarkodie, S. (2016). A review of renewable energy sources, sustainability issues and climate change mitigation. *Cogent Engineering* 3 (1): 1167990.

2 Thomas, A., Maiyelvaganan, K.R., Kamalakannan, S., and Prakash, M. (2019). Density functional theory studies on zeolitic imidazolate framework-8 and ionic liquid-based composite materials. *ACS Omega* 4 (27): 22655–22666.

3 Thomas, A., Kamalakannan, S., Cheriyan, A., and Prakash, M. (2023). Theoretical studies on the role of water in ionic liquids at ZIF (IL@ZIF) complex and its effect on selective CO_2 separation. *Materials Today Sustainability* 22: 100376.

4 Nocito, F. and Dibenedetto, A. (2020). Atmospheric CO_2 mitigation technologies: carbon capture utilization and storage. *Current Opinion in Green and Sustainable Chemistry* 21: 34–43.

5 Van Wagener, D.H. and Rochelle, G.T. (2011). Stripper configurations for CO_2 capture by aqueous monoethanolamine and piperazine. *Energy Procedia* 4: 1323–1330.

6 Li, K., Yu, H., Feron, P. et al. (2015). Technical and energy performance of an advanced, aqueous ammonia-based CO_2 capture technology for a 500

MW coal-fired power station. *Environmental Science & Technology* 49 (16): 10243–10252.

7 Li, K., Cousins, A., Yu, H. et al. (2016). Systematic study of aqueous monoethanolamine-based CO_2 capture process: model development and process improvement. *Energy Science & Engineering* 4 (1): 23–39.

8 Liu, H., Chen, Y., Lee, J. et al. (2022). Ammonia-mediated CO_2 capture and direct electroreduction to formate. *ACS Energy Letters* 7 (12): 4483–4489.

9 Luis, P. (2016). Use of monoethanolamine (MEA) for CO_2 capture in a global scenario: Consequences and alternatives. *Desalination* 380: 93–99.

10 Lei, Z., Chen, B., Koo, Y.M., and Macfarlane, D.R. (2017). Introduction: ionic liquids. *Chemical Reviews* 117 (10): 6633–6635.

11 Boeck, G. (2019). Paul Walden (1863–1957): the man behind the Walden inversion, the Walden rule, the Ostwald-Walden-Bredig rule and Ionic liquids. *ChemTexts* 5 (1): 1–7.

12 Kaur, G., Kumar, H., and Singla, M. (2022). Diverse applications of ionic liquids: a comprehensive review. *Journal of Molecular Liquids* 351: 118556.

13 Fortunati, A., Risplendi, F., Re Fiorentin, M. et al. (2023). Understanding the role of imidazolium-based ionic liquids in the electrochemical CO_2 reduction reaction. *Communications Chemistry* 6 (1): 1–13.

14 Ramdin, M., De Loos, T.W., and Vlugt, T.J.H. (2012). State-of-the-art of CO_2 capture with ionic liquids. *Industrial and Engineering Chemistry Research* 51 (24): 8149–8177.

15 Aghaie, M., Rezaei, N., and Zendehboudi, S. (2018). A systematic review on CO_2 capture with ionic liquids: current status and future prospects. *Renewable and Sustainable Energy Reviews* 96: 502–525.

16 Zhao, H. and Baker, G.A. (2023). Functionalized ionic liquids for CO_2 capture under ambient pressure. *Green Chemistry Letters and Reviews* 16 (1): 2149280.

17 Gao, H., Bai, L., Han, J. et al. (2018). Functionalized ionic liquid membranes for CO_2 separation. *Chemical Communications* 54 (90): 12671–12685.

18 Zhu, X., Chen, Z., and Ai, H. (2020). Amine-functionalized ionic liquids for CO_2 capture. *Journal of Molecular Modeling* 26 (345): 1–12.

19 Zhang, Y., Wu, Z., Chen, S. et al. (2013). CO_2 capture by imidazolate-based ionic liquids: effect of functionalized cation and dication. *Industrial and Engineering Chemistry Research* 52 (18): 6069–6075.

20 Liu, X., Sun, L., and Deng, W.Q. (2018). Theoretical investigation of CO_2 adsorption and dissociation on low index surfaces of transition metals. *Journal of Physical Chemistry C* 122 (15): 8306–8314.

21 Zhang, X., Guo, S.X., Gandionco, K.A. et al. (2020). Electrocatalytic carbon dioxide reduction: from fundamental principles to catalyst design. *Materials Today Advances* 7: 100074.

22 Ma, W., He, X., Wang, W. et al. (2021). Electrocatalytic reduction of $CO_2$and CO to multi-carbon compounds over Cu-based catalysts. *Chemical Society Reviews* 50 (23): 12897–12914.

23 Zhu, Q., Murphy, C.J., and Baker, L.R. (2022). Opportunities for electrocatalytic CO_2 reduction enabled by surface ligands. *Journal of the American Chemical Society* 144 (7): 2829–2840.

24 Luo, X.Y., Lv, X.Y., Shi, G.L. et al. (2019). Designing amino-based ionic liquids for improved carbon capture: one amine binds two CO_2. *AICHE Journal* 65 (1): 230–238.

25 Cho, M., Park, J., Yavuz, C.T., and Jung, Y. (2018). A catalytic role of surface silanol groups in CO_2 capture on the amine-anchored silica support. *Physical Chemistry Chemical Physics* 20 (17): 12149–12156.

26 Pan, M. and Wang, C. (2020). Ionic liquids for chemisorption of CO_2. *Materials Carbon Capture* 297–315.

27 Medina-Ramos, J., Zhang, W., Yoon, K. et al. (2018). Cathodic corrosion at the bismuth-ionic liquid electrolyte interface under conditions for CO_2 reduction. *Chemistry of Materials* 30 (7): 2362–2373.

28 Chernyshova, I.V. and Ponnurangam, S. (2019). Activation of CO_2 at the electrode-electrolyte interface by a co-adsorbed cation and an electric field. *Physical Chemistry Chemical Physics* 21 (17): 8797–8807.

29 Niemann, T., Li, H., Warr, G.G. et al. (2019). Influence of hydrogen bonding between ions of like charge on the ionic liquid interfacial structure at a mica surface. *Journal of Physical Chemistry Letters* 10 (23): 7368–7373.

30 Parameswari, A., Soujanya, Y., and Sastry, G.N. (2019). Functionalized rutile TiO_2 (110) as a sorbent to capture CO_2 through noncovalent interactions: a computational investigation. *Journal of Physical Chemistry C* 123 (6): 3491–3504.

31 M. J. Frisch, Trucks, G.W., Schlegel, H.B. et al. (2016) Gaussian 16, Wallingford CT.

32 Boys, S.F. and Bernardi, F. (2002). The calculation of small molecular interactions by the differences of separate total energies. Some procedures with reduced errors. *Molecular Physics* 19: 553–556.

33 Marsman, M. and Kresse, G. (2006). Relaxed core projector-augmented-wave method. *The Journal of Chemical Physics* 125 (10): 104101.

34 Kresse, G. and Furthmüller, J. (1996). Efficient iterative schemes for ab initio total-energy calculations using a plane-wave basis set. *Physical Review B: Condensed Matter and Materials Physics* 54 (16): 11169–11186.

35 Perdew, J.P., Burke, K., and Ernzerhof, M. (1996). Generalized gradient approximation made simple. *Physical Review Letters* 77: 3865–3868.

36 Perdew, J.P., Chevary, J.A., Vosko, S.H. et al. (1993). Atoms, molecules, solids, and surfaces: applications of the generalized gradient approximation for exchange and correlation. *Physical Review B* 48 (7): 4978.

37 Blöchl, P.E. (1994). Projector augmented-wave method. *Physical Review B* 50 (24): 17953–17979.

38 Joubert, D. (1999). From ultrasoft pseudopotentials to the projector augmented-wave method. *Physical Review B: Condensed Matter and Materials Physics* 59 (3): 1758–1775.

39 Pack, J.D. and Monkhorst, H.J. (1977). Special points for Brillouin-zone integrations—a reply. *Physical Review B* 16 (4): 1748–1749.

40 Henkelman, G., Arnaldsson, A., and Jónsson, H. (2006). A fast and robust algorithm for Bader decomposition of charge density. *Computational Materials Science* 36 (3): 354–360.

41 Tang, W., Sanville, E., and Henkelman, G. (2009). A grid-based Bader analysis algorithm without lattice bias. *Journal of Physics. Condensed Matter* 21 (8): 84204.

42 Henkelman, G., Uberuaga, B.P., and Jónsson, H.A. (2000). Climbing image nudged elastic band method for finding saddle points and minimum energy paths. *The Journal of Chemical Physics* 113: 9901–9904.

43 Henkelman, G. and Jónsson, H. (2000). Improved tangent estimate in the nudged elastic band method for finding minimum energy paths and saddle points. *The Journal of Chemical Physics* 113 (22): 9978–9985.

44 Yang, Z.Z., Zhao, Y.N., and He, L.N. (2011). CO_2 chemistry: task-specific ionic liquids for CO_2 capture/activation and subsequent conversion. *RSC Advances* 1 (4): 545–567.

45 Kamalakannan, S., Rudharachari Maiyelvaganan, K., Palanisamy, K. et al. (2022). Carbon dioxide adsorption and activation on ionic liquid decorated Au(111) surface: a DFT study. *Chemosphere* 286 (P1): 131612.

46 Kamalakannan, S., Maiyelvaganan, K.R., and Prakash, M. (2022). Role of functional groups in an ionic liquid decorated Au(111) surface for CO_2 capture and activation: a first principle approach. *Journal of the Electrochemical Society* 169 (5): 56524.

47 Yuksel Orhan, O. (2021). Effects of various anions and cations in ionic liquids on CO_2 capture. *Journal of Molecular Liquids* 333: 115981.

48 Shen, Z.-Z., Lang, S.-Y., Liu, R.-Z. et al. (2023). Revealing the CO_2 conversion at electrode/electrolyte interfaces in Li–CO_2 batteries via nanoscale visualization methods. *Angewandte Chemie, International Edition* 1–10.

49 Zhu, S., Guo, L., Li, P. et al. (2019). A computational study on linear and bent adsorption of CO_2 on different surfaces for its photoreduction. *Catalysis Today* 335: 278–285.

50 Li, F. and Tang, Q. (2019). Understanding the role of functional groups of thiolate ligands in electrochemical CO_2 reduction over Au(111) from first-principles. *Journal of Materials Chemistry A* 7 (34): 19872–19880.

10

Mixed Metal Oxides as Solid Base Catalysts: Fundamentals and Their Catalytic Performance

Naveen Kumar[1], *Sauraj*[2], *and Naveen Chandra Joshi*[3]

[1] *Sanatan Dharm College Muzaffarnagar, Department of Chemistry, Muzaffarnagar, Uttar Pradesh 251001, India*
[2] *Gachon University, College of Pharmacy, Incheon, South Korea*
[3] *Uttaranchal University, Division of Research & Innovation, Dehradun 248007, India*

10.1 Introduction

Catalysis plays an important role in chemical, physical, and biological sciences. A catalyst can be defined as a compound that accelerates or slows down the rate of a reaction and remains unchanged at the end of the reaction. Catalyst changes the activation energy of a chemical reaction without affecting the thermodynamic equilibrium. Catalysts are important in the refining of petroleum, the production of synthetic fibers and plastics, the processing of food, and the production of drugs and pharmaceutical compounds. They contribute to reducing air pollution through the development of environmentally friendly catalyst technology and new energy production processes. It is considered a pillar of green chemistry that protects our environment. Catalysts used in various processes are classified into two types: solid acid catalysts and solid base catalysts. Acid–base interactions between reactants and catalysts cause chemical reactions. Catalysts act simultaneously as an acid and a base on the reactants in acid- and base-catalyzed reactions. Many materials are amenable to solid-based catalysts if properly pretreated, and many important reactions are catalyzed by the use of these solid-based catalysts [1, 2]. Heterogeneous solid-based catalysts have attracted great attention in recent years for various industrial applications [3, 4] due to the ease of catalyst separation, reusability, more specific surface area, high concentration of basic sites, and thermal stability. Traditional liquid-based catalysts are generally associated with corrosion problems, high toxicity, catalyst waste, difficulty in product separation, and the production of other undesirable side products. Solid-based catalysts can overcome all these problems by offering several advantages that are listed here [5].

(a) Reactions catalyzed by liquid bases require neutralization of the base, whereas reactions catalyzed by solid bases do not require a neutralization step. Neutralization generates waste, which can be minimized by using solid base catalysts.

Solid Base Catalysts: Synthesis, Characterization, and Applications, First Edition.
Edited by Ravi Tomar, K.K. Pant, and Ramesh Chandra.

(b) Catalyst can be easily separated from the product and used catalyst can be recycled.
(c) Possibility to carry out high temperature reactions as opposed to catalytic reactions with liquid bases. Typically, solvents are used that can evaporate unless the system is operated below its boiling point. The use of flow reactor systems increases the productivity of the process.
(d) High productivity and efficiency when used in flow reactor systems and does not require the use of base-resistant materials in the reactor system.
(e) Possession of bifunctional catalytic properties that allow the coexistence of acidic and basic sites on the surface of solid base catalysts, which is not possible in the case of liquid base catalysts.
(f) There are some reactions that are unique to solid base catalysts, such as heterolytic cleavage of H_2 to H^+ and H^- was not observed in liquid bases, but was observed in solid bases and in the first step of base-catalyzed hydrogenation of conjugated dienes.

Considering the advantages of heterogeneous catalysts, the development of heterogeneous catalysts with high selectivity is always an interesting field. In recent years, various types of materials/nanomaterials have been used as solid-based heterogeneous catalysts, such as metal oxides, clays, zeolites, aluminosilicates, supported alkali oxides, or alkaline earth oxides, making the catalytic system benign [6–8]. Among all these, the main focus of this chapter is the fundamentals and applications of mixed metal oxides (MMOs) as solid base catalysts. MMO catalysts generally refer to those consisting of two or more metal oxides. The original idea for producing MMO catalysts was to increase the base or acid strength, increase the surface area, and increase the stability of these catalysts compared to single metal oxides.

10.2 Why Mixed Metal Oxides (MMOs)?

MMOs play a very important role in academic and industrial research due to their large surface area, acid–base and redox properties, and they represent the largest family of catalysts in heterogeneous catalysis. MMOs have numerous industrial applications in areas such as ceramics, electronics, nuclear research, and catalysis. They have been found to be applicable to a variety of reactions such as hydrocarbon oxidation, transfer hydrogenation, rearrangement reactions, isomerization, and hydrogenolysis [9]. This may be obvious considering that almost all metal oxide catalysts used in practice are mixed oxides. Here we introduce some of the important applications. Lucarelli et al. prepared binary and ternary mixed oxides of Mo, Nb, Sb, and V for the selective oxidation of propane [10]. Mesoporous Co_3O_4 and Au/Co_3O_4 catalysts were used for the low-temperature oxidation of trace amounts of ethylene by Chun Yan et al. Here, a conversion of 76% was achieved at 0 °C by promoting surface active oxygen species and ethylene oxidation [11]. For water–gas shift (WGS) reaction, Mg(Al)O catalysts containing MMOs on Cu nanoparticles were reported by Li et al. Among the prepared catalysts, the 30% Cu/Mg_2Al catalyst showed the largest Cu surface area and highest WGS activity [12]. In particular, some MMOs are superior to their constituent oxides in terms of catalytic activity in various reactions. This may be due to the increased number of active acidic or basic sites and increased surface area, which shortens the reaction time and increases the reaction yield or conversion of reactants [13]. For example, Rechi Siqueira et al. obtained a copper MMO from a hydrotalcite precursor and used it to convert ethanol to 1-butanol with high selectivity [14]. In one example, Kulkarni et al. prepared $La_{1-x}Sr_xFeO_3$ perovskites for catalytic liquid-phase transfer hydrogenation of aromatic nitro compounds. These perovskites exhibit excellent catalytic activity for liquid-phase catalytic transfer hydrogenation. The activity of $La_{0.8}Sr_{0.2}FeO_3$ was particularly remarkable, resulting in quantitative conversion of nitrobenzene to aniline. The activation energy values of $LaFeO_3$, $La_{0.8}Sr_{0.2}FeO_3$, and $SrFeO_3$ are 88.9, 73.9, and 84.4 kJ mol^{-1}, respectively, following the order $LaFeO_3 > SrFeO_3 > La_{0.8}Sr_{0.2}FeO_3$ [15]. Transesterification is the most important reaction for biodiesel production, and many MMOs are believed to have higher catalytic activity than pure oxides to produce biodiesel from this reaction. For biodiesel production by transesterification of palm oil, for example, Wong et al. prepared a mixed oxide catalyst with CaO supported on CeO_2. A 95% yield was obtained with a reaction temperature of 65 °C, a molar ratio of 12 : 1, and a catalyst loading of 5% by weight over four hours. This catalyst showed superior durability compared to commercially available CaO when reused seven times. However, its activity decreased significantly after sixth run due to CaO leaching [16]. Therefore, MMO has recently attracted much attention as an active catalyst and support for various reactions due to its superiority compared to single oxide materials.

10.3 Mixed Metal Oxides (MMOs)

MMOs are based on at least two metal or metalloid elements that must be closely combined when the material is synthesized. Catalysis uses different types of MMOs such as zeolites and zeotypes, layered double hydroxides (LDHs) and layered silicates, spinels, and heteropolyanions. Oxides can be binary, ternary, and quaternary depending on the number of different metal cations present. They can be further classified according to whether they are crystalline or amorphous. If the oxide is crystalline, the crystal structure is essentially determined by the size (ionic radius) and valence (or charge) of the metal cation (M) and the composition of the oxide. The ionic radius determines the arrangement to achieve close packing, and the valence determines the stoichiometry (ratio of M to O numbers) to maintain electrical neutrality. For example, if the tolerance factor t is between 0.75 and 1.0, the perovskite structure ABO_3 is obtained. The tolerance factor t defined as a function of ionic radii of A, B, and O is given as follows (Eq. 10.1):

$$t = \frac{r_A + r_O}{\sqrt{2} \cdot (r_B + r_O)} \tag{10.1}$$

Table 10.1 Typical structures of mixed oxides.

Structure	Name	General formula
Perovskite type	Perovskite	$A^{2+}B^{4+}O_3$
		$A^{3+}B^{3+}O_3$
	Brownmillerite	$A_2{}^{2+}B_2{}^{3+}O_4$
Spinel type	Spinel	$A^{2+}B_2{}^{3+}O_4$
	Magnetoplumbite	$A^{+}B_{12}{}^{3+}O_{19}$
	β-Alumina	$A^{3+}B_{11}{}^{3+}O_{18}$
Fluorite type	Scheelite	$A^{2+}B^{6+}O_4$
		$A^{3+}B^{5+}O_4$
	Pyrochlore	$A_2{}^{3+}B_2{}^{4+}O_7$
Corundum type	Ilmenite	$A^{3+}B^{3+}O_3$
β-Tungsten type	Garnet	$A_3B_2B'_3O_{12}$

where r_A, r_B, and r_O are ionic radii of A, B, and O, respectively. Typical structures of mixed oxides are provided in Table 10.1 [17–19]. The main advantage of the perovskite structure is that it can replace A cations (producing the $A_{1-x}A'_xBO_{3\pm\delta}$ composition) or B cations (producing the $AB_{1-y}B'_yO_{3\pm\delta}$ composition). For example, in $LaCoO_3$ and $LaNiO_3$, which have perovskite structures, Co^{3+} and Ni^{3+} are stabilized. Moreover, according to Eq. (10.2), partial introduction of Sr^{2+} instead of La^{3+} in these compounds results in the formation of anomalous Co^{4+} or Ni^{4+}. This procedure is called "valence control." Here, the higher valence instability is compensated by the stability of the perovskite structure. Tetravalent Co^{4+} and Ni^{4+} were observed in $BaCoO_3$, $SrCoO_{3-\delta}$, $La_{1-x}Sr_xCoO_{3-\delta}$, and $La_{1-x}Sr_xNiO_{3-\delta}$. According to Eq. (10.3), oxygen vacancies are likely to occur, while Co^{4+} partially reverts to Co^{3+} [18–20]. However, oxygen-deficient perovskites are also important because their oxygen storage capacity and redox properties make them very useful for overall oxidation reactions. The purpose of substitution at both positions is the same, i.e. to promote catalytic activity.

$$(1-x)\,La^{3+}Co^{3+}O_3 + xSr^{2+}Co^{4+}O_3 \rightarrow La_{1-x}Sr_xCo^{3+}_{1-x}Co^{4+}_xO_3 \tag{10.2}$$

$$La_{1-x}Sr_xCo^{3+}_{1-x}Co^{4+}_xO_3 \rightarrow La_{1-x}Sr_xCo^{3+}_{1-x+2\delta}Co^{4+}_{x-2\delta}O_{3-\delta} + \left(\frac{\delta}{2}\right)O_2 \tag{10.3}$$

10.4 Synthesis Aspect of MMO Catalysts

There are various methods for producing MMO, including coprecipitation, sol–gel, impregnation, kneading, hydrothermal, and solution combustion methods. For mixed oxides, coprecipitation is the most common technique. In this method, two metal salts are mixed in solution and forced to precipitate with an alkaline solution

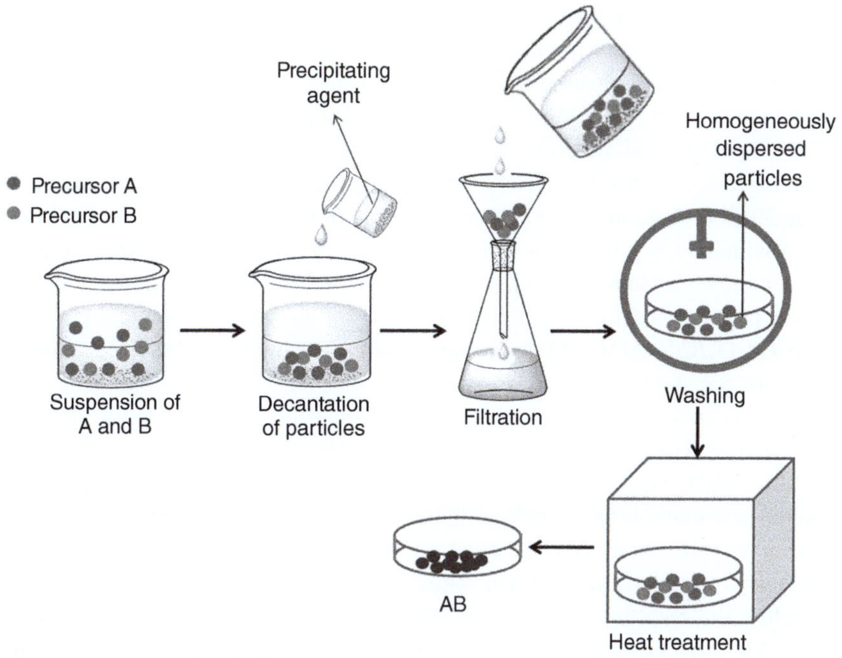

Figure 10.1 Representation of coprecipitation process for MMO catalyst synthesis.

at a specific pH (Figure 10.1). The resulting precipitate is calcined under a specific atmosphere depending on the catalyst used. The main concern of this method is to keep the pH constant during the precipitation process.

For example, $CeO_2/CuAlO_2$MMOs were prepared by a coprecipitation method using aluminum nitrate nonahydrate, cerium nitrate tetrahydrate, and copper nitrate hexahydrate as precursors. For this purpose, 1 M Na_2CO_3 solution was slowly added to the mixture of metal nitrate solutions while stirring until the precipitation was complete. The resulting mixture was stirred for an additional hour at 45 °C with constant stirring and then centrifuged to collect the precipitate. The samples were washed several times with deionized water and dried in an oven at 120 °C. This removed alkali metals and nitrate ions, after which it was calcined at 450 °C in the presence of an air flow. The samples were then calcined in a muffle furnace at 600 °C for two hours. The calcination converted the carbonates in the sample to oxides. Finally, the samples were annealed at 900 °C for six hours [21]. In the sol–gel method, metal precursor salts or their organic compounds are hydrolyzed in water or organic solvents at room temperature. However, insoluble salts require higher temperatures. Salinas et al. synthesized La_2O_3-ZrO_2 MMO catalyst using 80 wt% zirconium *n*-butoxide and lanthanum acetylacetonate as metal precursors. The gelation reaction was carried out under reflux at 80 °C for one hour in the presence of distilled water and 2-butanol, maintaining water: 2-butanol molar ratio of 1 : 4. The resulting gel was air-dried and calcined at 600 °C for six hours [22]. In some cases, alternative solvents are used to solubilize the precursor and obtain highly dispersed metals in solution, and either acids or bases are added to the hydrolysis solution to accelerate the peptization process. For example, Sivasamy et al. used a simple sol–gel method with nitrate–citrate route to synthesize Gd_2O_3/CdO MMO. In this method, solutions of calculated amounts of $CdN_2O_6{\cdot}4H_2O$ and $Gd(NO_3)_3$were stirred separately. $Gd(NO_3)_3$ was prepared by mixing Gd_2O_3 with 2 M HNO_3 and stirring well for 15 minutes, followed by adding 50 ml of 2 M citric acid solution. The pH of the solution was then adjusted to approximately 7 with liquid NH_3. The resulting solution was stirred at about 45 °C for 160 hours to form a gel, then the gel was decomposed at about 100 °C to obtain a powder, and finally sintered at 600 °C for six hours to obtain a polycrystalline Gd_2O_3/CdO-MMO [23]. In impregnation techniques, dissolved aqueous precursors are deposited onto a support (usually an oxide support). The solid support was contacted with a precursor solution containing the active elements of the catalyst. Tungsten–molybdenum catalyst ($W_{0.3}$-$Mo_{0.7}$/CaO, $W_{0.4}$-$Mo_{0.6}$/CaO, $W_{0.5}$-$Mo_{0.5}$/CaO, $W_{0.6}$-$Mo_{0.4}$/CaO, and $W_{0.7}$-$Mo_{0.3}$/CaO) was synthesized by wet impregnation method followed by calcination at a temperature of 650 °C. For this purpose, W_x-Mo_y from $H_3PO_4{\cdot}12WO_3{\cdot}xH_2O$ and $(NH_4)_6Mo_7O_{24}{\cdot}4H_2O$ ($x \geq 0.3$, $y \geq 0.7$) was dissolved in distilled water and stirred vigorously for one hour. Thereafter, the resulting solution was impregnated with powdered CaO to give a total of 5 g. Furthermore, the obtained slurry was dried in an oven at 110 °C for 24 hours. The catalyst was then calcined at 650 °C for four hours [24]. Recently, four MgO-SiO_2 catalysts (Mg_L-Si_L, Mg_L-Si_H, Mg_H-Si_L, and Mg_H-Si_H) were prepared by Szabó et al. WK was performed by stirring a mixture of MgO and SiO_2 (Mg/Si molar ratio equal to 2) in a small amount of water at room temperature for five hours. The water was then evaporated and the solid residue was dried at 120 °C overnight and then calcined at 550 °C for five hours [25]. In a hydrothermal process, a homogeneous aqueous solution of precursors and an alkali are exposed to high temperature and pressure for a certain period of time in an autoclave. ZnO·La_2O_3 (ZL) and the ternary MMO ZnO·La_2O_3·CeO_2 (ZLC) are synthesized using hydrothermal and solution combustion methods. In the hydrothermal method, $Zn(NO_3)_2{\cdot}6H_2O$ and $La(NO_3)_3{\cdot}6H_2O$ were dissolved in a small amount of distilled water, and 1 M NaOH was added dropwise to precipitate them. The precipitate was then transferred to a Teflon-lined stainless steel hydrothermal autoclave and stored in a preheated muffle furnace at 180 °C for 12 hours. The synthesized material was cooled, filtered, washed with water, and dried in a hot air oven. ZLC was also prepared according to the same scheme by adding zinc nitrate, lanthanum nitrate, and cerium nitrate as precursors and calcining the synthesized material at 900 °C [26] On the contrary, in the solution combustion method, $Zn(NO_3)_2{\cdot}6H_2O$ and $La(NO_3)_3{\cdot}6H_2O$ were placed in a crucible at a ratio of 1 : 2, and urea was added to this. A pasty mass was prepared by adding a few drops of water, and then the crucible was placed in a muffle furnace preheated to 500 °C for 15 minutes. The ternary MMO ZLC was prepared according to the same scheme by adding zinc nitrate, lanthanum nitrate, and cerium nitrate as starting materials in a ratio of 1 : 2 : 1 M. The resulting material was cooled, ground, and a portion was calcined at 900 °C for three hours [26].

10.5 Characterization Techniques for MMO Catalysts

MMOs are characterized by various spectroscopic and analytical techniques such as BET (Brunauer, Emmett, and Teller), X-ray diffraction (XRD), Fourier transform infrared (FTIR), Raman, X-ray photoelectron spectroscopy (XPS), thermogravimetric analysis–differential thermal analysis (TGA-DTA), scanning electron microscopy (SEM-EDS), transmission electron microscopy (TEM), and acidity and basicity measurements. To optimize catalytic processes, better control over catalyst properties is essential. Instrumentation technology plays a key role in solving this problem. For solid-based catalysts, fundamental properties such as the number of basic sites, their location, and base strength are of paramount importance. A new method has been developed to simultaneously measure the acidity and basicity of metal oxide catalysts using solid-state NMR techniques. In this method, three typical oxides, namely ZrO_2, Al_2O_3, and MgO, were characterized by ^{13}C and ^{31}P MAS NMR using CO_2 as the acidic probe molecule and trimethylphosphine as the basic probe molecule. To validate the simultaneous characterization method, the authors further investigated the acetone self-condensation reaction using in situ NMR techniques. The development of such technology arose from the fact that acidic and basic sites coexist on the surface of metal oxides and therefore interact [27]. Determining the surface area and pore distribution of the catalyst is important for understanding the range of possibilities for active metal dispersion. The larger the support surface, the higher the dispersion of the active metal. The size and number of pores determine the inner surface. Pore size also determines the accessibility of the reactants to the active sites and the ability of the products to diffuse back into the bulk liquid. Therefore, pore structure and surface area must be optimized to fully utilize the active sites of catalyst for a particular feedstock. To measure the surface area and pore size distribution of each of the samples, a well-known method, BET, is used. Chromium-based Cr/Ca and Cr/Zn mixed oxide catalysts were prepared using a precipitation process for transesterification of edible palm oil to fatty acid methyl esters. Structural properties such as surface area, pore volume, and average pore diameter of Cr/Ca oxide and Cr/Zn oxide were identified using a N_2 adsorption/desorption analyzer. Cr/Zn oxide has a larger surface area ($3.9897\,m^2\,g^{-1}$) compared to that of Cr/Ca oxide ($3.6915\,m^2\,g^{-1}$). However, Cr/Ca oxide indicates higher pore volume ($0.00684\,cm^3\,g^{-1}$) than Cr/Zn oxide ($0.006304\,cm^3\,g^{-1}$). Regarding the pore size, the pore size of Cr/Ca oxide is larger than that of Cr/Zn oxide, which is 93.919 Å and 80.113 Å, respectively. Since the average pore size of both catalysts is larger than 60 Å, triglycerides can diffuse and contact the active sites within the catalyst pores since the typical diameter of triglycerides is less than 60 Å. Based on these comparisons, it can be concluded that Cr/Ca oxide has a larger pore size despite having similar surface area and pore volume to Cr/Zn oxide. Therefore, the reactant molecules easily penetrate into the interior of the Cr/Ca oxide catalyst, and most of the active sites are used to promote the reaction [28]. For the nitroaldol condensation reaction, Bharali et al. synthesized high surface area MMOs (MgAl(O), NiMgAl(O), CoMgAl(O)) from LDHs. Using Quantachrome Instruments (Model: NOVA 1000e), they measured the BET surface area and the Barrett–Joyner–Halenda(BJH) method to measure the pore size and pore volume. NiMgAl(O) is observed to have a higher BET surface area (753 m^2 g^{-1}) and larger pore volume (279 cm^3 g^{-1}) compared to the mixed oxides MgAl(O) and CoMgAl(O). These MMOs were found to be more catalytically active for nitroaldol condensation reactions. The large BET surface area and large pore volume of NiMgAl(O) are better in catalytic activity than the other two mixed oxides, and NiMgAl(O) can be reused up to the fourth cycle without losing its activity properties [29].

XRD techniques are often used to characterize nanomaterials. This is useful for determining crystal structure, phase identification, and average crystallite size. A multicomponent metal oxide ($Al_{0.88}Fe_{0.67}Zn_{0.28}O_3$) was prepared to remove As(III) from aqueous solution. The formation of $Al_{0.88}Fe_{0.67}Zn_{0.28}O_3$ was confirmed by XRD using a PANalytical X-ray diffractometer equipped with a CuKa source (λ = 1.54187 Å) and operated at 30 mA and 40 kV. The XRD pattern of $Al_{0.88}Fe_{0.67}Zn_{0.28}O_3$ has a peak corresponding to the (211) plane at 2θ = 44.6489°, which corresponds to the standard file of $99Al_2O_3{\cdot}ZnO$. The peaks corresponding to planes (400) and (622) indicate the presence of Fe_2O_3. A peak from the (533) plane of Fe_3O_4 was also observed. Therefore, XRD analysis confirmed the formation of $Al_{0.88}Fe_{0.67}Zn_{0.28}O_3$ [30]. To characterize MMOs prepared by incorporating Li, K, Ca, Ba, Ce, and La metals into LDH, a Philips PW 3838 diffractometer was used to analyze monochromatic CuK radiation (=1.54 Å) in the 2θ range (10–80°). The scan speed was 1/4° minute^{-1}, and the step size was 0.02°. All XRD profiles show the two characteristic peaks for MgO (JCPDS 78-0430) and a third metal oxide {Li_2O_2, (JCPDS 74-0115), K_2O (JCPDS 77-2151), CaO (JCPDS 82-1691), BaO (JCPDS 74-1228), Ce_2O_3 (JCPDS 78-0484), La_2O_3 (JCPDS 83-1355) and spinel-type mixed oxide (MAl_2O_4)} were also observed [31]. To investigate the crystalline properties of mixed oxide catalysts composed of MgO and $MgAl_2O_4$ as periclase and spinel, respectively. An XRD instrument (MAC Science, MO_3XHF) was used with a CuKa radiation source (=1.54056 Å) at 40 kV and 30 mA and a scanning rate of 5° minute^{-1}. Peaks at 38.4°, 44.4°, 59.2°, 65.5° and 19.0°, 31.2°, 38.5°, 59.3°, 65.2°, respectively, indicated the presence of periclase and spinel (XRD peak at 34° for periclase MgO implies that the Mg^{2+} was replaced by Fe^{3+} and Al^{3+}, along with overlapping peaks of $MgAl_2O_4$) [32].

FTIR and Raman spectroscopiesare used to identify unknown materials and components in a mixture by recognizing the bonding among the atoms, ions, and molecules. A solid base catalyst containing $Ca_{12}Al_{14}O_{33}$ and CaO, for example, demonstrated a broad peak at around 3430 cm^{-1} due to O—H stretching vibration and a very minor absorption peak at around 1630 cm^{-1} due to O—H bending vibration of H_2O molecules absorbed from air. The presence of bands at 577 and 459 cm^{-1} indicated the stretching vibration of metal oxides Al—O and Ca—O, respectively [33]. FTIR spectroscopy was used to identify the functional groups present in MgAl, NiMgAl, and CoMgAl LDHs, as well as their corresponding MgAl(O), NiMgAl(O), and CoMgAl(O) MMO catalysts. The stretching and bending vibrations of the -OH group of LDH layers and the interlayer water molecule are responsible for the absorption bands at 3478.2 $cm^{-1}$and 1662.9 cm^{-1}. The peak at

1388.5 cm^{-1} is caused by asymmetric stretching of the CO_3^{2-} ion, and the broad peaks observed below 1000 cm^{-1}(500–800 cm^{-1}) are caused by M—O vibrations. While in MMOs, the bands observed at 558.6, 672.4, and 811.9 cm^{-1} could be attributed to metal–oxygen (M—O) bond vibration. Furthermore, the band at 1376.9 cm^{-1}is caused by carbonate, which may result from the adsorption of carbon dioxide from air on the surface of metal oxides [29]. Mitran and Marcu created Co(x)MgAlO mixed oxide catalysts with varying cobalt contents (1–20 wt%) by calcining LDH precursors at 1023 K. The Raman spectrum of Co(1)MgAlO revealed a band at 1090 cm^{-1} caused by residual carbonate species vibration. Al–O–Mg stretching vibrations in the mixed oxide with the periclase-like structure were assigned the 565 cm^{-1} band. The Raman spectrum of the Co(3)MgAlO compound revealed the presence of CoO crystalline phase at 695, 612, and 467 cm^{-1}. This phase was also thought to be responsible for the band at 303 cm^{-1}. The five Raman active modes of Co_3O_4 were systematically observed around 195, 490, 540, 620, and 695 cm^{-1} above $x = 3$. The slight shift of these bands from one sample to the next could reveal differences in spinel crystallinity. The band at around 590 cm^{-1} was attributed to Co—O stretching vibrations of widely dispersed cobalt species. In fact, this band and another at around 640 cm^{-1} could correspond to the symmetric and anti-symmetric stretching vibrations of Co—O—Al bonds, respectively [34].

XPS is a widely used characterization technique for obtaining quantitative and qualitative information on the elements present on the sample's surface, as well as the elements' electronic structure and chemical state. On a K-alpha XPS instrument with a monochromatic AlKa X-ray source, XPS analyses of CeO_2/Fe_2O_3, CeO_2/Mn_2O_3, and $CeO_2/Mn_2O_3/Fe_2O_3$ MMO nanocomposites were performed. Peaks for Ce, C, O, and Fe can be found in the XPS survey of CeO_2/Fe_2O_3. The C peak was due to the adventitious carbon. The peak for Mn 2p in CeO_2/Mn_2O_3 was present at 640 eV. Peaks corresponding to Ce, Mn, Fe, C, and O were clearly visible in $CeO_2/Mn_2O_3/Fe_2O_3$. The CeO_2/Fe_2O_3 HRXPS O 1s spectrum has three distinct components at 530.0, 532.3, and 534.0 eV, which were assigned to lattice oxygen (76.2%), hydroxyl groups (19.1%), and adsorbed water molecules (4.7%), respectively. Similar contributions for lattice oxygen, O–H, and water molecules were observed for CeO_2/Mn_2O_3 and $CeO_2/Mn_2O_3/Fe_2O_3$. The Ce $3d_{5/2}$ profiles were assigned peaks in the HRXPS Ce 3d spectrum of CeO_2/Fe_2O_3 at 898.0, 888.7, and 882.0 eV. The Ce $3d_{3/2}$ profiles have peaks at 916.4, 907.0, and 900.6 eV. The Ce $3d_{5/2}$ peaks were fitted to the oxidation states Ce^{3+} (28.1%) and Ce^{4+} (71.9%). Similar Ce 3d spectra were observed for CeO_2/Mn_2O_3 and $CeO_2/Mn_2O_3/Fe_2O_3$ with 24.1 and 25.5% Ce^{3+} ions, respectively. Peaks in the Fe 2p spectra of CeO_2/Fe_2O_3 and $CeO_2/Mn_2O_3/Fe_2O_3$ are at 710.0 (Fe $2p_{3/2}$) and 722.6 eV (Fe $2p_{1/2}$), respectively, with two shake-up satellites for Fe^{3+} ions in Fe_2O_3. CeO_2/Mn_2O_3 and $CeO_2/Mn_2O_3/Fe_2O_3$ Mn 2p spectra have two peaks at 640.0 (Mn $2p_{3/2}$) and 651.9 eV (Mn $2p_{1/2}$), with a spin-orbit splitting of 11.9 eV for Mn^{3+} in Mn_2O_3 [35]. Three Mg–Zr catalysts with varying Mg to Zr ratios were prepared, and the elemental composition was determined using an X-ray photoelectron spectrometer equipped with a dual anode and a non-monochromatized Al-K X-ray source ($h\nu = 1486.6$ eV) and operating at 12.5 kV and 16 mA. The B.E of O1s measured for Mg/Zr catalysts is between 531 and 532.1 eV. The B.E at 531 eV corresponds to metal oxide ions, and the high B.E around 532.1 eV is due to the presence of hydroxyl groups on the surface of MgO and/or ZrO_2. In addition to these signals, one sample shows another signal at 537 eV, which could be attributed to hydroxycarbonates. Zr^{4+} is represented by the Zr 3d signals at 182.6 and 184.9 eV. Finally, the presence of hydroxyl, hydroxyl carbonates, and Zr in tetragonal form increases the basicity of the catalysts [36].

TGA, differential scanning calorimetry (DSC), DTA, temperature programmed reduction (TPR), and temperature programmed desorption (TPD) techniques are primarily used to investigate thermal properties. TGA is a thermal analysis technique that measures sample weight as a function of temperature in order to determine the thermal stability of a catalyst sample. The DTA and DSC are used to calculate phase transitions at different temperatures, heating or cooling rates, and atmospheric effects. DSC is used to determine the difference in the rate of change of the sample's heat flow rate, whereas DTA analyzes the temperature difference with the temperature of the reference material. TPR and TPD are two other thermal characterization techniques for solid catalysts. The primary distinction between the two is that TPR includes surface processes, whereas TPD includes a bulk reaction. TPD exposes the pretreated sample to an increasing temperature at a constant rate while being swept by an inert gas such as nitrogen, helium, or argon. The gas that is thus chemisorbed is later desorbed from the sample's surface and can be detected. The observed peaks in TPD can be used to deduce the relationship between base or acid strength and desorption temperature or adsorption volume [2]. For example, the total basicity and basic site distributions in three Mg–Zr catalysts (designated MZ-1, MZ-2, and MZ-3) were measured using the TPD of CO_2 technique. Initially, the catalyst was treated with helium gas for two hours at 100 °C to remove absorbed impurities. Then, at a flow rate of 30 ml minute^{-1}, a mixture of 10% CO_2 and helium gas was passed through it. The main CO_2 desorption peaks in the TPD patterns are at 100 °C, 300 °C, and at a high temperature around 600 °C. CO_2 interaction with weak and strong basic sites resulted in the desorption peaks observed at 100 °C and 300 °C, respectively. The presence of much stronger basic sites, i.e. super basic sites, could explain the higher intensity desorption peak at 600 °C. These strong basic sites are most likely due to isolated O^{2-}ions located in a specific position on the mixed oxide surface. The presence of strong basic sites confirmed that the MZ-2 catalyst with a Mg/Zr ratio of 2 : 1 was extremely active [36]. The thermal properties of the MgAlFe catalysts were evaluated using TGA curves (Perkin–Elmer Pyris 1 series) with a heating rate of 10 K minute^{-1} from room temperature to 800 K under N_2 atmosphere (30 ml min^{-1}). The thermogram revealed a three-step thermal decomposition, with the first weight loss (5 wt%) caused by the elimination of physically adsorbed water at lower temperatures (400 K). The second weight loss is associated with the removal of chemically adsorbed water from the internal and external surfaces of the hydrotalcite structure and ranges between 400 and 500 K. This weight loss suggests that MgAlFe catalysts with Mg absorb moisture more easily, and absorbed water increases as Mg content increases. Furthermore, the third weight loss at 500–600 K is attributed to the combination of the decomposition of the brucite layer's interlayer carbonate anions and the dehydroxylation of the

OH groups in the hydrotalcite. The basicity of the MgAlFe catalysts was also investigated using the TPD with CO_2 and helium flow (10 K $minute^{-1}$, 30 ml $minute^{-1}$) at temperatures ranging from 373 to 1073 K. MgAlFe catalysts have higher CO_2 adsorption volumes than MgO, Al_2O_3, and Mg_3Al catalysts. It means that MgAlFe catalysts have a higher basicity and fatty acid methyl ester (FAME) conversion. Furthermore, the CO_2 adsorption volumes increase with increasing Mg content in the hydrotalcite structure, and thus basicity increases as Mg content in the catalysts increases [32]. The thermal stability of the MgO–ZrO_2 catalyst was studied using thermograms obtained using the TGA/DTA method. Both measurements were performed at a heating rate of 5 °C $minute^{-1}$ from 30 to 650 °C. The TG-DTA profiles revealed that the thermal decomposition in the differential thermal curve consists of two major exothermic peaks, indicating that the precursor hydroxides are thermally stable up to nearly 250 °C [37].

SEM and TEM are commonly used to study the morphology and chemical composition characteristics of nanomaterials. SEM images are created by detecting secondary and backscattered electrons, whereas TEM images are created by transmitting electrons. TEM has greater resolution than SEM and can resolve particles as small as 1 nm. Wang et al. used a scanning electron microscope to examine the morphology of a solid base catalyst containing $Ca_{12}Al_{14}O_{33}$ and CaO. Many large pores and cracks with micrometer sizes were observed among these agglomerates, resulting in improved catalyst activity [33]. Juma et al. investigated the morphology of CuO-NiO-ZnO mixed oxide nanoparticles using a JSM-7100F FE-SEM (field emission SEM). The SEM image depicts nanoparticles with spherical and rice-like shapes. The TEM image of CuO-NiO-ZnO mixed oxide nanoparticles shows nanoparticles ranging in shape and size from 15 to 40 nm [38]. Gupta et al. used a Hitachi S-4300 microscope to capture SEM images of CeO_2/Fe_2O_3, CeO_2/Mn_2O_3, and $CeO_2/Mn_2O_3/Fe_2O_3$ MMOs after ion-sputtering the samples with a gold–platinum alloy. It was discovered that nanospheres clustered together in all three MMOs nanocomposites. HR-TEM images were obtained at 100 nm resolutions on a JEM-2010F, JEOL, Japan microscope, revealing that the nanospheres were formed with polyhedral nanocrystals ranging in size from 10 to 20 nm [35]. Abdelrahman et al. investigated the surface morphologies of $Mn_{0.5}Zn_{0.5}Fe_2O_4/Fe_2O_3$ and $Fe_{0.5}Mn_{0.5}Co_2O_4/Fe_2O_3$ MMOs using a Quanta 250 FEG scanning electron microscope and a Talos F200iS transmission electron microscope. According to the SEM results, the surface of the $Fe_{0.5}Mn_{0.5}Co_2O_4/Fe_2O_3$was composed of cubic and irregular shapes with an average grain size of 1.71 m. Furthermore, $Mn_{0.5}Zn_{0.5}Fe_2O_4/Fe_2O_3$surface is composed of spherical shapes with an average grain size of 0.26 m. The TEM results, however, show that both $Fe_{0.5}Mn_{0.5}Co_2O_4/Fe_2O_3$and $Mn_{0.5}Zn_{0.5}Fe_2O_4/Fe_2O_3$MMOs had cubic and irregular shapes with average diameters of 100.27 and 84.29 nm, respectively [39].

10.6 Catalytic Applications of MMO Catalysts

Because of their superior thermal and chemical stability as compared to component oxides, MMOs have been widely used in a variety of organic conversions and

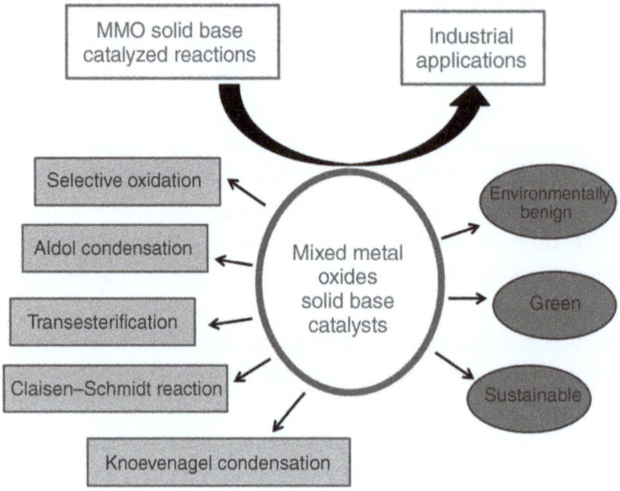

Figure 10.2 Applications of solid base mixed metal oxide catalysts.

catalytic applications. MMOs have been used in a wide range of reactions, including oxidation, isomerization, aldol condensation, transesterification, Knoevenagel, and Claisen–Schmidt reactions (Figure 10.2) [40]. Table 10.2 summarizes the applications of MMOs as solid base catalysts in industrially important reactions, organic synthesis, and green chemistry.

10.6.1 Applications in Industrially Important Reactions

MMOs have been used in a variety of industrially significant synthetic conversions, such as aldol condensation between substituted 4-alkylaldehyde and linear aldehydes, which produces a variety of fragrances. These fragrances are commonly found in the perfume and soap industries. Jasmine aldehyde (2-pentylcinnamylaldehyde) is a popular fragrance that is made by aldol condensation of benzaldehyde with heptanal. Tiler et al. created a variety of zinc-modified Mg–Al MMO catalysts with varying Mg:Al:Zn ratios (1.0 : 2.1 : 1.0–3.2 : 0.57 : 1.0) for use in aldol condensation of benzaldehyde with heptanal. The amount of zinc in the catalyst affects the reaction rate, and Mg-Al-Zn oxides produced 15–20% more jasmine aldehyde than pure Mg–Al oxide under the same conditions [62]. Vrbková et al. used sol–gel and impregnation methods to create a variety of MoO_3-SiO_2 and WO_3-SiO_2 MMOs. The prepared catalysts' catalytic activity was evaluated in three model reactions: aldol condensation (benzaldehyde with heptanal), acetalization (heptanal with ethylene glycol), and esterification (lauric acid with methanol). The as-prepared MMO catalysts demonstrated good catalytic performance, with aldol condensation achieving 78% heptanal conversion with 58% selectivity to jasmine aldehyde in 24 hours. In acetalization, more than 90% heptanal conversion with 90% selectivity to ethylene glycol acetal was observed in five hours. Furthermore, the sol–gel catalysts demonstrated 91–94% conversion with 100% selectivity to methyl laurate in a 24-hour esterification reaction [41]. Pseudoionones are important precursors

Table 10.2 Summary of various reactions catalyzed by mixed metal oxides as catalysts.

Catalysts	Preparation method	Reactions	Ref.
MoO_3-SiO_2/WO_3-SiO_2	Sol–gel/ wet impregnation	Aldol condensation (benzaldehyde with heptanal), Acetalization (heptanal with ethylene glycol), and Esterification (lauric acid with methanol)	[41]
$Mg_3Al_{1-a}Y_aO_x$	Co-precipitation	Synthesis of pseudoionones	[42]
La/MgO ZrO_2	Combustion	Synthesis of styrene carbonate	[43]
$Mg_2Al_{0.5}Ti_{0.5}$	Co-precipitation	Synthesis of styrene carbonate	[44]
CuCr-MMO	Co-precipitation	Synthesis of styrene oxide	[45]
CaO-NiO and CaO-Nd_2O_3	Co-precipitation	Transesterification of *Jatropha curcas* oil	[46]
M-CaO; M = ZnO, MnO_2, Fe_2O_3, and Al_2O_3	—	Transesterification of Jatropha and Karanja oils	[47]
Barium–calcium–zinc mixed oxide	Wet impregnation	Biodiesel production from microalgal oil	[48]
Sr–Ce bimetal oxide	Gel combustion	Transesterification of waste cooking oil	[49]
$NiMgO_x$	Co-precipitation	Green synthesis of glycerol carbonate	[50]
$SrTiO_3$, $YNbO_4$, and $BaCeO_3$	Sol–gel	Synthesis of glycerol carbonates	[51]
MMO of Ba and lanthanides (La, Ce, and Nd)	Co-precipitation	Synthesis of glycidol	[52]
$ZnAl_2O_4$	Co-precipitation	Synthesis of 4′-(arylmethylene)bis(3-methyl-1*H*-pyrazol-5-ol) derivatives	[53]
$Co_2Ni_1Al_1O_x$	Co-precipitation	Oxidation of ethylbenzene to acetophenone	[54] NK75
Al_2O_3@Co-Cu-Al mixed metal oxide	Co-precipitation	Oxidation of ethylbenzene to acetophenone	[55]
La-modified mesoporous Mg-Al mixed oxides	Co-precipitation followed by hydrothermal	Synthesis of dimethyl carbonate from methylcarbamate and methanol	[56]
ZnO(0.54)-Nb_2O_5(0.20)-TiO_2(0.26)	Sol–gel	Synthesis of dimethyl carbonate from urea methanolysis	[57]
K loaded La_2O_3-MgO	Hydrothermal method/wet impregnation	Synthesis of veratraldehyde	[58]
MgO, Al/Mg mixed oxide	Co-precipitation	Claisen–Schmidt condensation of acetophenones with 5-hydroxymethylfurfural	[59]
MgO/ZrO_2	Co-precipitation	Knoevenagel condensation of aldehydes with malononitrile and ethyl cyanoacetate	[60]
$ZnFe_2O_4$	—	Synthesis of 4*H*-pyran-3-carboxylate	[61]

for α- and β-ionones, which are widely used to synthesize vitamins A and E, as well as carotenoids and a variety of aroma chemicals. Pseudoionones are synthesized commercially through the cross-aldol condensation of citral and acetone. Wang et al. used the coprecipitation method to create rare earth element (Y, La, and Eu)-modified MgAl mixed oxides. The Y-modified MgAl mixed oxides catalyst demonstrated unexpectedly high catalytic activity for citral and acetone cross-aldol condensation. They achieved 100% conversion of citral with 85% selectivity to pseudoionone at 60 °C in three hours. $Mg_3Al_{1-a}Y_aO_x$ catalysts exhibited well-developed small flake morphology with high surface area and pore volume when the Y^{3+}/M^{3+} molar ratio was 0.1–0.2, resulting in an increasing number of exposed basic sites on the surface of $Mg_3Al_{1-a}Y_aO_x$ catalysts to further improve the catalyst activity for the aldol reaction [42].

It is a difficult task to use carbon dioxide for value-added chemical synthesis rather than capturing and storing it because it is an important key to a sustainable low-carbon society. Carbon dioxide will be one of the future sources of a variety of chemicals, including methanol, dimethyl ether, formic acid, formaldehyde, and syngas for the Fischer–Tropsch process. There is a lot of research going on for the utilization of CO_2 by heterogeneous catalytic processes in the bulk and specialty chemical industries. Styrene carbonate prepared from CO_2 has a wide range of applications as a solvent, reagent, and fuel additive, as well as a precursor in the synthesis of polymers, herbicides, and disinfectants. Cyclic carbonates are also widely used in the manufacture of personal care and cosmetic products [43]. According to Kulal et al., MMO catalysts are more effective than individual oxides for the synthesis of cyclic carbonates because MMOs exhibit acidity and basicity on their surface. The basicity of the reaction activates CO_2, whereas the acidity of the reaction aids in the activation of the epoxide molecule. Thus, the combination of acidity and basicity plays an important role in this reaction. Kulal et al. prepared several MMOs for CO_2 cycloaddition reactions with epoxides in this context. Mn-Ba and Sn-Ni MMOs were discovered to be effective solid catalysts and under optimized reaction conditions; 160 °C, 25 bar pressure, and CO_2: epoxide ratio of 1.6 : 1, Mn-Ba oxide (4.3 : 1) and Sn-Ni oxide (1.5 : 1) gave 96.0% and 90.2% yield for propylene carbonate, respectively [63]. Velpuri prepared a high surface area (45.8 m^2 g^{-1}) copper and zinc MMO catalyst (CuO-ZnO) for cycloaddition of CO_2 with various epoxides in one example. The CuO-ZnO catalyst demonstrated excellent catalytic performance, with cyclic carbonate yields exceeding 95% [64]. Tambe et al. synthesized styrene carbonate by cycloaddition of carbon dioxide to styrene oxide using various solid catalysts such as 0.1% Li/MgO, calcined hydrotalcite (CHT), 0.1% La/MgO ZrO_2 and La-Zr mixed oxide (1 : 3 La/Zr). Catalysts with both strong basic and weak acidic sites are used to carboxylate styrene oxide with carbon dioxide. The best catalyst was found to be La-Zr mixed oxide (1 : 3 La/Zr), which gave 90% conversion of styrene oxide with 100% selectivity of styrene carbonate in three hours at a moderate 130 °C temperature. Furthermore, the MMO catalyst was reused three times without losing activity [43]. By calcining Ti^{4+} doped Mg-Al LDHs for the cycloaddition reaction of styrene oxide with CO_2, $Mg_2Al_{1-x}Ti_x$-MMOs were obtained. It was discovered that Ti^{4+} doping raises the residual Na content from 1.3% to 2.3%, increasing the number of strongly basic sites (up to twice as many), and thus improving the catalytic performance of MMOs. $Mg_2Al_{0.5}Ti_{0.5}$-MMO converted styrene oxide slightly faster and with higher selectivity to styrene carbonate than Mg_2Al-MMO. Thus, increasing the number and strength of basic sites and improving the catalytic performance of LDH-derived MMOs in cycloaddition of CO_2 with styrene oxide via Ti^{4+} doping is an effective method [44].

Styrene oxide is a key intermediate in organic synthesis, as well as in the pharmaceutical and cosmetics industries. Wang et al. prepared a series of MMO materials $M1Mg_3Al_2$-MMO (M1 = Cu, Ni, Mn, Ti, Zn, Cr, Co, and Fe) for styrene epoxidation with H_2O_2. The acidity and basicity of the catalysts are important factors in their selectivity and stability. As a result, they altered the surface acid-basic properties of MMOs in order to improve catalytic performance in this reaction. CuCr-MMO catalyst demonstrated the highest styrene conversion of 82.8% with 79.7% selectivity to styrene oxide. The acid–base properties of the catalyst were modulated by the unique composition of the $CuCr_2O_4$ and CuO phases in CuCr-MMO. A suitable basicity was facilitated in particular to selectively transform H_2O_2 into HOO• radicals, resulting in a high conversion and selectivity to styrene oxide [45]. Because of the utility of styrene monomer-based polymers such as polystyrene (PS), polystyrene–butadiene rubber (SBR), styrene acrylonitrile (SAN), and polyacrylonitrilebutadienestyrene (ABS), styrene demand in the plastic industry has been increasing year after year. Jiang et al. investigated ZrO_2-based MMO catalysts for the industrially important dehydrogenation process of ethylbenzene to styrene monomer. The desired surface basicity, increased surface area, and CO_2 affinity of ZrO_2-based MMO catalysts resulted in higher ethylbenzene conversion with good selectivity to styrene monomer [65]. Sancheti and Yadav used the co-precipitation method to create a Mo/Zr MMO catalyst for highly selective styrene production. A 5% Mo/Zr catalyst with optimal acidity and basicity for better selectivity yields 19.53% conversion and 97.5% styrene selectivity at 853 K [66].

Vanillin is widely used in perfumes, food, cosmetics, pharmaceuticals, and other chemical industries. It is produced industrially through the bioconversion of various materials such as lignin, vanillyl alcohol, phenolic stilbenes, glucose, creosol, isoeugenol, and eugenol [67]. Lignin oxidative depolymerization is a promising route for converting lignin to vanillin. Lignin is discarded as cellulosic waste in biorefinery industries, so converting it into fine chemicals is economically valuable in the biorefinery process. Vanillyl alcohol is a lignin B-0-4 linkage that produces Vanillin via selective oxidation. Jeon et al. used the sol–gel method to create Cu-Mn mixed oxide catalysts with varying Cu/Mn ratios for the production of vanillin from lignin. The N_2-physisorption analyses revealed that $Cu_{1.5}Mn_{1.5}O_4$ had a high BET surface area, and the formation of the spinel structure improved the redox ability and oxygen mobility. As a result, when compared to pure Cu and Mn oxides, $Cu_{1.5}Mn_{1.5}O_4$ mixed oxide demonstrated superior catalytic performance for lignin oxidation to vanillin [68]. Saha and Hamid synthesized spinel structured Cu-Mn mixed oxide and demonstrated superior catalytic activity for the oxidation of vanillin alcohol to vanillin using air and H_2O_2. In the presence of a base, 94% conversion with 99% selectivity to vanillin was obtained using H_2O_2, whereas 91% conversion with 81% selectivity to vanillin was obtained in liquid phase aerobic oxidation in the absence of a base. Furthermore, the MMO catalyst as prepared demonstrated high stability in three subsequent oxidation reactions [69]. For isoeugenol and vanillyl alcohol oxidation to vanillin, Rahmanivahid et al. used $CuO/MgAl_2O_4$ (BET surface area = $20\,m^2\,g^{-1}$, pore volume $0.09\,cm^3\,g^{-1}$) and $CuO/MgFe_2O_4$ (BET surface area = $5\,m^2\,g^{-1}$, pore volume $0.03\,cm^3\,g^{-1}$) catalysts. For both reactions, the catalyst $CuO/MgAl_2O_4$ demonstrated superior catalytic performance. However, the separation of $CuO/MgFe_2O_4$ by the reaction mixture is easy due to the high magnetic strength of $CuO/MgFe_2O_4$ [67]. Jha et al. also used a solvothermal method to prepare a mixed oxide MnCo-MO catalyst that converted 62% vanillyl alcohol with 83% selectivity to vanillin in two hours. The higher activity of the MnCo-MO catalysts is due to a greater capacity for oxygen adsorption than a single metal oxide, as confirmed by temperature-programmed oxidation analysis [70].

Alternative energy resources are becoming increasingly important as fossil fuel sources deplete over time. Biodiesel is a popular potential alternative to fossil fuels due to the depletion of traditional fuels, the rising cost of petroleum, and the environmental problems caused by these fuels. Biodiesel is typically composed of fatty acid alkyl (methyl or ethyl) esters (FAME and FAEE) produced from fats or oils via a process known as transesterification and/or esterification in the presence of a homogeneous or heterogeneous catalyst [71, 72]. The heterogeneous catalytic process is preferable to the homogeneous one because solid catalysts can be recovered and reused. In this order, as an improved method, the use of solid MMO heterogeneous catalysts has received significant attention in recent years for the production of biodiesel, and a comprehensive overview can be found in review articles published in scientific literature [73–75]. CaO has demonstrated promising potential as a solid base catalyst in the transesterification reaction. Furthermore, several research groups reported that the catalytic property of pure CaO could be improved by preparing mixed oxides and improving transesterification processes. Teo et al., for example, synthesized CaO-NiO and CaO-Nd_2O_3 mixed oxides via co-precipitation method for biodiesel production from *Jatropha curcas* oil. CO_2-TPD analysis revealed that when compared to pure oxides, the mixed CaO-NiO and CaO-Nd_2O_3 catalysts have the highest basic site. As a result, both CaO-NiO and CaO-Nd_2O_3 catalysts demonstrated high activity, with the highest FAME yields of 86.3% and 82.2% achieved over CaO-NiO and CaO-Nd_2O_3 catalysts, respectively, under optimum reaction conditions [46]. *Jatropha curcas* plant is gaining popularity as a biodiesel feedstock. Because of the presence of phorbol ester as toxic constituents in *J. curcas* oil, it is not suitable for nutrition; however, without detoxification, it can be used as an energy or fuel source. To make biodiesel from *J. curcas* oil, a mixed CaO-La_2O_3 catalyst with different Ca/La molar ratios was used. The higher the CaO loading amount, the higher the FAME yield, and under optimized reaction conditions at 65 °C, 4% catalyst ($CaLa_4$) dose with a 24 : 1 molar ratio of methanol to Jatropha oil, the catalyst produced 86.51% biodiesel. The presence of strong basic strength and a high amount of basicity on the calcium-based mixed oxide catalyst could explain the high activity of the $CaLa_4$

catalyst [76]. Joshi et al. used MMOs (M-CaO; M = ZnO, MnO_2, Fe_2O_3, and Al_2O_3) derived from waste chicken egg shells (i.e. CaO(cesp)) for transesterification of Jatropha and Karanja oils. Metal oxide-impregnated catalysts outperformed neat CaO(cesp) in terms of activity. Experiments with the Hammett indicator and BET revealed that the Zn-CaO(csep) catalyst had higher basic strength and surface area than other metal impregnated catalysts and neat CaO(cesp). Thus, the best catalyst was found to be ZnO-CaO(cesp), with maximum conversion of 98.2% for Jatropha and 95.8% for Karanja biodiesels obtained with Zn-impregnated CaO(cesp) under optimized reaction conditions [47]. $CaTiO_3$, $CaMnO_3$, $CaZrO_3$, and $Ca_2Fe_2O_5$ calcium-containing perovskites were used in one case to methanolyze sunflower oil. The catalytic activity of pure perovskite phases was low or non-existent at 60 °C; however, catalytic activity of perovskites containing a small amount of CaO was obtained at 60 °C because the presence of CaO phase increased the basic strength. In two hours, the $CaTiO_3$/CaO and $Ca_2Fe_2O_5$/CaO catalysts produced FAME yields of 98.1% and 72.7%, respectively. Therefore, their catalytic activities are most likely caused by the presence of CaO phase. In contrast to low or no activity at 60 °C, methanolysis at 165 °C using pure perovskites as catalyst resulted in a FAME yield of more than 90% with $CaTiO_3$, $CaMnO_3$, $Ca_2Fe_2O_5$, and $CaZrO_3$, respectively, after three hours. This finding suggests that pure perovskites could be effective biodiesel catalysts, but only at higher temperatures [77]. Singh et al. recently prepared a barium-impregnated CaOZnO mixedoxide catalyst for biodiesel synthesis from *Spirulina* oil using a wet impregnation method. Under optimized conditions (catalyst = 2.5 wt%, methanol-to-oil molar ratio = 1 : 18, temperature = 65 °C, time = 120 minutes), high 98.94% FAME conversion was obtained from microalgal oil [48]. Banerjee et al. used gel combustion to create Sr-Ce-based MMOs for transesterification of waste cooking oil. At pH 7.0, the enhanced activity of the catalyst with a Sr-Ce atomic ratio of 3 : 1 was investigated, which gave a maximum conversion of 99.5% under optimized conditions (catalyst dose = 2.0 wt%, oil-to-methanol ratio = 1 : 14, reaction time = 120 minutes, reaction temperature = 65 °C, and stirring speed = 700 rpm). The obtained biodiesel sample in this reaction was compared to diesel fuel in terms of calorific value, cetane number, flash point, pour point, cloud point, density, and ash content, and the results were quite comparable and within the limits of ASTM standards. In a nutshell, the synthesized biodiesel could be easily used in any existing diesel engine as a favorable substitute and eco-friendly fuel [49]. As a result, the long-term activity of MMO base catalysts will help to reduce biodiesel production costs on an industrial scale.

Although biodiesel has provided a safe and renewable alternative to fossil fuels, there are a few major challenges, such as the unavoidable excess of crude glycerol produced as a by-product of biodiesel production. The biodiesel industry is one of the fastest growing in Europe, North America, and South Asia, producing massive amounts of by-products such as glycerol, soap, and methanol. Glycerol accounts for roughly one-tenth of biodiesel production, and its valorization is critical. Glycerol is a versatile compound with numerous applications in cosmetics, pharmaceutical preparations, and the pharmaceutical and food industries. The efficient conversion of glycerol to cyclic glycerol carbonates, alcohols, acrolein, lipids, lactic acid, polyglycerols, and polyols adds significant market value to the oversupplied glycerol. Glycerol carbonate is the most eco-friendly of these chemicals and is used in a wide range of applications in the cosmetics, paints, and lithium-ion battery industries as an excellent chemical solvent or membrane coating in gas separation systems [50, 78]. Marimuthu et al. synthesized Cu-containing Mg/Al-based mixed oxides that were found to be highly efficient for the transesterification of glycerol with dimethyl carbonate (DMC) to produce glycerol carbonates. CO_2-TPD analysis confirms that the catalysts had strong surface basic sites, resulting in 96.4% glycerol conversion with 91.2% glycerol carbonate yield. The increased catalytic performance was attributed primarily to the fine-tuned Cu molar ratio, strong surface basicity, and large surface area [79]. Poolwong et al. recently used the non-hydrolytic sol–gel process to synthesize perovskite-based MMOs (PBMMOs) such as $YNbO_4$. TPD and BET analysis revealed that the as-synthesized PBMMOs contain a large number of critical strong basic sites and have larger surface areas. As the basicity of the catalyst increased, the catalytic transesterification of DMC with glycerol of as-synthesized PBMMOs ($SrTiO_3$-200, $YNbO_4$-200, and $BaCeO_3$-200) followed a linear trend. Under optimized reaction conditions, the $SrTiO_3$-200 catalyst demonstrated high robustness (99% glycerol carbonate yield) and could be reused with at least 95% glycerol carbonate yield after five catalytic cycles [51]. Liu et al. created a Ca-Mg-Al MMO catalyst from waste steel slag through acidolysis, coprecipitation, and calcination at various temperatures. Because of the large pore size and proper alkalinity, the catalyst calcined at 600 °C had excellent catalytic performance in the transesterification reaction of glycerol and DMC. Under optimal reaction conditions (catalyst = 3 wt%, molar ratio of DMC to glycerol = 3 : 1, temperature = 75 °C), the catalyst achieves 98.3% glycerol conversion and 96.2% glycerol carbonate yield in 90 minutes. Furthermore, the hydrotalcite-type mixed oxide crystal structure enhanced the anti-leaching ability of active sites, and the glycerol conversion and glycerol carbonate yield remained above 90% after five cycles [80]. Dalma et al. recently synthesized double-layer hydroxide and used coprecipitation to incorporate third transition metal ions (Cu, Ni, or Zn) in addition to Mg and Al ions in these LDHs. MMOs generated by LDHs through thermal decomposition and used as a catalyst in the conversion of glycerol to glycerol carbonate. Under solvent-free and mild reaction conditions, the as-synthesized mixed oxides demonstrated high catalytic activity with an 80% yield of glycerol carbonate. Because of their high surface area, strong basic site, and good thermal stability, the addition of Cu, Ni, or Zn to the matrix of Mg and Al improved their catalytic performance. Furthermore, the catalyst could be reused up to four times without losing significant catalytic activity [81]. Glycidol is also a key raw material in the production of polyglycerols, glycidyl ethers, and pharmaceuticals, as well as in the perfume and cosmetics industries. By co-precipitation, single metal oxide and MMO catalysts of Ba and lanthanides (La, Ce, and Nd) were prepared for one pot glycidol synthesis by transesterification of glycerol with DMC. Among the single metal oxides, the BaO with the highest basicity as measured by CO_2-TPD demonstrated the greatest conversion of glycerol (71%) with selective formation of glycerol carbonate. With 80% selectivity and 98% glycerol conversion, mixed oxides of Ba and lanthanides (La and Ce) were discovered to be the best catalyst for glycerol transesterification to glycerol carbonate followed by decarboxylation to tandem synthesis of glycidol. Although BaO alone exhibited the highest basicity as measured by CO_2 TPD, tuning basicity through the incorporation of CeO_2-induced oxygen vacancies in the cerium oxide material resulted in the formation of glycidol [52]. In one case, glycerol is used in the synthesis of 2-methylpyrazine via vapor-phase dehydrocyclization with ethylene diamine over ZnO-$ZnCr_2O_4$ mixed oxide. The mixed oxides were created by precipitating corresponding Zn-Cr hydrotalcite precursors at pH 7 and 9 and then calcining in air at 450 °C. TEM and XPS analysis revealed the formation of small (24 nm) particles in Zn-Cr-O, with Zn enrichment at the near-surface region at pH 9. As a result of their small-sized Zn-Cr-O particles and Zn-enriched Zn-Cr-O surface, these catalysts were highly efficient in the conversion of bioglycerol to 2-methylpyrazine [82].

10.6.2 Applications in Organic Synthesis

Many heterocyclic compounds are used as starting materials in the synthesis of various drugs. Nitrogen-containing heterocyclic compounds such as pyrazole, pyrazine, quinoxaline, and their derivatives, for example, play an important role in the design of a variety of anti-inflammatory, antibiotic, anticancer, and antiviral drugs. For example, 4,4′-(arylmethylene)bis(3-methyl-1*H*-pyrazol-5-ol) derivatives were synthesized using $ZnAl_2O_4$ as an MMO basic catalyst in a three-component cyclocondensation reaction of hydrazine hydrate, ethyl acetoacetate, and aldehydes. The products were obtained in high yields in few minutes using this method [83]. Teodorescu et al. used a mechanochemical method to synthesize MMOs mo-ZnAl, mo-NiAl, and mo-CoAl (M^{2+}/Al = three molar ratio) for 2-methylpyrazine synthesis. TPR analysis revealed that the prepared MMOs are thermally stable and that the catalytic sites can be regenerated after being subjected to reaction conditions. At 400 °C, the solid yield of 2-methylpyrazine was close to 90%, and the results after the fifth reaction cycle showed no significant changes in conversion compared to the initial value [84]. Ajaikumar and Pandurangan loaded various weight percentages of ZrO_2/M_xO_y (where M = Al, Ga, In, and La) MMOs on Si-MCM-41 mesoporous molecular sieves. The materials' TEM and SEM images show fine dispersion of metal oxides on the host Si-MCM-41, and the catalysts were highly efficient for the synthesis of quinoxaline derivatives via liquid-phase condensation of 1,2-diamine with 1,2-diketones at room temperature. Various reaction parameters were optimized to improve product yield, with 17% ZrO_2/4% Ga_2O_3/MCM-41 yielding the highest yield at lower catalyst concentration and reaction time. The activity of the catalyst was not significantly affected after three to four reuses, and the reaction proceeded smoothly with a yield of 97–94% [85]. Tetrazole derivatives are well-known heterocycle compounds that are used in a variety of fields, including biochemistry, pharmacology, and medicine. The CuO-NiO-ZnO MMO-based catalyst was synthesized using *Rheum ribes* leaves, and its catalytic activity and selectivity for the regioselective synthesis of aminotetrazole derivatives (1-aryl-5-amino-1*H*-tetrazoles) were investigated. Under ultrasonic irradiation at 70–80 °C, 0.03 g catalyst yielded a high reaction yield. Furthermore, the prepared MMO could be recovered and reused multiple times before becoming significantly deactivated [86]. Flavonoids are one of the most important classes of natural products, and they can be found in a wide variety of fruits and vegetables. These are phenolic fused rings that have been shown to have anticancer, antioxidant, anti-inflammatory, anticarcinogenic, and antibacterial properties. According to Khosravani Farahani et al., cobalt-copper spinel ferrite MNPs ($Cu_{1-x}Co_xFe_2O_4$) catalysts were discovered to be highly efficient nanocatalysts for the synthesis of various flavanones antioxidants. Furthermore, the catalyst was separated using a simple magnetic extraction method and reused six times with no noticeable loss of activity. The prepared catalyst's high activity was attributed to the cooperative activation of the carbonyl group by both Cu and Co through a synergistic catalytic effect that facilitates the Michael addition of the -OH group to the, -unsaturated ketone [87]. A critical chemical transformation in the organic production process is the selective oxidation of alkanes and alkyl aromatics to produce highvalue-added alcohols, ketones, aldehydes, carboxylic acids, and other chemical intermediates. Ji et al. created Co-Ni-Al-O MMOs using Co-Ni-Al-LDHs ($Co_xNi_yAl_1$-LDHs) as precursors for the selective oxidation of ethylbenzene to acetophenone. Due to the catalyst's mesoporous and sheet-like nanostructure, the prepared catalyst demonstrated a high catalytic performance for the oxidation of ethylbenzene to acetophenone, with ethylbenzene conversion reaching 80.0% and acetophenone selectivity reaching 88.9% over the $Co_2Ni_1Al_1O_x$ catalyst under optimized reaction conditions. Here, the synergy between the highly dispersed active site Co and promoter Ni species greatly enhances the catalytic activity. Moreover, the catalyst shows good stability and can be recycled at least four times without significant deactivation [54]. In one example, a hierarchical flower-like Al_2O_3@Co-Cu-Al MMO catalyst was synthesized using Co-Cu-Al LDH as a precursor. These MMO catalysts showed excellent catalytic performance in the selective liquid-phase oxidation of ethylbenzene to acetophenone. Incorporation of copper in MMO maximizes the synergistic effect between Co and Cu species and significantly improves the catalytic performance of the catalyst. The as-formed Al_2O_3@CoCuAl MMO catalyst with a Cu/(Co + Cu) molar ratio of 0.25 showed the best catalytic performance with a high conversion of 92.8% and acetophenone selectivity of 89.4% [55]. Bankar et al. used coprecipitation followed by hydrothermal method to create Cu dope-based cubic ZrO_2 MMO catalysts Cu_xZrO_{100-x} for the selective oxidation of aromatics active methyl, alcohol, and amine groups. Under milder reaction conditions, the Cu_xZrO_{100-x}catalyst demonstrated excellent liquid-phase oxidation properties even with a low Cu loading (i.e. 2 wt%) and high conversion of active methyl groups, alcohols, and amines with high selectivities to corresponding ketones, aldehydes, and amides [88]. Partial oxidation of saturated hydrocarbons to valuable products is an important organic reaction. In this context, Jian et al. developed a hydrotalcite-derived Cu-MgAlO-MMO catalyst for selective oxidation of cyclohexane to KA oil. Here, the catalytic reaction results show that 8.3% conversion of cyclohexane and 82.9% selectivity to KA oil were achieved using 9% Cu-MgAlO catalyst at 150 °C and 0.6 MPa oxygen pressure for two hours [89].

Because of their excellent properties and commercial applications, much effort has recently been devoted to the development of new synthesis methods based on

CO_2as a feedstock for organic carbonates. DMC is a nonpolluting building block that can replace dimethyl sulfate, phosgene, chloromethane, and methyl chloroformate as a methylation, carbonylation, esterification, or ester interchange reagent in organic synthesis. Because of its low toxicity, DMC is also used as a green solvent and food flavoring agent [56]. In comparison to traditional synthetic methods, direct synthesis of DMC from CO_2 and methanol was a more cost-effective and eco-friendly route that not only avoided the use of toxic substances such as CO and phosgene but also solved the problem of CO_2 mitigation and methanol excess capacity. In this context, Li et al. synthesized a series of Fe-Zr mixed oxides with varying Fe contents using the sol–gel method for direct synthesis of DMC from CO_2 and methanol. Under reaction conditions of 110 °C and 12 MPa, the mixed oxide $Fe_{0.7}Zr_{0.3}O_y$ catalyst showed better DMC yield of 0.44 mmol g cat^{-1} and DMC selectivity of 100% than pure Fe_2O_3 and ZrO_2 [90]. DMC was also synthesized by direct conversion of CO_2 with methanol in the presence of cerium–calcium MMO catalyst. Three catalysts (Ce_3-Ca_1, Ce_1-Ca_1, and Ce_1-Ca_3) with different Ce/Ca molar ratios were synthesized by surfactant template method. The BET surface areas and average pore diameters of these catalysts are 102, 82, 109 m^2 g^{-1} and 9, 12, and 5 nm, respectively. CO_2 and NH_3 TPD analysis showed that Ce_1-Ca_1 catalyst has a high density of acidic and basic centers, and therefore Ce_1-Ca_1 catalyst exhibited the highest catalytic activity for DMC synthesis by direct conversion of CO_2 [91].

Wang et al. synthesized DMC from methyl carbamate (MC) and methanol using La-modified mesoporous Mg-Al mixed oxides (HTC-La) derived from hydrotalcite-like precursors (Mg^{2+}: Al^{3+}: La^{3+} = 3 : 1: x, where x varies from 0 to 1.0). The structure and surface basic properties of the HTC-La catalyst were shown to be dependent on the amount of La additive, with a DMC yield of 54.3% and a high DMC selectivity of 80.9% achieved when x was tuned to 0.5 under optimized reaction conditions [56]. One of the eco-friendly and cost-effective methods for DMC production is urea alcoholysis (or methanolysis). With its plentiful, lowcost, and relatively green reactants (urea and methanol), the urea alcoholysis method has piqued the interest of both researchers and industry. Mandooie et al. recently reviewed zinc-based MMO catalysts for DMC synthesis via urea alcoholysis over a 20-year period. They concluded that the highest DMC production yields were achieved by zinc-based MMO catalysts with ZnO/CaO binary metal oxide and Zn/Ce/La ternary metal oxide catalysts [92]. Asghari and Ghiaci used a modified sol–gel method to create a series of catalysts with different molar ratios of ZnO, Nb_2O_5, and TiO_2 for DMC via urea methanolysis. ZnO and Nb_2O_5 incorporation into the TiO_2 matrix improves catalyst stability and reusability. Furthermore, the TiO_2 matrix effectively prevented metal leaching into the reaction medium. Among the tested catalysts, ZnO(0.54)-Nb_2O_5(0.20)-TiO_2(0.26) demonstrated the best catalytic performance under optimal conditions, with a DMC yield of 39.1% [57].

10.6.3 Applications in Green Chemistry

It is widely acknowledged that there is a growing need in the chemical industry for more eco-friendly processes. This trend has been dubbed "green chemistry" or "sustainable technology." Green chemistry deals with synthesis procedures in accordance with its classic 12 principles, thereby contributing to the sustainability of chemical processes, making reagents and products less hazardous/toxic, causing less damage to the environment and human health, lowering the risk of global overheating, and making more rational use of natural resources and agricultural waste [93]. Two useful indicators of a chemical process's potential environmental acceptability are as follows: (i) Sheldon defines the environmental factor (E) as the weight of waste divided by the weight of product. It is used to assess the amount of waste generated during a synthesis process or the environmental impact of chemical processes. The E factor is the actual amount of waste produced during the process, which is defined as everything other than the desired product. There is one exception: water is typically excluded from the E factor. A higher E factor results in more waste and, as a result, a greater negative environmental impact. Zero is the ideal E factor [94]. (ii) Atomic efficiency: This relates to the formula weight of the desired final product and the sum of the formula weights of all precursors [94]. Different control methods can be applied at different steps of the chemical process in material production to minimize risks to the environment, for example, the use of selected alternative precursors, solvents, catalysts and reagents, target product modification, and real-time process monitoring. To minimize the risk to the environment, keeping in mind the concerns of the factors, we report here some examples of MMOs used as base catalysts in green chemical processes.

Molleti and Yadav reported a green process for producing veratraldehyde by O-alkylation of vanillin with an eco-friendly reagent, DMC, on a K-loaded La_2O_3-MgO MMO catalyst. The addition of K to La_2O_3-MgO increases the basic sites, and a 2 wt% K/La_2O_3-MgO catalyst provides 96% conversion with 0.03 g cm^{-3} catalyst loading at 160 °C in two hours. The catalyst was recovered and reused four times with consistent activity. Scale-up experiments were also conducted to determine the viability of the current method, and 92% conversion with 88% yield of desired product was discovered [58]. Green synthesis of α-methylcinnamaldehyde (2-methyl-3-phenyl-propen-2-al) by Claisen–Schmidt condensation of benzaldehyde with propanal using Mg-Zr mixed oxide as base catalyst was reported by Wagh et al. They investigated the activity and selectivity of Mg-Zr mixed oxide on HMS at various loadings. It was discovered that a 20% (w/w) Mg-Zr (2 : 1) mixed oxide/HMS catalyst with a selectivity of 92% for benzaldehyde conversion was highly active. The activation energy for this Claisen–Schmidt reaction was 9.54 kcal mol^{-1}, and it follows second-order kinetics. Furthermore, the catalyst demonstrated excellent selectivity and reusability [95]. Arias et al. used MgO, Al/Mg mixed oxide (HTc), and a hydrated Al/Mg mixed oxide (HTr) catalysts to synthesize furanochalcones via Claisen–Schmidt condensation of acetophenones with 5-hydroxymethylfurfural (HMF). Lewis basic sites exist in Al/Mg (HTc), whereas Bronsted basic sites exist in hydrated Al/Mg mixed oxide (HTr), and both play an important role in Claisen–Schmidt condensation. Furthermore, in the absence of a solvent, these catalysts demonstrated high selectivity in Claisen–Schmidt condensation [59]. Gawande and Jayaram used an eco-friendly MgO/ZrO_2 base catalyst for the solvent-free Knoevenagel condensation of several aldehydes with ethyl cyanoacetate and malononitrile at 60 °C. The indicator method revealed that the MgO/ZrO_2 catalyst had strong basic sites, which increased the catalyst's activity toward the Knoevenagel condensation reaction, resulting in good yield (65–98%) [60] (Scheme 10.1). The reaction's E factor and atom economy are depicted below.

$$\text{Atom economy} = \text{m. wt. of the desired product/m. wt. of all products}$$
$$= 242/260 \times 100 = 93.07\%$$

$$\text{E} = \text{sum of m. wt. of the by products/m. wt. of the desired product}$$
$$= 18/242 = 0.07$$

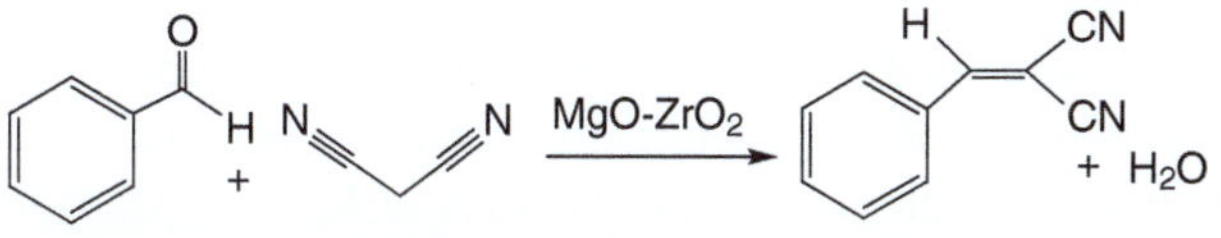

Scheme 10.1 Knoevenagel condensation of benzaldehyde with malononitrile.

The theoretical atom efficiency and Efactor are 93.07% and 0.074%, respectively. Under the given reaction conditions, the experimental atom efficiency for the 10 mmol scale reaction is 88.07%, and the Efactor is 0.078 [13]. Pradhan et al. created a greener and more potent heterogeneous Ni/Mg-based mixed oxide catalyst ($NiMgO_x$) for the green synthesis of glycerol carbonate from used cooking oil. They used co-precipitation to create a series of catalysts with varying Mg/Ni molar ratios (1 : 1, 2 : 1, and 3 : 1). The 3 : 1 MgNi catalyst was discovered to be the best catalyst for maximum glycerol conversion (97%). The reaction is catalyzed by a base, so the higher the base strength of the catalyst, the higher the activity. Here, the environmental impact of transesterification reactions using $NiMgO_x$ catalysts was calculated using the green chemistry "Efactor" metric (Eq. 10.4).

$$\text{E} - \text{factor} = \frac{\text{Mass of waste (mg)}}{\text{Mass of product (mg)}} \tag{10.4}$$

In the transesterification reaction of glycerol to glycerol carbonate, methanol is the only easily recyclable waste product, and the low Efactor value of 0.074 in this analysis confirms that it generates the least amount of waste. Moreover, the catalyst is very stable up to its fifth cycle and the high TOF and low Efactor values indicate that the catalyst is efficient and the reaction process is eco-friendly in nature [50].

Imines serve as starting materials and intermediates in a variety of organic transformation reactions, including the synthesis of bio-active nitrogen-containing compounds and nitrogen heterocycles [96]. One of the most convenient methods for imine synthesis is aerobic amine oxidation, and MMO-based catalytic systems have been reported as capable catalysts. A Cu/Al MMO catalyst, for example, was synthesized using a modified inverse micelle method (surface area: 140 $m^2\,g^{-1}$ and average pore size: 9.63 nm). The synthesized MMO efficiently catalyzes the aerobic oxidative coupling of amines to imines under mild, solvent-free, eco-friendly conditions with atmospheric air as the sole oxidant. The catalytic activity of a prepared MMO catalyst was investigated for a variety of amine substrates, with conversions as high as >99% and excellent selectivity [97]. The one-pot green approach employing various mixed oxides as catalysts provides a broad platform for the design and development of novel heterocyclic compounds with promising properties and features. Dhale et al. recently introduced $Mg_xCu_{0.3}Zn_{0.7-x}La_yFe_{2-y}O_4$ (x = 0.1, 0.2, 0.3, 0.4, 0.5; and y = 0.0, 0.025, 0.050, 0.075, 0.10) MMO catalysts for one-pot synthesis of biologically and pharmaceutically important quinoxalines (90–97% yield) and 1,5-benzodiazepines derivatives (89–98%) under mild reaction conditions. Furthermore, the catalyst can be reused up to five times without losing significant catalytic activity. As a result, these catalysts can be studied further and used in industrial applications for the preparation of quinoxaline and 1,5-benzodiazepine derivatives [98]. Lingaiah and colleagues reported the synthesis of biologically and medicinally important 1,2,4,5-tetrasubstituted imidazole derivatives via four-component condensation of benzil or benzoin, aldehydes, amines, and ammonium acetate using a mixed oxide of $K_5CoW_{12}O_{40}H_2O$ as a catalyst. The catalyst demonstrated good to excellent yields of the desired product in a short reaction time with good reusable activity under solvent-free conditions. As a result of the reusability, atom economy, fast and eco-friendly protocols, the method for the synthesis of 1,2,4,5-tetrasubstituted imidazoles is simple, economical, benign, and user-friendly [99]. Naeimi et al. reported a one-pot synthesis of pyrido[2,3-d:6,5-d′]dipyrimidine derivatives via four-component condensation reactions of aromatic aldehydes, 2-thiobarbituric acid, and NH_4OAc in water as a green solvent using $CuFe_2O_4$ mixed oxide as catalyst. In short reaction times and with simple workup, the desired products were obtained in excellent yields of up to 98%. As a result, this microwave-oriented heterocyclic synthesis with a green solvent and catalyst, as well as the environmental acceptability and recyclability of the catalyst, may be an ideal method for the synthesis of pyrido[2,3-d:6,5-d′] dipyrimidine derivatives [100]. 4*H*-Pyran derivatives are the fundamental building blocks of many natural products and the foundation of valuable biologically active compounds. Thus, the 4*H*-pyran framework served as a valuable lead pharmacophore and sparked a lot of interest in the synthesis of its derivatives. $ZnFe_2O_4$ nano catalyst was used in this order for the one-pot synthesis of 4*H*-pyran-3-carboxylate (1,4-DHPs) from 4*H*-pyran-3-carboxamide in water at room temperature. As the reaction requires, the $ZnFe_2O_4$nanocatalyst provides both acidic (Fe^{3+}) and basic (O_2) functionalities. The functional group interconversion from amide of 4*H*-pyran-3-carboxamide to ester leads to 4*H*-pyran-3-carboxylate in this reaction. As a result of its simplicity, cost-effectiveness, eco-friendliness, and ease of scaling up for large-scale synthesis, this method is considered green [61]. The fusion of coumarin to pyrimidine/isoxazoles/pyrazole rings to form polycyclic fused compounds may result in improved pharmaceutical applications. Bhagi et al. designed a one-pot synthesis route of various fused heterocycles (benzopyranones and benzoxazinones) using $CaAl_2O_4$ as a catalyst in this context, eliminating the use of toxic compounds and solvents. In this method, the reactants are completely converted into products without the formation of side products in a shorter reaction time [101]. Styrene is an industrially important chemical that serves as a precursor to a variety of resins, plastics, rubbers, and copolymers. One of the major challenges in green chemistry is the development of a suitable catalyst that provides high ethyl benzene conversion and styrene selectivity, while remaining active over time. Negi et al. prepared $Mn_xTi_{1-x}O_2$ catalyst by solution combustion and used it for oxidative dehydrogenation (ODH) of ethyl benzene with O_2 or air in this order. The catalyst is both stable and active over a long period of time, and it has demonstrated a high and sustainable yield of styrene at low temperatures. Since this is a more economical and safer approach for styrene synthesis, especially in terms of stability and yield, it can be adopted as a green chemistry approach using air or O_2as the oxidizing agent [102].

10.6.4 Applications in Environmental Catalysis

Catalysis for clean air such as automotive exhaust catalysts, desulfurization, wastewater treatment, and volatile organic compound (VOC) removal are important applications for heterogeneous catalysts. MMOs have applications in environmental catalysis where VOCs are further oxidized to less hazardous compounds. VOCs are commonly released from poorly controlled emissions from the chemical industry and are responsible for many health risks to humans [1, 103]. Castao et al. synthesized Co-Mn-Mg-Al mixed oxides to eliminate VOCs using auto-combustion and co-precipitation techniques. The oxides' catalytic behavior was investigated in the total oxidation of a mixture of 250 ppm toluene and 250 ppm 2-propanol. The catalysts demonstrated good activity in the oxidation of VOCs mixtures due to their ability to favor redox processes in the catalyst structure and good oxygen mobility [104]. Toluene is a VOC that is commonly used in industry and pollutes the environment. Dong et al. investigated the catalytic oxidation of toluene to CO_2 and H_2O over a $CoMn_2O_4$ catalyst. T_{50} and T_{90} values for $CoMn_2O_4$ catalyst were reported as 202 °C and 210 °C, respectively, which are significantly higher than those for Co_3O_4, MnO_x, and Co_3O_4/MnO_x. After the 700 -minute stability test (2.0 vol%), the $CoMn_2O_4$ showed 100% conversion of toluene at 220 °C in the presence or absence of water vapor. The structural differences between the spinel $CoMn_2O_4$ and the MMO Co_3O_4/MnO_x lead to different mobilities and contents of active oxygen species in the samples, which play an important role in the catalytic activity. In addition, $CoMn_2O_4$ has a large surface area, rich cation vacancies, and high mobility of oxygen species, resulting in high activity in toluene oxidation [105]. Layered transition metal oxide (Co, Ni, Cu, Zn)-doped δ-MnO_2 materials have been synthesized for catalytic total oxidation of toluene using an eco-friendly, very fast, and scalable self-propagated flaming technique. At 300 °C, the CuO-δ-MnO_2 with a Mn/Cu molar ratio of 10 : 1 demonstrated the best catalytic performance for complete toluene removal. CuO modification increased catalytic activity due to the highly dispersed CuO phase, as well as the rich content of lattice oxygen with its mobility and higher specific surface area. Thus, by allowing the facile and energy-saving fabrication of large quantities of layered CuO-δ-MnO_2 catalyst, the self-propagated flaming technique may provide a method for developing highly efficient catalysts for the complete removal of volatile organic compounds (De-VOCs) [106]. Bisphenol A (BPA) is a compound with low volatility in the environment and often used as a raw material in the production of epoxy and polycarbonate resins. The U.S. Environmental Protection Agency (EPA) reports that more than one million pounds of BPA is released into the environment annually and that its xenoestrogenic activity can affect the growth, reproduction, and development of aquatic organisms. Li et al. prepared CoMnAl-MMO catalyst with spinel structure from LDH. This catalyst showed high catalytic activity in activating potassium peroxymonosulfate (PMS) for BPA degradation. The rate of BPA degradation was affected by pH, temperature, and other factors, and it was discovered that 10 $mg\,l^{-1}$ of BPA was completely degraded in 90 minutes at 25 °C with 0.02 $g\,l^{-1}$ of CoMnAl-MMO and 0.15 $g\,l^{-1}$ of oxone. Furthermore, even after three recyclings, the catalyst demonstrated good recyclability. The degradation pathway of BPA was studied using LC-MS and GC-MS and involved hydroxylation of BPA to form hydroxyl-BPA and cleavage of benzene rings to form aliphatic compounds. The degradation process here follows first-order kinetics with activation energy of 76.83 $kJ\,mol^{-1}$. As a result, this MMO/oxone system was ideal for treating neutral or alkaline wastewater [107]. Ibrahim et al. synthesized MFe_2O_4 (M = Zn, Co, Mn) spinel using a sol–gel hydrothermal method, followed by six hours of heating at 573 K in air, and used it as a catalyst for nitroarenes reduction. They reported that a $MnFe_2O_4$ catalyst converted 4-nitrophenol (4-NP), 2,4,6 tri-nitrophenol (2,4,6-NP), and 4-nitroaniline (4-NA) into their respective amino derivatives in 270 seconds. The supercatalytic reduction performance of $MnFe_2O_4$ was due to the increase in mesoporous character due to the increase in pore volume and pore radius and the spacing of electron charge transfer between Mn^{2+}-Mn^{3+} and Fe^{2+}-Fe^{3+} ion pairs [108]. At room temperature, magnetically separable heterogeneous catalysts Fe_3O_4 and Fe_3O_4-CeO_2 were synthesized using the coprecipitation method for the degradation of catechol in the presence of H_2O_2. The 15% Fe_3O_4-CeO_2 nanocomposite has the highest activity for catechol degradation, which is primarily due to its large surface area and small particle size. Catechol degradation was monitored using a UV–vis spectrophotometer, and the formation of degraded products such as 2-hydroxy-1,4-benzoquinone, acetic acid, β-ketoadipic acid, glyoxylic acid, and gluconic acid was identified using liquid chromatography mass spectrometry (LC-MS) studies. The Fe_3O_4-CeO_2 catalyst was recovered from the reaction mixture using an external magnet and used for five consecutive cycles without loss of catalytic activity under eco-friendly conditions. From the standpoint of green chemistry, the catalyst's efficiency in degradation and ease of separation show that it is promising for the removal of catechol from aqueous solutions [109].

10.7 Challenges and Future Scope of Mixed Metal Oxides

MMOs are the heart of heterogeneous catalysis, and various concepts are used to improve MMO catalytic activity. One of the most difficult tasks is to achieve 100% selectivity of the desired product molecules in uni- or multi-path reactions (dubbed "green chemistry"). As a result, there are additional opportunities for the development of new MMO catalysts that exhibit both activity and selectivity while maintaining the economic viability of their preparation method. Synthetic

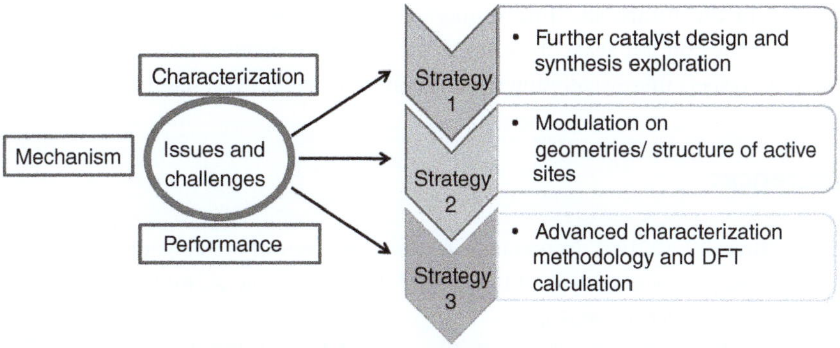

Figure 10.3 Schematic representation of the challenges and development of solid base mixed metal oxide catalysts.

methods for producing monodisperse metal oxide nanoparticles with controlled shape, oxidation states, and composition have been developed in catalyst synthesis nanoscience; these nanoparticles can be used as selective catalysts because chemical selectivity appears to be dependent on all of these experimental parameters. Furthermore, the complete catalytic and deactivation mechanism is not well understood, and characterization and identification of active basic sites are difficult due to the complexity of a heterogeneous reaction and basic site structure deactivation. For this, a combination of advanced characterization methodology and DFT(density functional theory) calculation is required to characterize the intrinsic active site structure of catalysts, which would provide a relatively fundamental understanding of catalyst fine structure and structure–activity correlation. DFT calculations are a necessary supplement to understanding structure–activity correlation and heterogeneous base-catalyzed reaction mechanisms, and they can provide a reasonable explanation or even a prospect for experimental observations. Catalysts with three metals (trimetallic oxides) have yet to be fully explored, and the field is in its infancy. As a result, there is plenty of room for further research into trimetallic oxides and their applications. MMOs are also being investigated in the context of the energy crisis, where they are being used as catalysts for the conversion of waste to energy in the form of fuels. Thus, the use of MMOs in waste valorization and refinery is one of the hottest topics that need to be researched further in order to develop new methodologies in green chemistry prospects. Figure 10.3 depicts the challenges and development of MMO catalysts.

10.8 Conclusion

This chapter discusses the significance of MMOs as solid base catalysts in a variety of organic and industrial applications, as well as in green chemistry. This chapter discusses the various benefits of MMO catalysts, their preparation methods, and characterization techniques in brief. MMO catalyst can successfully catalyze a variety of reactions in various chemical industries and can be recovered and reused multiple times with no discernible loss in catalytic activity. Thus, MMO will be modified in the future and find numerous applications as catalysts and catalyst carriers in the coming years. To summarize, the more complex catalysis science of MMO catalysts is finally catching up to metal catalyst catalysis science.

References

1 Védrine, J.C. (2019a). Importance, features and uses of metal oxide catalysts in heterogeneous catalysis. 40 (11): 1627–1636. Available at: https://doi.org/10.1016/S1872-2067(18)63162-6.

2 Jambhulkar, D.K., Ugwekar, R.P., Bhanvase, B.A. et al. (2022). A review on solid base heterogeneous catalysts: preparation, characterization and applications. *Chemical Engineering Communications* Available at: https://doi.org/10.1080/00986445.2020.1864623.

3 Tomar, R., Kundra, P., Sharma, J. et al. (2024). *Letters in Organic Chemistry* 21 (1): 16–28.

4 Tomar, R., Jain, S., Yadav, P. et al. (2022). *Current Organic Synthesis* 19 (3): 414–425.

5 Hattori, H. (2014). Applied catalysis A: general. *Applied Catalysis A: General'* [Preprint] Available at: https://doi.org/10.1016/j.apcata.2014.10.060.

6 Arya, K. and Tomar, R. (2013). *Research on Chemical Intermediates* 41 (6): 3389–3400.

7 Yoshio, O. and Hideshi, H. (2020). *Solid Base Catalysis*, vol. 101. Springer Science & Business Media.

8 Kumar, L., Verma, N., Tomar, R. et al. (2023). *Dalton Transactions* 52 (10): 3006–3015.

9 Burange, A.S. and Gawande, M.B. (2016). Role of mixed metal oxides in heterogeneous catalysis. *Encyclopedia of Inorganic and Bioinorganic Chemistry* 1–19. Available at: https://doi.org/10.1002/9781119951438.eibc2458.

10 Lucarelli, C., Moggi, P., Cavani, F. et al. (2007). Sol-gel synthesis and characterization of transition metal based mixed oxides and their application as catalysts in selective oxidation of propane. *Applied Catalysis A: General* 325 (2): 244–250. Available at: https://doi.org/10.1016/j.apcata.2007.02.034.

11 Chun Yan, M., Mu, Z., Li, J.J. et al. (2010). Mesoporous CO_3O_4 and AU/CO_3O_4 catalysts for low-temperature oxidation of trace ethylene. *Journal of the American Chemical Society* 132 (8): 2608–2613. Available at: https://doi.org/10.1021/ja906274t.

12 Li, D., Cai, Y., Chen, C. et al. (2016). Magnesium-aluminum mixed metal oxide supported copper nanoparticles as catalysts for water-gas shift reaction. *Fuel* 184: 382–389. Available at: https://doi.org/10.1016/j.fuel.2016.06.131.

13 Gawande, M.B., Pandey, R.K., and Jayaram, R.V. (2012). Role of mixed metal oxides in catalysis science—versatile applications in organic synthesis. *Catalysis Science and Technology* 2 (6): 1113–1125. Available at: https://doi.org/10.1039/c2cy00490a.

14 Rechi Siqueira, M., Perrone, O.M., Metzker, G. et al. (2019). Highly selective 1-butanol obtained from ethanol catalyzed by mixed metal oxides: reaction optimization and catalyst structure behavior. *Molecular Catalysis* 476 (July): 110516. Available at: https://doi.org/10.1016/j.mcat.2019.110516.

15 Kulkarni, A.S. and Jayaram, R.V. (2004). Liquid phase catalytic transfer hydrogenation of aromatic nitro compounds on $La_{1-x}Sr_xFeO_3$ perovskites prepared by microwave irradiation. *Journal of Molecular Catalysis A: Chemical* 223 (1–2): 107–110. Available at: https://doi.org/10.1016/j.molcata.2003.12.042.

16 Wong, Y.C., Tan, Y.P., Taufiq-Yap, Y.H. et al. (2015). Biodiesel production via transesterification of palm oil by using CaO-CeO_2 mixed oxide catalysts. *Fuel* 162: 288–293. Available at: https://doi.org/10.1016/j.fuel.2015.09.012.

17 Carrier, X., Royer, S., and Marceau, E. (2018). Synthesis of metal oxide catalysts. *Metal Oxides in Heterogeneous Catalysis* Available at: https://doi.org/10.1016/B978-0-12-811631-9.00002-8.

18 Misono, M. (2013a). Catalysis of perovskite and related mixed oxides. *Studies in Surface Science and Catalysis* 176: 67–95. Available at: https://doi.org/10.1016/B978-0-444-53833-8.00003-X.

19 Misono, M. (2013b). Chemistry and catalysis of mixed oxides. *Studies in Surface Science and Catalysis* 176: 25–65. Available at: https://doi.org/10.1016/B978-0-444-53833-8.00002-8.

20 Nitadori, T., Ichiki, T., and Misono, M. (1988). Catalytic properties of perovskite-type mixed oxides (ABO3) consisting of rare earth and 3d transition metals. The roles of the A- and B-site ions. *Bulletin of the Chemical Society of Japan* 621–626. Available at: https://doi.org/10.1246/bcsj.61.621.

21 Abdus, S.,.M., Ahmed, T., Awal, M.R. et al. (2015). Structure and photoluminescence studies of $CeO_2 \cdot CuAlO_2$ mixed metal oxide fabricated by co-precipitation method. *Spectrochimica Acta Part A: Molecular and Biomolecular Spectroscopy* 135: 466–471. Available at: https://doi.org/10.1016/j.saa.2014.07.030.

22 Salinas, D., Sepúlveda, C., Escalona, N. et al. (2018). Sol–gel La_2O_3–ZrO_2 mixed oxide catalysts for biodiesel production. *Journal of Energy Chemistry* 27 (2): 565–572. Available at: https://doi.org/10.1016/j.jechem.2017.11.003.

23 Sivasamy, R., Venugopal, P., and Mosquera, E. (2020). Synthesis of Gd_2O_3/CdO composite by sol-gel method: structural, morphological, optical, electrochemical and magnetic studies. *Vacuum* 175: 109255. Available at: https://doi.org/10.1016/j.vacuum.2020.109255.

24 Mansir, N., Teo, S.H., Ibrahim, M.L. et al. (2017). Synthesis and application of waste egg shell derived CaO supported W-Mo mixed oxide catalysts for FAME production from waste cooking oil: effect of stoichiometry. *Energy Conversion and Management* 151 (June): 216–226. Available at: https://doi.org/10.1016/j.enconman.2017.08.069.

25 Szabó, B., Novodárszki, G., Lónyi, F. et al. (2022). Texture and morphology-directed activity of magnesia-silica mixed oxide catalysts of ethanol-to-butadiene reaction. *Journal of Molecular Structure* 1259: 132764. Available at: https://doi.org/10.1016/j.molstruc.2022.132764.

26 Kumari, V., Kumar, N., Yadav, S. et al. (2019). Materials today: proceedings novel mixed metal oxide ($ZnO.La_2O_3.CeO_2$) synthesized via hydrothermal and solution combustion process—a comparative study and their photocatalytic properties. *Materials Today: Proceedings* 19: 18–21. Available at: https://doi.org/10.1016/j.matpr.2019.07.748.

27 Fu, Y., Zhang, L., Yue, B. et al. (2018). Simultaneous characterization of solid acidity and basicity of metal oxide catalysts via solid state NMR technique. *The Journal of Physical Chemistry* 1–36. Available at: https://doi.org/10.1021/acs.jpcc.8b06827.

28 Bashah, N.A.A., Luin, A., Jalaluddin, I.A. et al. (2019). Characteristics of chromium based mixed oxide catalyst in biodiesel production characteristics of chromium based mixed oxide catalyst in biodiesel production. *Journal of Physics: Conference Series* 1349: 1–7. Available at: https://doi.org/10.1088/1742-6596/1349/1/012143.

29 Bharalia, D., Devi, R., Bharali, P. et al. (2014). Synthesis of high surface area mixed metal oxide from NiMgAl LDH precursor for nitro-aldol condensation reaction. *New Journal of Chemistry* 39 (1): 1–8. Available at: https://doi.org/10.1039/b000000x.

30 Subhan, M.A., Monim, S.A., Bhuiyan, M.B.R. et al. (2015). Synthesis, characterization of a multi-component metal oxide. *Open Journal of Inorganic Chemistry* 1 (2): 9. Available at: https://doi.org/10.4236/ojic.2011.12002.

31 Balsamo, N., Mendieta, S., Oliva, M. et al. (2012). Synthesis and characterization of metal mixed oxides from layered double hydroxides. 1: 506–513. Available at: https://doi.org/10.1016/j.mspro.2012.06.068.

32 Wang, S., Wang, Y.B., Dai, Y.M. et al. (2012). Applied catalysis A: general preparation and characterization of hydrotalcite-like compounds containing transition metal as a solid base catalyst for the transesterification. *Applied Catalysis A: General* 439–440 (3): 135–141. Available at: https://doi.org/10.1016/j.apcata.2012.06.049.

33 Wang, B., Li, S., Tian, S. et al. (2013). A new solid base catalyst for the transesterification of rapeseed oil to biodiesel with methanol. *Fuel* 104: 698–703. Available at: https://doi.org/10.1016/j.fuel.2012.08.034.

34 Mitran, G. and Marcu, I. (2012). Oxidative dehydrogenation of propane over cobalt-containing mixed oxides obtained from LDH precursors. *Applied Catalysis A: General* 417-418: 153–162. Available at: https://doi.org/10.1016/j.apcata.2011.12.038.

35 Gupta, N.K., Bae, J. and Kim, K.S. (2021) 'A novel one-step synthesis of Ce/Mn/Fe mixed metal oxide nanocomposites for oxidative removal of hydrogen sulfide at room temperature', pp. 26739–26749. Available at: https://doi.org/10.1039/d1ra03309c.

36 Sree, R., Babu, N.S., Prasad, P.S.S. et al. (2008). Transesterification of edible and non-edible oils over basic solid Mg/Zr catalysts. *Fuel Processing Technology* 90 (1): 152–157. Available at: https://doi.org/10.1016/j.fuproc.2008.08.008.

37 Gawande, M.B., Branco, P.S., Parghi, K. et al. (2011). Synthesis and characterization of versatile MgO-ZrO_2 mixed metal oxide nanoparticles and their applications. *Catalysis Science and Technology* 1 (9): 1653–1664. Available at: https://doi.org/10.1039/c1cy00259g.

38 Juma, A.O., Arbab, E.A.A., Muiva, C.M. et al. (2017). Synthesis and characterization of CuO-NiO-ZnO mixed metal oxide nanocomposite. *Journal of Alloys and Compounds* 723: 866–872. Available at: https://doi.org/10.1016/j.jallcom.2017.06.288.

39 Abdelrahman, E.A. and Al-Farraj, E.S. (2022). Facile synthesis and characterizations of mixed metal oxide nanoparticles for the efficient photocatalytic degradation of rhodamine B and Congo red dyes. *Nanomaterials* 12 (22): Available at: https://doi.org/10.3390/nano12223992.

40 Bing, W. and Wei, M. (2019). Recent advances for solid basic catalysts: structure design and catalytic performance. *Journal of Solid State Chemistry* 269: 184–194. Available at: https://doi.org/10.1016/j.jssc.2018.09.023.

41 Vrbkovál, E., Šteflová, B., Sekerová, L. et al. (2020). Contribution to MoO_3–SiO_2 and WO_3–SiO_2 utilization—active catalysts in jasmine aldehyde, 2-hexyl-1,3-dioxolane and methyllaurate synthesis. *Reaction Kinetics, Mechanisms and Catalysis (2020)* 129: 645–658. Available at: https://doi.org/10.1007/s11144-020-01727-6.

42 Wang, Z., Lu, G., Guo, Y. et al. (2015). Effect of one-pot rehydration process on surface basicity and catalytic activity of Mg. *ACS Sustainable Chemistry & Engineering* 1–11. Available at: https://doi.org/10.1021/acssuschemeng.5b01533.

43 Tambe, P.R. and Yadav, G.D. (2018). Heterogeneous cycloaddition of styrene oxide with carbon dioxide for synthesis of styrene carbonate using reusable lanthanum—zirconium mixed oxide as catalyst. *Clean Technologies and Environmental Policy* 20 (2): 345–356. Available at: https://doi.org/10.1007/s10098-017-1475-1.

44 Zhou, S., Zhou, B., and Wang, L. (2021). Effects of Ti4+ doping on residual Na contents and catalytic properties of MgAl-LDH-derived MMOs in cycloaddition of CO_2 to styrene oxide. *Catalysis Communications* 148 (September 2020): 106174. Available at: https://doi.org/10.1016/j.catcom.2020.106174.

45 Wang, Q., Liang, X., Bi, R. et al. (2019). Highly efficient CuCr-MMO catalyst for a base-free styrene epoxidation with H_2O_2 as the oxidant: synergistic effect between Cu and Cr. *Dalton Transactions* 48: 16402–16411. Available at: https://doi.org/10.1039/c9dt02357g.

46 Teo, S.H., Rashid, U., and Taufiq-Yap, Y.H. (2014). Biodiesel production from crude *Jatropha curcas* oil using calcium based mixed oxide catalysts. *Fuel* 136: 244–252. Available at: https://doi.org/10.1016/j.fuel.2014.07.062.

47 Joshi, G., Rawat, D.S., Lamba, B.Y. et al. (2015). Transesterification of Jatropha and Karanja oils by using waste egg shell derived calcium based mixed metal oxides. *Energy Conversion and Management* 96: 258–267. Available at: https://doi.org/10.1016/j.enconman.2015.02.061.

48 Singh, R., Kumar, A., and Chandra Sharma, Y. (2019). Biodiesel production from microalgal oil using barium-calcium-zinc mixed oxide base catalyst: optimization and kinetic studies. *Energy and Fuels* 33 (2): 1175–1184. Available at: https://doi.org/10.1021/acs.energyfuels.8b03461.

49 Banerjee, S., Sahani, S., and Chandra Sharma, Y. (2019). Process dynamic investigations and emission analyses of biodiesel produced using Sr–Ce mixed metal oxide heterogeneous catalyst. *Journal of Environmental Management* 248 (February): 109218. Available at: https://doi.org/10.1016/j.jenvman.2019.06.119.

50 Pradhan, G. and Chandra Sharma, Y. (2020). Studies on green synthesis of glycerol carbonate from waste cooking oil derived glycerol over an economically viable $NiMgO_x$ heterogeneous solid base catalyst. *Journal of Cleaner Production* 264: 121258. Available at: https://doi.org/10.1016/j.jclepro.2020.121258.

51 Poolwong, J., Del Gobbo, S., and D'Elia, V. (2021). Transesterification of dimethyl carbonate with glycerol by perovskite-based mixed metal oxide nanoparticles for the atom-efficient production of glycerol carbonate. *Journal of Industrial and Engineering Chemistry* 104: 43–60. Available at: https://doi.org/10.1016/j.jiec.2021.08.008.

52 Kondawar, S.E., Patil, C.R., and Rode, C.V. (2017). Tandem synthesis of glycidol via transesterification of glycerol with DMC over ba-mixed metal oxide catalysts. *ACS Sustainable Chemistry and Engineering* 5 (2): 1763–1774. Available at: https://doi.org/10.1021/acssuschemeng.6b02520.

53 Paquin, F., Rivnay, J., Salleo, A. et al. (2015). Multi-phase semicrystalline microstructures drive exciton dissociation in neat plastic semiconductors. *Journal of Materials Chemistry C* 3: 10715–10722. Available at: https://doi.org/10.1039/b000000x.

54 Ji, D., Xi, N., Li, G. et al. (2021). Hydrotalcite-based Co_xNi_yAl1Ox mixed oxide as a highly efficient catalyst for selective ethylbenzene oxidation. *Molecular Catalysis* 508 (April): 111579. Available at: https://doi.org/10.1016/j.mcat.2021.111579.

55 Xie, R., Fan, G., Yang, L. et al. (2016). Hierarchical flower-like Co-Cu mixed metal oxide microspheres as highly efficient catalysts for selective oxidation of ethylbenzene. *Chemical Engineering Journal* 288: 169–178. Available at: https://doi.org/10.1016/j.cej.2015.12.004.

56 Wang, D., Zhang, X., Ma, J. et al. (2016). La-modified mesoporous Mg-Al mixed oxides: effective and stable base catalysts for the synthesis of dimethyl carbonate from methyl carbamate and methanol Catalysis. *Science and Technology* 6 (5): 1530–1545. Available at: https://doi.org/10.1039/c5cy01712b.

57 Asghari, S. and Ghiaci, M. (2020). Dimethyl carbonate synthesis from urea methanolysis over ZnO-Nb_2O_5-TiO_2 mixed oxide catalysts. *Industrial and Engineering Chemistry Research* Available at: https://doi.org/10.1021/acs.iecr.9b05637.

58 Molleti, J. and Yadav, G.D. (2017). Green synthesis of veratraldehyde using potassium promoted lanthanum-magnesium mixed oxide catalyst. *Organic Process Research and Development* 21 (7): 1012–1020. Available at: https://doi.org/10.1021/acs.oprd.7b00127.

59 Arias, K.S., Climent, M.J., Corma, A. et al. (2016). Chemicals from biomass: synthesis of biologically active furanochalcones by Claisen–Schmidt condensation of biomass-derived 5-hydroxymethylfurfural (HMF) with acetophenones.

Topics in Catalysis 59 (13–14): 1257–1265. Available at: https://doi.org/10.1007/s11244-016-0646-3.

60 Gawande, M.B. and Jayaram, R.V. (2006). A novel catalyst for the Knoevenagel condensation of aldehydes with malononitrile and ethyl cyanoacetate under solvent free conditions. *Catalysis Communications* 7 (12): 931–935. Available at: https://doi.org/10.1016/j.catcom.2006.03.008.

61 Das, P., Dutta, A., Bhaumik, A. et al. (2014). Heterogeneous ditopic $ZnFe_2O_4$ catalyzed synthesis of 4H-pyrans: further conversion to 1,4-DHPs and report of functional group interconversion from amide to ester. *Green Chemistry* 16 (3): 1426–1435. Available at: https://doi.org/10.1039/c3gc42095g.

62 Tišler, Z., Vrbková, E., Kocík, J. et al. (2018). Aldol condensation of benzaldehyde and heptanal over zinc modified mixed Mg/Al oxides. *Catalysis Letters* 148 (7): 2042–2057. Available at: https://doi.org/10.1007/s10562-018-2416-2.

63 Kulal, N., Vasista, V., and Shanbhag, G.V. (2019). Identification and tuning of active sites in selected mixed metal oxide catalysts for cyclic carbonate synthesis from epoxides and CO_2. *Journal of CO2 Utilization* 33 (February): 434–444. Available at: https://doi.org/10.1016/j.jcou.2019.07.018.

64 Velpuri, V.R. (2021). High yield room temperature conversion of carbon dioxide into cyclic carbonates catalyzed by mixed metal oxide (CuO–ZnO) nano-flakes/micro-flakes (Cozi-nmf). *Applied Organometallic Chemistry* 35 (6): 1–11. Available at: https://doi.org/10.1002/aoc.6224.

65 Jiang, N., Burri, A., and Eon, S. (2016). Ethylbenzene to styrene over ZrO_2-based mixed metal oxide catalysts with CO_2 as soft oxidant. *Chinese Journal of Catalysis* 37 (1): 3–15. Available at: https://doi.org/10.1016/S1872-2067(15)60901-9.

66 Sancheti, S.V. and Yadav, G.D. (2021). Highly selective production of styrene by non-oxidative dehydrogenation of ethylbenzene over molybdenum-zirconium mixed oxide catalyst in fixed bed reactor: activity, stability and kinetics. *Catalysis Communications* 154 (March): 106307. Available at: https://doi.org/10.1016/j.catcom.2021.106307.

67 Rahmanivahid, B., Dios, M.P.-d., Haghighi, M. et al. (2019). Mechanochemical synthesis of $CuO/MgAl_2O_4$ and $MgFe_2O_4$ spinels for vanillin production from isoeugenol and vanillyl alcohol. *Molecules* 24: 1–18.

68 Jeon, W., Jeon, W., Choi, I.H. et al. (2020). Alkaline wet oxidation of lignin over Cu-Mn mixed oxide catalysts for production of vanillin. *Catalysis Today* 352: 95–103. Available at: https://doi.org/10.1016/j.cattod.2019.12.037.

69 Saha, S. and Hamid, S.B.A. (2016). Nanosized spinel Cu-Mn mixed oxide catalyst prepared via solvent evaporation method for liquid phase oxidation of vanillyl alcohol using air and H_2O_2. *RSC Advances* 6 (98): 1–27. Available at: https://doi.org/10.1039/C6RA21221B.

70 Jha, A., Patil, K.R., and Rode, C.V. (2013). Mixed Co–Mn oxide-catalysed selective aerobic oxidation of vanillyl alcohol to vanillin in base-free conditions. *ChemPlusChem* 78: 1384–1392. Available at: https://doi.org/10.1002/cplu.201300247.

71 Mahdavi, V. and Monajemi, A. (2014). Optimization of operational conditions for biodiesel production from cottonseed oil on CaO–MgO/Al_2O_3 solid base catalysts. *Journal of the Taiwan Institute of Chemical Engineers* 45 (5): 2286–2292. Available at: https://doi.org/10.1016/J.JTICE.2014.04.020.

72 Maroa, S. and Inambao, F. (2021). A review of sustainable biodiesel production using biomass derived heterogeneous catalysts. *Engineering in Life Sciences* 21 (12): 790–824. Available at: https://doi.org/10.1002/elsc.202100025.

73 Chang, F., Zhou, Q., Pan, H. et al. (2014). Solid mixed-metal-oxide catalysts for biodiesel production: a review. *Energy Technology* 2 (11): 865–873. Available at: https://doi.org/10.1002/ente.201402089.

74 Gabriel, R., De Carvalho, S.H.V., da Silva Duarte, J.L. et al. (2022). Mixed metal oxides derived from layered double hydroxide as catalysts for biodiesel production. *Applied Catalysis A: General* 630 (October 2021): 1–17. Available at: https://doi.org/10.1016/j.apcata.2021.118470.

75 Zul, N.A., Ganesan, S., Hamidon, T.S. et al. (2021). A review on the utilization of calcium oxide as a base catalyst in biodiesel production. *Journal of Environmental Chemical Engineering* 9 (4): 105741. Available at: https://doi.org/10.1016/j.jece.2021.105741.

76 Taufiq-Yap, Y.H., Teo, S.H., Rashid, U. et al. (2014). Transesterification of *Jatropha curcas* crude oil to biodiesel on calcium lanthanum mixed oxide catalyst: effect of stoichiometric composition. *Energy Conversion and Management* 88: 1290–1296. Available at: https://doi.org/10.1016/j.enconman.2013.12.075.

77 Kesić, Ž., Lukić, I., Zdujić, M. et al. (2016). Assessment of $CaTiO_3$, $CaMnO_3$, $CaZrO_3$ and $Ca_2Fe_2O_5$ perovskites as heterogeneous base catalysts for biodiesel synthesis. *Fuel Processing Technology* 143: 162–168. Available at: https://doi.org/10.1016/j.fuproc.2015.11.018.

78 Kaur, A. and Ali, A. (2020). Lithium zirconate as a selective and cost-effective mixed metal oxide catalyst for glycerol carbonate production. *Industrial and Engineering Chemistry Research* 59 (7): 2667–2679. Available at: https://doi.org/10.1021/acs.iecr.9b05747.

79 Marimuthu, M., Marimuthu, P., Ashok Kumar, S.K. et al. (2018). Tuning the basicity of Cu-based mixed oxide catalysts towards the efficient conversion of glycerol to glycerol carbonate. *MolecularCatalysis* 460 (August): 53–62. Available at: https://doi.org/10.1016/j.mcat.2018.09.002.

80 Liu, G., Yang, J., and Xu, X. (2020). Synthesis of hydrotalcite-type mixed oxide catalysts from waste steel slag for transesterification of glycerol and dimethyl carbonate. *Scientific Reports* 10 (1): 1–14. Available at: https://doi.org/10.1038/s41598-020-67357-z.

81 Dalma, A., Nancy, B., Griselda, E. et al. (2022). Mixed oxides from calcined layered double hydroxides for glycerol carbonate production to contribute to the biodiesel economy. *Frontiers in Chemical Engineering* 4 (October): 1–10. Available at: https://doi.org/10.3389/fceng.2022.1027152.

82 Venugopal, A., Sarkari, R., Kumar, S.N. et al. (2014). Effective utilization of glycerol for the synthesis of 2-methylpyrazine over ZnO-$ZnCr_2O_4$ catalyst. *Journal of Chemical Sciences* 126 (2): 387–393. Available at: https://doi.org/10.1007/s12039-014-0585-0.

83 Safaei-Ghomi, J., Khojastehbakht-Koopaei, B., and Shahbazi-Alavi, H. (2014). Pseudo five-component process for the synthesis of 4,4′-(arylmethylene) bis(3-methyl-1H-pyrazol-5-ol) derivatives using $ZnAl_2O_4$ nanoparticles in aqueous media. *RSCAdvances* 4 (86): 46106–46113. Available at: https://doi.org/10.1039/C4RA07584F.

84 Teodorescu, F., Slabu, A.I., Pavel, O.D. et al. (2020). A comparative study on the catalytic activity of ZnAl, NiAl, and CoAl mixed oxides derived from LDH obtained by mechanochemical method in the synthesis of 2-methylpyrazine. *Catalysis Communications* 133: 105829. Available at: https://doi.org/10.1016/j.catcom.2019.105829.

85 Ajaikumar, S. and Pandurangan, A. (2009). Efficient synthesis of quinoxaline derivatives over ZrO_2/MxOy (M = Al, Ga, In and La) mixed metal oxides supported on MCM-41 mesoporous molecular sieves. *Applied Catalysis A: General* 357 (2): 184–192. Available at: https://doi.org/10.1016/j.apcata.2009.01.021.

86 Atarod, M., Safari, J., and Tebyanian, H. (2020). Ultrasound irradiation and green synthesized CuO-NiO-ZnO mixed metal oxide: an efficient sono/nano-catalytic system toward a regioselective synthesis of 1-aryl-5-amino-1H-tetrazoles. *Synthetic Communications* 50 (13): 1993–2006. Available at: https://doi.org/10.1080/00397911.2020.1761396.

87 Khosravani Farahani, M., Fareghi-Alamdari, R., and Kiasat, A.R. (2018). Cooperative activation in the synthesis of flavanone antioxidants using a simple and highly efficient magnetically recoverable nano-Cu-$CoFe_2O_4$ catalyst. *Polycyclic Aromatic Compounds* 38 (5): 464–478. Available at: https://doi.org/10.1080/10406638.2016.1255233.

88 Bankar, B.D., Advani, J.H., and Biradar, A.V. (2021). Cubic $Cu_xZrO100_x$ as an efficient and selective catalyst for the oxidation of aromatics active methyl, alcohol, and amine groups. *Polyhedron* 200: 115129. Available at: https://doi.org/10.1016/j.poly.2021.115129.

89 Jian, J., Yang, D., Liu, P. et al. (2022). Solvent-free partial oxidation of cyclohexane to KA oil over hydrotalcite-derived Cu-MgAlO mixed metal oxides. *Chinese Journal of Chemical Engineering* 42: 269–276. Available at: https://doi.org/10.1016/j.cjche.2021.04.020.

90 Li, A., Pu, Y., Li, F. et al. (2017). Synthesis of dimethyl carbonate from methanol and CO_2 over Fe-Zr mixed oxides. *Journal of CO2 Utilization* 19: 33–39. Available at: https://doi.org/10.1016/j.jcou.2017.02.016.

91 Kumar, P., Srivastava, V.C., Gläser, R. et al. (2017). Active ceria-calcium oxide catalysts for dimethyl carbonate synthesis by conversion of CO_2. *Powder Technology* 309: 13–21. Available at: https://doi.org/10.1016/j.powtec.2016.12.016.

92 Mandooie, M., Rahimi, M., Nikravesh, G. et al. (2023). A comprehensive review on zinc-based mixed metal oxide catalysts for dimethyl carbonate synthesis via urea alcoholysis process. *Journal of Industrial and Engineering Chemistry* 117: 70–84. Available at: https://doi.org/10.1016/j.jiec.2022.10.042.

93 Kharissova, O.V., Kharisov, B.I., González, C.M.O. et al. (2019). Greener synthesis of chemical compounds and materials. *Royal Society Open Science* 6 (11): 191378. Available at: https://doi.org/10.1098/rsos.191378.

94 Sheldon, R.A. (2018). Metrics of green chemistry and sustainability: past, present, and future. *ACS Sustainable Chemistry and Engineering* 6 (1): 32–48. Available at: https://doi.org/10.1021/acssuschemeng.7b03505.

95 Wagh, D.P. and Yadav, G.D. (2018). Green synthesis of A-methylcinnamaldehyde via Claisen-Schmidt condensation of benzaldehyde with propanal over Mg–Zr mixed oxide supported on HMS. *MolecularCatalysis* 459 (August): 119–128. Available at: https://doi.org/10.1016/j.mcat.2018.08.014.

96 Tomar, R., Ebitani, K., and Chandra, R. (2019). *ChemistrySelect* 4 (12): 3577–3581.

97 Dissanayake, D., Achola, L.A., Kerns, P. et al. (2019). Aerobic oxidative coupling of amines to imines by mesoporous copper aluminum mixed metal oxides via generation of reactive oxygen species (ROS). *Applied Catalysis B: Environmental* 249: 32–41. Available at: https://doi.org/10.1016/j.apcatb.2019.02.037.

98 Dhale, L.A., Pathan, S.K., Deshmukh, S. et al. (2023). $Mg_xCu_{0.3}Zn_{0.7\text{-}x}La_yFe_{2\text{-}y}O_4$ magnetic mixed metal oxide nanocatalyst: synthesis, characterization and application for one-pot N-heterocycle synthesis. *Polycyclic Aromatic Compounds* 0 (0): 1–16. Available at: https://doi.org/10.1080/10406638.2023.2257845.

99 Nagarapu, L., Apuri, S., and Kantevari, S. (2007). Potassium dodecatugstocobaltate trihydrate ($K_5CoW_{12}O_{40}·3H_2O$): a mild and efficient reusable catalyst for the one-pot synthesis of 1,2,4,5-tetrasubstituted imidazoles under conventional heating and microwave irradiation. *Journal of Molecular Catalysis A: Chemical* 266 (1–2): 104–108. Available at: https://doi.org/10.1016/j.molcata.2006.10.056.

100 Naeimi, H., Didar, A., and Rashid, Z. (2017). Microwave-assisted synthesis of pyrido-dipyrimidines using magnetically $CuFe_2O_4$ nanoparticles as an efficient, reusable, and powerful catalyst in water. *Journal of the Iranian Chemical Society* 14 (2): 377–385. Available at: https://doi.org/10.1007/s13738-016-0986-8.

101 Bhagi, A.K., Kirsch, W., Vinay, K. et al. (2022). One pot synthesis of benzopyranones and benzoxazinones catalyzed by MMO. *Indian Journal of Chemistry (IJC)* 61 (11): 1173–1179. Available at: https://doi.org/10.56042/ijc.v61i11.65837.

102 Negi, S.S., Venugopalan, A.T., Raja, T. et al. (2014). A green chemistry approach to styrene from ethylbenzene and air on $Mn_xTi_{1\text{-}x}O_2$ catalyst. *RSCAdvances* 4 (100): 57087–57097. Available at: https://doi.org/10.1039/c4ra11814f.

103 Védrine, J.C. (2019b). Metal oxides in heterogeneous oxidation catalysis: state of the art and challenges for a more sustainable world. *ChemSusChem* 12 (3): 577–588. Available at: https://doi.org/10.1002/cssc.201802248.

104 Castaño, M.H., Molina, R., and Moreno, S. (2015). Oxygen storage capacity and oxygen mobility of Co-Mn-Mg-Al mixed oxides and their relation in the VOC oxidation reaction. *Catalysts* 5 (2): 905–925. Available at: https://doi.org/10.3390/catal5020905.

105 Dong, C., Qu, Z., Qin, Y. et al. (2019). Revealing the highly catalytic performance of spinel $CoMn_2O_4$ for toluene oxidation: involvement and replenishment of oxygen species using in situ designed-TP techniques. *ACS Catalysis* 9 (8): 6698–6710. Available at: https://doi.org/10.1021/acscatal.9b01324.

106 Li, L., Luo, J., Liu, Y. et al. (2017). Self-propagated flaming synthesis of highly active layered CuO-δ-MnO_2 hybrid composites for catalytic total oxidation of toluene pollutant. *ACS Applied Materials and Interfaces* 9 (26): 21798–21808. Available at: https://doi.org/10.1021/acsami.7b04380.

107 Li, W., Wu, P., Zhu, Y. et al. (2015). Catalytic degradation of bisphenol A by CoMnAl mixed metal oxides catalyzed peroxymonosulfate: performance and mechanism. *Chemical Engineering Journal* 279: 93–102. Available at: https://doi.org/10.1016/j.cej.2015.05.001.

108 Ibrahim, I., Ali, I.O., Salama, T.M. et al. (2016). Synthesis of magnetically recyclable spinel ferrite (MFe2O4, M=Zn, Co, Mn) nanocrystals engineered by sol gel-hydrothermal technology: high catalytic performances for nitroarenes reduction. *Applied Catalysis B: Environmental* 181: 389–402. Available at: https://doi.org/10.1016/j.apcatb.2015.08.005.

109 Gogoi, A., Navgire, M., Sarma, K.C. et al. (2017). Fe_3O_4-CeO_2 metal oxide nanocomposite as a Fenton-like heterogeneous catalyst for degradation of catechol. *Chemical Engineering Journal* 311: 153–162. Available at: https://doi.org/10.1016/j.cej.2016.11.086.

11

Recent Advances in Conversion of Carbon Dioxide into Value-Added Product over the Solid Base Catalyst

Rajan Singh and Kamal K. Pant

Indian Institute of Technology Delhi, Department of Chemical Engineering, Hauz Khas, New Delhi, 110016, India

11.1 Introduction

Fossil fuel reserves such as coal, petroleum, and natural gas have been extensively exploited to fulfill the energy demand of the growing world economy and increasing population [1, 2]. Along with the valuable energy, the combustion of these fossil fuels produces CO_2, contributing to the anthropogenic sources of CO_2 emissions. There has been a continuous rise in CO_2 emissions since the industrial revolution. Currently, the atmospheric concentration of CO_2 is around 405 ppm compared to 280 ppm at the preindustrial level [3]. In 2019, the global CO_2 emission was about 39.152×10^3 million tons of CO_2 equivalent. In the year 2020, there was a significant dip in CO_2 emissions due to the partial and full lockdown around the globe because of the COVID-19 pandemic. However, the emissions have now reached pre-COVID levels and are expected to rise further. The concentration of atmospheric CO_2 is projected to reach 570 ppm with a temperature rise of 1.9 °C by the end of the twenty-first century [4]. The rising CO_2 concentration has adversely affected our environment by increasing global average temperature, climate change, and melting of glaciers [5]. Besides, a 32% rise in ocean acidity is also reported [6]. Therefore, the global research community is compelled to search for and develop new technologies to mitigate CO_2 emissions. Three possible strategies to reduce CO_2 emission includes (i) the development of energy-efficient systems from which maximum energy is extracted with the minimum quantity of fuel, (ii) CO_2 capture and storage, and (iii) CO_2 capture and utilization of the captured CO_2 as feed to produce value-added chemicals and fuels also termed as carbon capture utilization and storage (CCUS) (Figure 11.1) [7, 8].

The first strategy targets designing energy-efficient technologies to reduce overall fuel consumption, including promotion of less carbon-intensive renewable energy sources such as solar, wind, and geothermal energy. The continuous and economical energy production from these renewable energy resources has not been feasible until now, and ongoing efforts are required to make this viable. The second strategy is

Solid Base Catalysts: Synthesis, Characterization, and Applications, First Edition.
Edited by Ravi Tomar, K.K. Pant, and Ramesh Chandra.

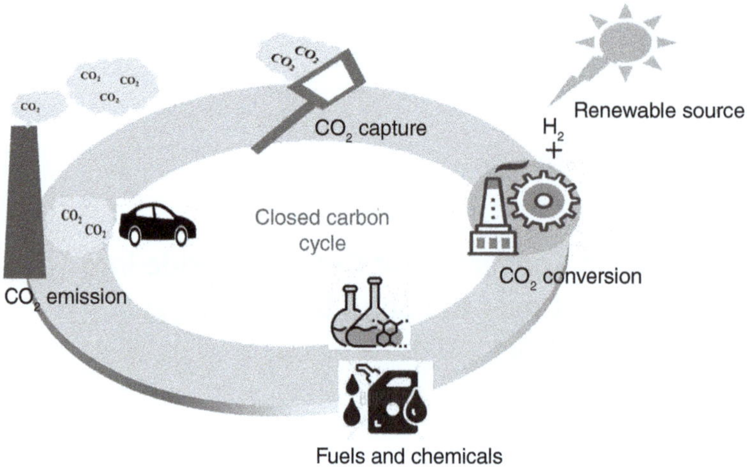

Figure 11.1 Closed carbon cycle.

CO_2 capturing and sequestration (CCS). The CCS process can be further subdivided into three steps: (i) capture, (ii) transportation, and (iii) storage. In general, CO_2 is captured from the point sources using different technologies such as chemical adsorption on liquid (generally amine), adsorption on solids, and membrane separation. The choice of CO_2 capturing methodology depends upon CO_2 concentration, capturing medium, temperature, and pressure [9]. Subsequently, the captured CO_2 is separated from adsorbate and transported for storage via ships and pipelines. CO_2 is pumped deep into the ocean, saline formations, aquifers, or depleted oil/gas wells under high pressure. The CCS technology is promising for the effective mitigation of CO_2 emissions. However, the major drawbacks are the economics of the CCS process and the need for continuous storage tank monitoring [10]. These drawbacks can be overcome by utilizing the captured CO_2 (CCU) as feed for synthesizing value-added chemicals. Using CO_2 as industrial feedstock alleviates the increasing CO_2 emissions and challenges researchers to explore new technological improvements in the present catalytic and industrial practices. Currently, CO_2 is used as feedstock in the industrial production of urea and its derivatives, such as melamine and urea resins. Further, CO_2 can be used as a feed to produce many industrially relevant chemicals such as methanol (MeOH), dimethyl ether (DME), olefins, and methane. The details of possible CO_2 hydrogenation products and the documented catalytic systems is shown in Figure 11.2 [11].

This text documents a concise examination of the thermodynamics and catalysts involved in the thermochemical conversion of CO_2 to products such as methane, MeOH, DME, and light olefins. It highlights the key catalysts that facilitate these transformations and discusses the energy requirements and efficiencies of these processes. Additionally, a thorough overview of the mechanisms driving these reactions is provided, shedding light on the steps and conditions necessary for effective conversion. This analysis aims to encapsulate the essential aspects of CO_2 conversion technology, emphasizing both the chemical principles and the innovative catalysts that make these processes viable.

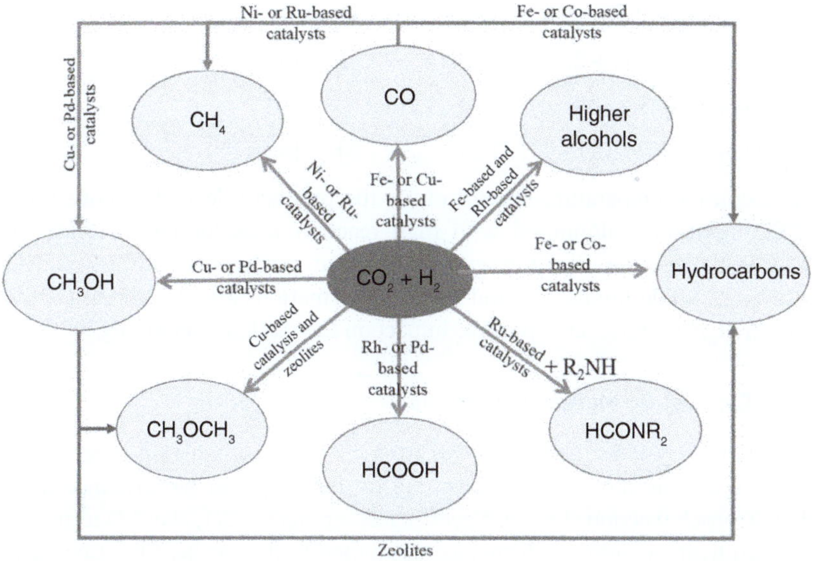

Figure 11.2 Routes for CO_2 hydrogenation to different products. Source: Ref. [11]/with permission of Elsevier.

11.2 CO_2 Hydrogenation to Methane

The CO_2 methanation reaction, also known as Sabatier reaction, was first documented in 1902 by Sabatier and Senderens with Ni-based catalyst. This reaction is recognized as one of the promising routes for CO_2 hydrogenation and finds practical application in ammonia industry and fuel cells. Further, the National Aeronautics and Space Administration (NASA) has suggested incorporating the methanation reaction into the closed cycle life support systems to recover oxygen. Specifically, CO_2 methanation has garnered growing attention within the realm of power-to-gas (PtG) technologies, primarily due to the preexisting infrastructure for methane storage and transportation. Several pilot plants have integrated CO_2 methanation into PtG processes, and the majority of industrial instances are notably concentrated in Germany [12, 13].

11.2.1 Thermodynamics of CO_2 Hydrogenation to Methane

The methanation reaction is exothermic in nature, with low temperature being thermodynamically favorable for higher methane yields. The thermodynamic stability of CO_2 poses a challenge as it requires higher temperature of activation. Thus, the Sabatier reaction (reaction (11.1)) is mainly operated at temperatures between 200 and 400 °C. The side reactions at the same reaction condition are namely reverse water–gas shift reaction (reaction (11.2)), carbon formation reaction (reaction (11.3)), Boudouard reaction (reaction (11.4)), and CO_2 methane reforming (reaction (11.5)).

$$CO_2 + 4H_2 \leftrightarrow CH_4 + 2H_2O \quad \Delta H_{298K} = -165\ \text{kJ mol}^{-1} \tag{11.1}$$

$$CO_2 + H_2 \leftrightarrow CO + 2H_2O \ \Delta H_{298K} = 41.2\ \text{kJ mol}^{-1} \quad (11.2)$$

$$CO_2 + 2H_2 \leftrightarrow C + 2H_2O \ \Delta H_{298K} = -90.1\ \text{kJ mol}^{-1} \quad (11.3)$$

$$CO_2 + C \leftrightarrow 2CO \ \Delta H_{298K} = 172.4\ \text{kJ mol}^{-1} \quad (11.4)$$

$$CO_2 + CH_4 \leftrightarrow 2CO + 2H_2O \ \Delta H_{298K} = -247.3\ \text{kJ mol}^{-1} \quad (11.5)$$

The operating temperature is key to tune the product selectivity; above 600 °C the CO formation is dominant with high reaction rates for reverse water–gas shift (RWGS) reaction, Boudouard reaction and methane reforming reaction (Figure 11.3), while the carbon and methane formations are disfavored. Further, the higher H_2/CO_2 ratio has a positive impact on the methane selectivity.

11.2.2 Catalyst for Methane Synthesis

Many catalytic systems have been reported for the methanation reaction. Ni, Fe, and Co are the most used metals, due to their superior performance in the Fischer–Tropsch reaction [14, 15]. Nevertheless, opting for CO_2 over CO resulted in heightened hydrogenation performance, attributed to the weaker CO_2–Co adsorption in contrast to CO–Co. This weaker adsorption led to the formation of CH_4 instead of long-chain hydrocarbons. Additionally, the higher cost of Co compared to Ni has impeded its widespread adoption for industrial-scale methane production. While pure Fe alone has not demonstrated comparable CO_2 methanation performance, its combination with other metals, such as Co and Ni, has the potential to enhance process yield by improving CH_4 selectivity. The formation of Fe-Co alloys in Fe-Co/alumina systems increased methane selectivity by reducing aluminate formation compared to Co/alumina – Ni-based catalyst. Nevertheless, the metal's performance is significantly influenced by the interaction and properties of the metal–support interface. CeO_2 emerges as the superior support for Ni when compared to yttria-stabilized zirconia (YSZ), TiO_2, Al_2O_3, and MgO, demonstrating a remarkable 100% selectivity to methane. Ceria oxide proves to be highly effective in providing substantial oxygen quantities due to its facile oxygen exchange characteristics, thereby enhancing the CO_2 hydrogenation activity. Moreover, ZrO_2–CeO_2 supports exhibit an impressive 99% selectivity to CH_4.

Noble metals such as Rh, Ru, Pd, and Pt supported on various oxides have been reported as very active methanation catalysts even at low temperatures and low metal loading. Ru and Rh emerge as the most active catalysts in CO_2 methanation, demonstrating significantly higher turnover numbers (approximately 20 times) than other noble metals when supported on alumina. When used as a dopant, Ru proves capable of facilitating the reduction of CeO_2 at lower temperatures compared to undoped Ce-oxide, resulting in around 50% CO_2 conversion. Furthermore, oxides based on Ru and Rh exhibit notable methane selectivity, reaching about 99%, in contrast to Pt or Pd, which tend to be more selective toward CO via RWGS. Additionally, noble metals show increased resistance to deactivation across a broad temperature range, attributed to fewer carbon deposits. However, the limited availability and high cost of noble metals have hindered their widespread application in CO_2 methanation [12, 13].

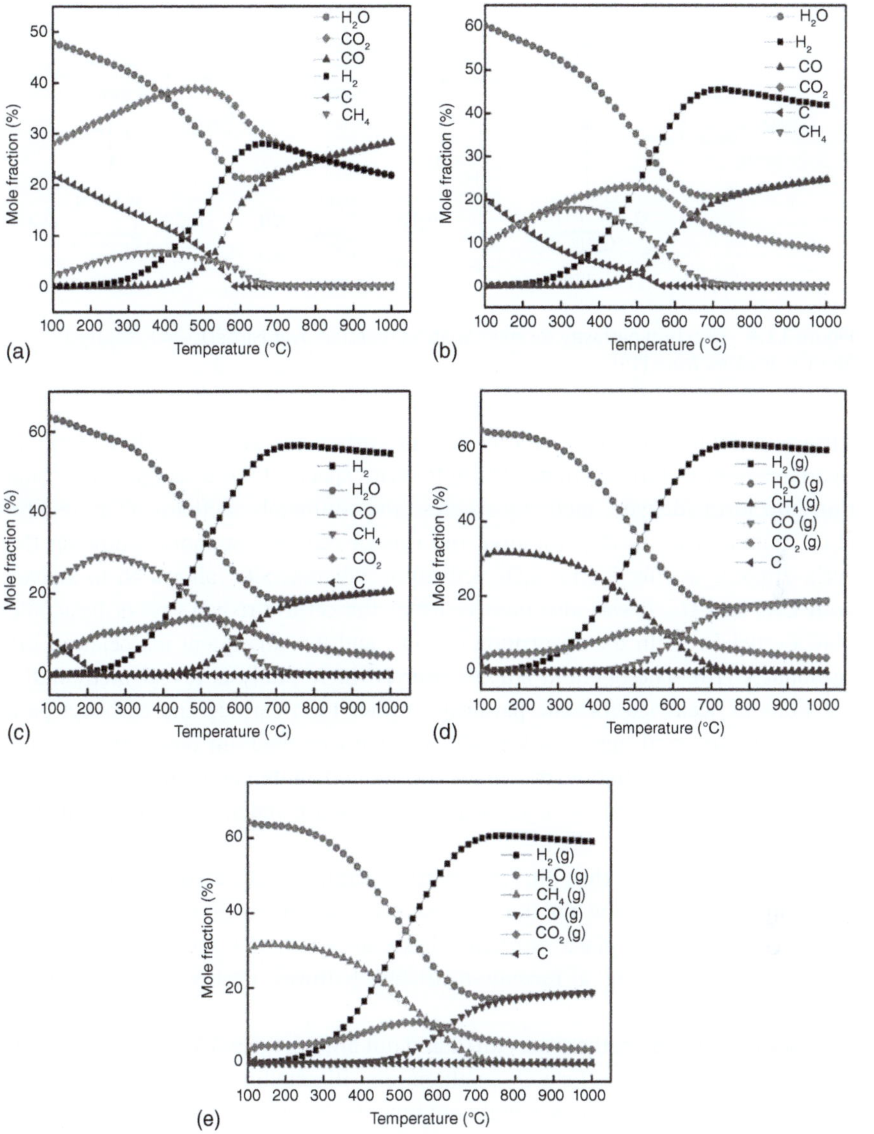

Figure 11.3 The equilibrium composition of components as a function of reaction temperature in the CO_2 hydrogenation reaction with varied H_2/CO_2 ratios. (a) $H_2/CO_2 = 1/1$, (b) $H_2/CO_2 = 2/1$, (c) $H_2/CO_2 = 3/1$, (d) $H_2/CO_2 = 3.5/1$, (e) $H_2/CO_2 = 4/1$. $P = 1$ bar.

11.2.3 Proposed Reaction Pathways for CO_2 Hydrogenation to Methane

Ashok et al. demonstrated that methanation proceeds via the formate pathway on Ni/CZ (CeO_2–ZrO_2). Ni plays a role in dissociating H_2 and transferring the resulting hydrogen atoms to the Ce species. CO_2 can adsorb on the catalyst

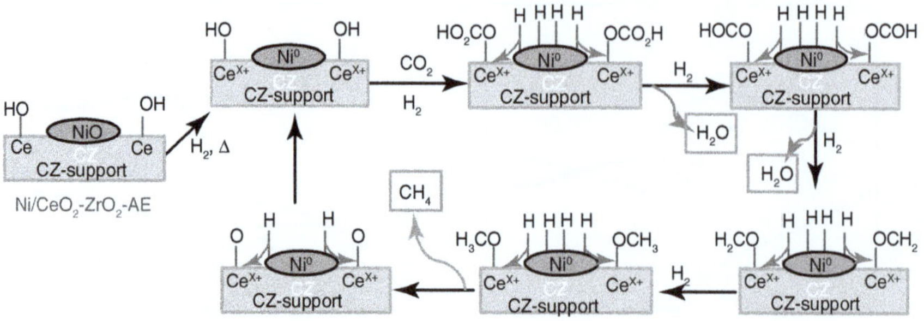

Figure 11.4 Reaction pathway for methanation reaction over Ni/CeO_2-ZrO_2 catalyst. Source: Adapted from [14].

surface in the form of mono- and bi-dentate carbonates, subsequently undergoing hydrogenation to form formate. These formate species then undergo transformation into formaldehyde, methoxy species, and ultimately methane. (Figure 11.4) [14]. Alanda et al. further explored the role of different supports, showing that ceria–zirconia is effective for CO_2 activation, whereas SiO_2 allows Ni to activate both CO_2 and H_2. It was also observed that the ceria–zirconia's weak basic sites play a crucial role in the adsorption of CO_2, and its subsequent dissociation into carbonates [16]. The operand studies over Ni/SiO_2 catalysts reveal methanation reactions to occur via formate pathways, rather than carbide and carboxyl pathways [17]. Pan et al. reported $Ce_{0.5}Zr_{0.5}O_2$, with its medium basic sites, notably improves CO_2 methanation by enhancing monodentate carbonate formation. In contrast, basic sites over γ-Al_2O_3 lead to less active bidentate formates in CO_2 hydrogenation [18].

Jia et al. suggested that Ni/ZrO_2 could facilitate CO_2 methanation through two pathways depending on the catalyst preparation method. Specifically, CO_2 hydrogenation takes place through dissociative or CO-intermediated routes over Ni/ZrO_2–P, while a formate-mediated pathway occurs over Ni/ZrO_2–C (Figure 11.5) [19].

Rhodium (Rh) and ruthenium (Ru) stand out among the noble metals as particularly effective active sites for CO_2 methanation. Bimetallic versions, incorporating more abundant non-precious metal components, have garnered significant attention. In a study by Martin et al., the CO_2 hydrogenation over ceria-supported Ni and Rh catalysts was investigated at atmospheric pressure. Activated CO_2 on the Ce^{3+} sites underwent a reaction with the dissociated hydrogen over Ni and Rh, resulting in formate formation. The study identified the presence of large Ni nanoparticles (6 nm) and small Rh nanoparticles (4 nm) on CeO_2 as support, leading to a weaker Ni-support interaction. Consequently, Rh/CeO_2 exhibited higher activity toward CH_4 formation. Introducing potassium (K) to Rh/Al_2O_3 and RhNi/Al_2O_3 catalysts results in the formation of an active intermediate conducive to carbon monoxide (CO) production. Specifically, K modifies the Rh sites, enhancing CO desorption and simultaneously boosting the catalyst's ability to dissociate hydrogen (H_2) [20] (Figure 11.6).

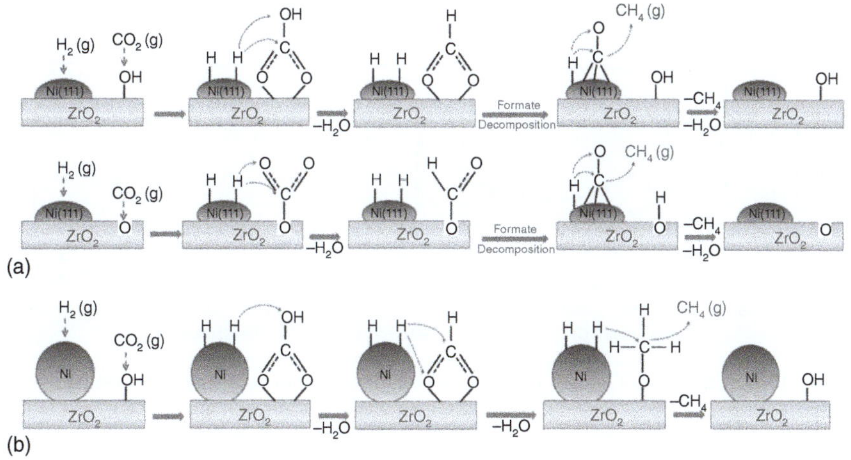

Figure 11.5 Methanation reaction pathways over Ni/ZrO_2-C catalyst. Source: Jia et al. [19]/ with permission of Elsevier. (a) CO route (b) formate route.

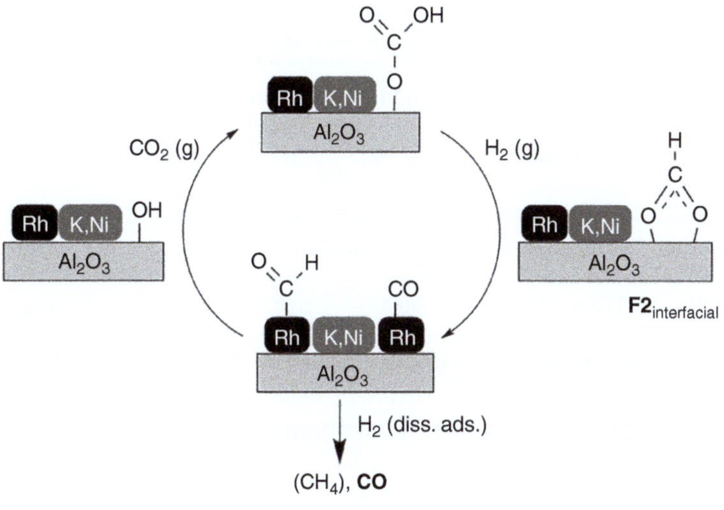

Figure 11.6 Methanation reaction pathways over RhNi/Al_2O_3 catalyst. Source: Heyl et al. [20]/with permission of American Chemical Society.

11.3 CO_2 Hydrogenation to Methanol

11.3.1 Thermodynamics of CO_2 Hydrogenation to Methanol

CO_2 is a highly inert linear molecule with a carbon atom combined with two oxygen atoms by a covalent bond having 116.3 pm bond length with 1072 kJ mol^{-1} dissociation energy [21]. The carbon atom of CO_2 is in +4 oxidation state, which is the most stable configuration. Thus, thermodynamically, MeOH synthesis is an energy-driven

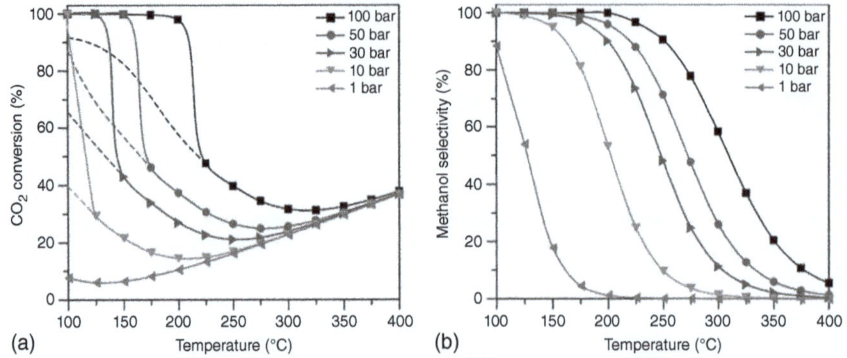

Figure 11.7 Equilibrium (a) CO_2 conversion and (b) methanol selectivity at various pressure [23].

process and requires 228 kJ of energy and six electrons to convert one mole of CO_2 to MeOH [22].

Following are the main reactions occurring during MeOH synthesis.

The first reaction is MeOH synthesis from CO_2 and hydrogen (reaction (11.6))

$$CO_2 + 3H_2 \rightarrow CH_3OH + H_2O \ \Delta H^0_{f298K} = -49.58\frac{kJ}{mol} \ \Delta G^0_f = 3.82\frac{kJ}{mol} \quad (11.6)$$

The RWGS reaction that produces CO and H_2O (reaction (11.7))

$$CO_2 + H_2 \rightarrow CO + H_2O \ \Delta H^0_{f298K} = 48.96\frac{kJ}{mol} \ \Delta G^0_f = 55.262\frac{kJ}{mol} \quad (11.7)$$

The two-step CO_2 hydrogenation can also occur; the first CO_2 is converted to CO via RWGS and then hydrogenated to MeOH (reaction (11.8)).

$$CO + 2H_2 \rightarrow CH_3OH \ \Delta H^0_{f298K} = -90.45\frac{kJ}{mol} \ \Delta G^0_f = -24.612\frac{kJ}{mol} \quad (11.8)$$

The MeOH synthesis by CO_2 and CO hydrogenation is exothermic, whereas RWGS is an endothermic reaction [23]. Therefore, high pressure and low temperature shift the equilibrium toward MeOH synthesis according to the Le-Chataliers principle (Figure 11.7). In addition, a high molar ratio of H_2/CO_2 is favorable for MeOH synthesis both thermodynamically as well as kinetically [23]. The equilibrium MeOH yield at 230 °C and 50 bar is around 30%. Therefore, efficient recycling of unreacted feed gases is required after the separation of products [24]. Moreover, CO_2 reduction to CO (RWGS) is very facile under similar reaction conditions. The RWGS reaction is endothermic; thus, high temperature favors CO production. The RWGS reaction lowers MeOH yield and increases water production, promoting catalyst deactivation.

11.3.2 Catalytic System for CO_2 Hydrogenation to Methanol

The Cu-ZnO-Al_2O_3 catalyst system is widely used in the commercial synthesis of MeOH from synthesis gas (syngas), which is a mixture of CO, CO_2 (>5%), and hydrogen H_2. The typical operating conditions for MeOH synthesis using this catalyst system are within the temperature range of 200–300 °C and at pressures between 50 and 100 bar. The reaction pathways for CO and CO_2 hydrogenation to MeOH on Cu–ZnO–Al_2O_3 catalysts share many similarities, primarily because both processes involve the activation and hydrogenation of carbon oxides in the presence of hydrogen. The Cu component of the catalyst is known for its excellent ability to catalyze the hydrogenation reactions, while ZnO acts as a structural promoter that enhances the dispersion of Cu and contributes to the overall catalytic activity and selectivity toward MeOH. Al_2O_3 serves as a support material that provides structural stability and affects the interaction between Cu and ZnO, further influencing the catalyst's performance.

To enhance the performance of the Cu–ZnO catalyst system in MeOH synthesis, research has focused on incorporating various promoters to tweak the catalyst's surface characteristics, such as its basicity and hydrophobicity. These adjustments are pivotal for increasing the reaction's speed and directing it more selectively toward MeOH production while minimizing by-products. Promoters such as K and Ba, chosen for their basic properties, improve CO_2 adsorption, thus facilitating its conversion. Additionally, the use of rare earth metal oxides (La, Ce, Nd, Pr) and transition metal oxides (ZrO_2, TiO_2) has been explored for their beneficial effects on the catalytic system. Noble metals such as Pd, Pt, and Au are also employed as promoters, further refining the catalyst's performance by modifying its electronic and structural properties, leading to improved efficiency and selectivity in MeOH synthesis.

Palladium-based catalysts are frequently used for hydrogenation reactions, especially for the selective reduction of double and triple-bond carbon [25–27]. Dissociative adsorption of H_2 over Pd^0 makes it a hydrogen source during the hydrogenation reaction. The activity of Pd-based catalysts is strongly influenced by the nature of support and its preparation method. In addition, the formation of bimetallic sites with metal oxides such as ZnO, Ga_2O_3, and In_2O_3 during in situ H_2 reduction facilitates the performance of Pd-based catalytic systems.

Jiang et al. studied the synergistic effect of Pd–Cu bimetallic catalyst supported over amorphous SiO_2, SBA-15, MCM-41, and MSU-f. The formation of PdCu and $PdCu_3$ alloy phases on reduction facilitated MeOH synthesis over (Pd/(Pd + Cu), 0.25–0.34) Pd–Cu/amorphous silica catalyst [28]. Besides, Lin et al. reported the effect of supports (TiO_2, ZrO_2, CeO_2, Al_2O_3, and SiO_2) on the physicochemical properties of the Pd–Cu bimetallic catalysts [29]. The TiO_2-supported Pd–Cu shows the best catalytic performance with $S_{CH_3OH} = 25.70\%$ and $X_{CO_2} = 16.40\%$ at 250 °C and 41 bar, due to formation of $PdCu_3$ bimetallic sites on H_2 pre-treatment. Whereas, for Pd–Cu/CeO_2 catalyst the $PdCu_3$ alloy re-crystalizes, adversely affecting the catalytic activity. In contrast, theoretical studies confirm stepped PdCu (111) surface with unsaturated Pd atoms is more active for MeOH synthesis than flat $PdCu_3$ (111) facets [30]. Moreover, in situ DRIFTS (diffuse reflectance infrared Fourier transform spectroscopy) reveals CO_2 adsorbs as carbonyl and formate species over Pd–Cu bimetallic catalyst surface. The surface concentration of the reaction intermediates increased over bimetallic sites compared to monometallic sites implying superior catalytic activity due to formation of bimetallic catalysts [28, 30–32].

The hydrogen spillover on Pd^0 surfaces facilitated the formation of Pd–Zn alloy by the reduction of ZnO around the Pd edges. This alloying alters the nature of Pd by charge transfer to fill its d-band [33]. Thus, the d-band shifts below with fermi level such as Cu. At higher reduction temperatures Pd–Zn alloy and Zn rich β-Pd–Zn alloy is formed, increasing the concentration of oxygen vacancies on the ZnO surface [34]. Liang et al. reported Pd–Zn bimetallic catalyst supported over multi-walled carbon nanotubes (MWCNTs) [35]. The increase in surface concentration of Pd–Zn sites promoted H_2 adsorption, improving catalytic activity. Ota et al. derived Pd–Zn bimetallic catalyst from reductive decomposition of HTLs precursor [36]. Homogeneous distribution of Pd^{2+} cation interacting with ZnO periphery was formed after catalyst pre-treatment. Liao et al. incorporated CdSe to ZnO support, and the formation of type II hetero junction helped in ZnO reduction [37]. Thus, a highly dispersed Zn-enriched nanocore-shell bimetallic Pd@Zn alloy particle was synthesized. The same group later synthesized Pd–Zn bimetallic catalyst derived after pyrolysis of Pd nanoparticle confined in zeolitic imidazolate framework (ZIF-8). The MeOH yield of 0.65 g gcat^{-1} h^{-1} at 270 °C, 45 bar was reported, with a TOF of 972 h^{-1} [38]. Bahruji et al. prepared Pd–Zn catalyst by sol immobilization method controlling the size distribution of Pd–Zn NPs. The MeOH selectivity decreased from 60% for 3 nm average diameter to 20% at 7 nm (250 °C and 20 bar) [39].

Gallium oxide-supported Pd catalyst has been used extensively for steam reforming and MeOH dehydrogenation reactions. Fujitani et al. were the first to report Pd–Ga_2O_3 catalyst for CO_2 hydrogenation to MeOH [40]. The improvement in the catalytic performance of Pd–Ga_2O_3 catalyst was attributed to the formation of Pd^{n+} ($0 < n < 2$) species stabilized by Ga_xO_y layer on the Pd surface [41]. The β-Ga_2O_3 supported Pd-Ga catalyst with high Pd_2Ga concentration shows better MeOH selectivity compared to α-Ga_2O_3, α-β-Ga_2O_3 supported Pd-Ga catalyst [42].

The experimental and theoretical investigations confirm the catalytic activity of the In_2O_3 catalyst for CO_2 hydrogenation to MeOH. The In_2O_3 catalyst is highly selective to MeOH synthesis reactions with low CO_2 conversion due to the absence of sufficient atomic hydrogen on the In_2O_3 surface. The DFT (density functional theory) study over Pd-doped In_2O_3 catalyst predicts improvement in catalytic activity, as Pd and bimetallic Pd–In interfacial sites are active sites for adsorption of H_2 and CO_2, respectively [43]. In contrast, experimental studies report poor MeOH yield for Pd–In bimetallic catalysts due to the strong interaction between Pd and In_2O_3. The formation of Pd–In bimetallic sites alters the nature of interfacial sites, resulting in poor H_2 chemisorption as compared to monometallic Pd catalyst [44]. Therefore, correct tuning between Pd and In_2O_3 is required to improve the catalytic performance of the Pd/In_2O_3 catalyst. Qingfeng et al. used a peptide template for the synthesis of highly dispersed Pd NPs (3.6 nm) on In_2O_3. The Pd–In bimetallic surface formed after reduction displayed improved hydrogen adsorption and preserved the density of the oxygen vacancies on the In_2O_3 surface [45]. Further, Pd–In bimetallic catalyst supported on SBA-15 improved Pd and In dispersion. The metal–support interaction facilitated the catalytic performance of the Pd/In_2O_3/SBA-15 catalyst with S_{CH_3OH} = 83.90% and X_{CO_2} = 12.60% at 260 °C and 50 bar [46].

Au supported over ZnO, TiO_2, Al_2O_3, CeO_2, and ZrO_2 is active for MeOH synthesis. High MeOH selectivity on Au/ZnO catalyst than Au/ZrO_2, Au/TiO_2, and Au/Al_2O_3 catalyst implies the influence of metal–support interaction. The catalytic activity of Au/ZnO catalyst improves with CO co-feeding due to the formation of oxygen vacancy over the ZnO surface [47]. Moreover, a higher MeOH formation rate over Au/CeO_2 (4.1×10^{-6} mol s^{-1} g Au^{-1} at 250 °C) catalyst than Au/Al_2O_3, Au/TiO_2, Au/Fe_2O_3, and Au/ZnO [48] is attributed to the presence of oxygen vacancies [49]. However, the Au/CeO_2 catalyst deactivates fast due to either over-reduction of support or adsorption of active sites by carbonate species [49]. The theoretical investigations demonstrated that CO_2 activation over Au/CeO_2 catalyst requires both metallic Au and Ce^{3+}. In addition, the CeO_2/TiO_2 interfacial sites on Au/CeO_2/TiO_2 catalyst activated Au nanoparticles by charge redistribution, enhancing CO_2 activation [50]. Further, the formation of reactive $Au^{\delta+}$-In_2O_{3-x} interfacial sites over Au/In_2O_3 catalyst facilitates MeOH synthesis [51].

Men et al. prepared highly dispersed Pt NPs by cold-plasma/peptide-assembly (CPPA) method. The catalytic activity was tested in DBD (dielectric barrier discharge) cold-plasma reactor, and 37.0% CO_2 conversion with 62.6% MeOH selectivity was achieved at atmospheric pressure and 30 °C [52]. Besides Pt_3Co octopods and carbon-supported Pt_4Co nanorods are reported to be active for low-temperature MeOH synthesis [53]. Moreover, the interaction between Pt and reduced MoOx promoted selective CH_3OH synthesis over Pt/MoOx/TiO_2 catalyst with 73% yield (150 °C, pCO_2 = 10 bar, pH_2 = 50 bar) [54].

The catalytic activity of Ag-based catalysts is low compared to Cu-based catalysts [55]. However, higher MeOH selectivity is reported for Ag/ZrO_2 than Cu/ZrO_2 at low temperatures (<227 °C) [56]. Whereas Ag/ZrO_2 exhibits higher catalytic activity with low MeOH selectivity than Ag/ZnO. This is due to the stability of Ag^+ on t-ZrO_2 crystal phase. Nevertheless, for Ag/ZnO/ZrO_2 system both activity and selectivity improved remarkably [57]. The first principle calculations over Ag (111)/ZnO (0001) surfaces show that d-band states of Ag exceed the fermi level on addition of ZnO (0001) to Ag (111). In addition, CO_2 binding on supported Ag (111) is also significantly enhanced. Therefore, the metal–support interaction facilitates the catalytic performance of Ag/ZnO catalyst [58]. Rh also promotes H_2 adsorption and spillover [59]. The formation of Rh–In bimetallic sites aids in the catalytic performance of Rh/In_2O_3 catalyst [60], 17.1% CO_2 conversion with 56.1% MeOH selectivity and 0.5448g_{MeOH} h^{-1} g_{cat}^{-1} STY was reported under 300 °C, 50 bar [61].

The DFT studies on defective [62] and non-defective [63] In_2O_3(110) facets confirm their catalytic activity for MeOH synthesis by CO_2 hydrogenation with cyclic creation and annihilation of oxygen vacancies. The cleavage of carbon–oxygen O bond, over the Cu surface occurs by dissociation of reaction intermediates [62]. Whereas, over In_2O_3 C—O bond cleavage is achieved by interaction of O atom of the reaction intermediates to fill the oxygen vacancy. The DFT investigation was experimentally confirmed with In_2O_3 catalyst, 54.9% MeOH selectivity with 3.69 mol h^{-1} kg cat^{-1} MeOH production rate at 330 °C and 40 bar [64]. Recently theoretical and experimental investigations confirm hexagonal In_2O_3 with (104) facets exhibits higher activity and MeOH selectivity than (110) facets rich cubic In_2O_3 [65].

DFT calculation over defective ZrO_2 supported In_2O_3 (110) reveals presence of ZrO_2 under In_2O_3 over layer inhibits RWGS reaction and enhances the concentration of oxygen vacancies on In_2O_3 surface [66]. Therefore, 99.8% MeOH selectivity was observed at 300 °C and 50 bar over ZrO_2 supported In_2O_3 catalyst. The In_2O_3/ZrO_2 catalyst was stable for 1000 h and no change in In_2O_3 crystallite size and surface area [67]. The strongly adsorbed CO_2 and H_2 facilitated MeOH synthesis over In_2O_3/ZrO_2 catalyst. However, too strong adsorption of reactants over In_2O_3/ZrO_2 catalyst decreases MeOH yield [64]. Moreover, formation of InOOH in presence of H_2O improved the catalytic activity of In_2O_3/ZrO_2 catalyst [68]. The higher oxygen vacancy concentration on m-ZrO_2 facilitates catalytic performance of In_2O_3/m-ZrO_2 compared to In_2O_3/t-ZrO_2 catalyst [65]. Besides hexagonal-phase In_2O_3 (h-In_2O_3) supported on Al_2O_3/Al-fiber matrixes is also reported as active catalyst. The h-In_2O_3/Al_2O_3/Al-fiber catalyst was successful in achieving 4.4% CO_2 conversion and 67.6% MeOH selectivity (325 °C and 40 bar) [69]. In addition, incorporation of Ga ions into In_2O_3 lattice increases the crystallite dispersion, BET surface area, and desorption. This enhanced MeOH yield for Ga_x/In_xO_{n-2} catalyst [70]. Chou et al. correlated increase in MeOH selectivity to the improvement in surface basicity on La and Y promoted In_2O_3 [71].

Li et al. reported formation of Mn–Co bimetallic with manganese oxide NPs supported on cobalt oxide synthesized by capillary inclusion method [72]. MnO is the active site for CO_2 reduction to CO, subsequently, CO is transformed to MeOH over CoO. Mn–Co spinel catalyst prepared by modified sol–gel inverse micelle method has also been reported active for CO_2 hydrogenation reaction. The physicochemical characterization of catalyst confirmed incorporation of Mn into the lattice of Co_3O_4 spinel, MeOH selectivity of 29.8% was reported at 250 °C and 20 bar [73].

Le and coworkers reported around 10% single-pass CO_2 conversion with >85% MeOH selectivity over ZnO–ZrO_2 catalysts. The superior catalytic performance of ZnO–ZrO_2 catalyst was attributed to the synergistic effects between Zr and Zn, where CO_2 adsorbs on the basic sites and unsaturated Zr whereas dissociative adsorption of hydrogen occurs over ZnO [74]. Moreover, Cd_xZrO_y and Ga_xZrO_y solid solutions are also active for MeOH synthesis, the synergistic effect facilitated heterolytic dissociative H_2 activation and promoted MeOH synthesis [75].

11.3.3 Reaction Pathways for Methanol Synthesis

The molecular level interactions between the reactants and catalyst surface provide us pivotal information for designing of highly active and stable catalyst. Many researchers revealed the mechanistic insights of MeOH synthesis reaction by conducting in situ experiments, modeling, and single crystal DFT. However, the research community has not reached to consensus and many questions are active for debate. These questions include nature and type of active site for CO_2 adsorption to initiate the hydrogenation reaction, the reaction pathways, intermediates, and carbon source (CO or CO_2). Here, we briefly present discussion on all the key aspects related to the mechanism of CO_2 hydrogenation.

CO_2 adsorbs very weakly over bare Cu^0 and CuO surface [76] thus, the catalytic activity for CO_2 hydrogenation reaction is poor. Literature reports a significant role of synergistic effects between Cu^0 and Cu^+ sites present in the catalyst. The synergistic effect at Cu–ZnO interfacial sites due to metal–support interaction improves the MeOH yield. Willinger et al. [77] reported the presence of metastable layer of "graphitic-like" ZnO over Cu/ZnO catalysts after reduction of $CuO/ZnO/Al_2O_3$ catalyst. In addition, ZnO acts as reservoir for atomic hydrogen and helps in spillover of hydrogen during hydrogenation reaction [78]. Furthermore, theoretical and experimental studies reveal that Cu/ZnO interfacial sites are responsible for promotional effect on C=O bond activation [79].

Moreover, formation of Zn–Cu alloy due to partial reduction of ZnO during reduction is also reported as active site [80]. The incorporation of Zn atoms onto Cu steps significantly decreases the energy barrier of MeOH synthesis thus enhancing the catalytic performance of $Cu/ZnO/Al_2O_3$ catalyst [81]. The migration of Zn atoms in the Cu surface promotes the activity of Cu–ZnO catalyst and correlation between Zn coverage and activity for MeOH synthesis reaction is experimentally confirmed [80]. Later The DFT and KMC (kinetic Monte Carlo) simulations over ZnCu (211) and Cu/ZnO revealed surface oxidation of ZnCu (211) occurs under the reaction conditions. The transformation of Zn to ZnO during reaction indicates pivotal role of Cu–ZnO interfacial sites and synergistic effects for the catalytic activity [82]. DFT calculations over Zn/Cu, ZnO/Cu, and ZnOH/Cu surfaces revealed ZnOH/Cu as the most active phase for MeOH synthesis [83].

Adsorption of reactants on the active sites of the catalyst is the initial step for the reaction. Hydrogen adsorb dissociatively over the bare Cu° metal surface. Whereas, CO_2 adsorption can be either dissociative over bare Cu° surface in co-adsorbed CO and O (such as hydrogen) or over pre-absorbed hydrogen by Eley–Rideal mechanism [76]. Atomic oxygen co-adsorbed with CO on bare Cu surface poisons the surface and inhibits further adsorption of CO_2 [84]. In addition, CO_2 is reported to adsorb over Cu/ZrO_2 interfacial surface in bent structure with two oxygen atoms binding with Zr^{4+} Lewis acidic sites and the carbon atom interacts with the copper particle surface [85]. Moreover, the oxygen vacancies on the surface of In_2O_3 and CeO_2 act as active sites for CO_2 adsorption. After successful adsorption of CO_2, the reaction is known to follow either formate (HCOO) pathway, hydrocarboxyl (COOH) pathway, or RWGS+ CO hydrogenation pathway based on the species formed on hydrogenation of adsorbed CO_2. Formate (HCOO) pathway has been prominently suggested in literature compared to hydrocarboxyl and RWGS + CO pathways.

11.3.3.1 HCOO Pathway

The DFT calculations on Cu (111) and Cu nanoparticle surface suggested MeOH synthesis to occur via formate (HCOO) pathway [86]. The formate species binds to the Cu nanoparticle in a bidentate configuration through two O atoms, further carbon of formate species was hydrogenated to H_2CO_2 in the proposed reaction mechanism. H-guided dissociation of H_2CO_2 forms H_2CO and OH, followed by hydrogenation of H_2CO to H_3CO. Hydrogenation of formate and dioxomethylene species was the rate-determining step (RDS) for both Cu (111) and Cu nanoparticles. Larmier et al. proposed formate pathway for MeOH synthesis over Cu/ZrO_2 surface. The formate species was hydrogenated to acetal (H_2COO) like species with two oxygen atoms bonding to the two Zr^{4+} atoms. The third C—H bond was formed in a concerted SN2-like step, while the C—OH bond was broken, forming methoxy species. Subsequently, methoxy species undergo protonation by hydrogen transfer forming MeOH [85]. Wang et al. revealed the role of oxygen vacancy and Cu/ZrO_2 interfacial sites for CO_2 adsorption and activation [87]. Whereas, Copéret and coworkers reported Lewis acidic sites on ZrO_2 are pivotal to initiate the reaction rather than oxygen vacancy on ZrO_2 surface. The interfacial sites between Zr^{4+} and Cu were the active site for MeOH synthesis [88].

In contrast, Garbow et al. [89] observed MeOH synthesis on Cu(111) to occur via formate pathway with HCOO, HCOOH, H_2COOH, H_2CO, CH_3O_2, and CH_3O intermediates. They reported hydrogenation of O atom of formate species was more favorable compared to hydrogenation of carbon atom. Therefore, formate species was hydrogenated to formic acid (HCOOH) followed by hydrogenation to H_2COOH, which dissociates to CH_2O and OH species. The CH_2O was subsequently hydrogenated to MeOH via methoxy intermediate. Again, Kattel et al. confirmed MeOH formation occurred via formate (HCOO) pathway with HCOOH, H_2COOH, and CH_3O as surface intermediates over Cu/ZnO [82]. In addition, Behrens et al. [90] also proposed similar HCOO pathways over stepped Cu(211) surface decorated with Zn atoms. The interfacial sites of Cu steps and ZnO_x improve adsorption of oxygen-bound intermediates and enhance the catalytic performance of Cu–ZnO catalyst. Moreover, HCOO pathways proposed by DFT studies on overstepped PdCu (111) with energetically favorable HCOOH, H_2COOH, and CH_3O surface intermediates [30].

In_2O_3 oxide is extensively used as support for tri-reforming of methane and CO_2 methanation reaction. The CO_2 adsorbs on In_2O_3(110) surface on the surface oxygen and forms carbonate species. Whereas, dissociative adsorption of H_2 occurs with one H combining with surface oxygen forming hydroxy and second H forming hydride over In. MeOH synthesis over In_2O_3(110) facets follows HCOO pathways [63]. The formate species is then subsequently hydrogenated to bi-H_2CO, mono-H_2CO, CH_3O, and then MeOH. The oxygen vacancy on In_2O_3 surface helps in CO_2 adsorption, activation and dissociation of reaction intermediates to follow MeOH synthesis pathways. The high concentration of oxygen vacancies on In_2O_3 surface inhibits dissociation of H_2 and protonation of H_3CO. Thus, MeOH yield over In_2O_3 (110) surface can be tuned by optimizing surface oxygen atoms and surface oxygen vacancies [91].

Pérez-Ramírez and coworkers reported superior activity of In_2O_3 (111) than In_2O_3 (110) facet for MeOH synthesis reaction via formate pathways. The oxygen atom of the formate (HCOO) species was hydrogenated to CHOOH surface species, followed by hydrogenation of carbon to CH_2OOH and then oxygen to $CH_2(OH)_2$. The $CH_2(OH)_2$ species dissociates to CH_2O and H_2O followed by CH_2O hydrogenation to methoxy species and then MeOH [92]. The proposed elementary reaction mechanism steps over In_2O_3(111) and In_2O_3(110) facets are different. Thus, both In_2O_3 facets and oxygen vacancies influence the MeOH synthesis mechanism. The reaction proceeds through formate pathway followed by HCOOH, H_2COOH, CH_2O, and CH_3O intermediates on both the PdIn (110) and PdIn (211) surfaces. However, RDS on PdIn (110) and PdIn (211) surfaces was hydrogenation of HCOOH to H_2COOH and dissociation of H_2COOH to CH_2O and OH, respectively [93].

11.3.3.2 COOH Pathways

The COOH route is also termed as trans-COOH pathway. Zheo et al. included the effect of water in DFT calculations over Cu(111) and proposed COOH pathway for MeOH synthesis [94]. In the presence of adsorbed water-derived species, the energy barrier for hydrogenation of formate to H_2COO increased significantly implying formate is a stable spectator under the reaction conditions. Instead, hydrogenation of adsorbed CO_2 to hydrocarboxyl (trans-COOH) is energetically more favorable in the presence of water. The adsorbed CO_2 initially hydrogenated via hydrogen transfer mechanism from H_2O and formed COOH. COOH is then hydrogenated to dihydroxycarbene (COHOH), which subsequently dissociates to OH and COH species. Further MeOH is formed via the hydrogenation COH to HCOH and H_2COH intermediates. Moreover, experimental studies over Cu/SiO_2 also confirm that presence of water (0.04–0.5 mol%) significantly increases rate of MeOH synthesis. The transient surface reaction experimentally proved that CO_2 hydrogenation follows hydro-carboxyl (COOH) pathway for MeOH synthesis and formate species is co-adsorbed spectator [95]. In addition, N_2O or O_2 pre-treated Cu catalyst facilitated MeOH synthesis, signifying the role of adsorbed surface oxygen or water-derived species in the reaction initiation by changing the bidentate formate to mono-dentate configuration [96]. Fang et al. reported trans-COOH reaction pathways over $PdCu_3$(111) with t-COOH, t,t-COHOH, t,c-COHOH, c,c-COHOH, COH, HCOH, and H_2COH reaction intermediates [97] (Figure 11.8).

11.3.3.3 RWGS+ CO Hydrogenation Pathways

Yang et al. performed DFT calculations on the transition metal doped Cu (111) and reported MeOH synthesis to occur via RWGS + CO hydrogenation. The adsorbed CO_2 was first hydrogenated to carboxyl (COOH), which then dissociates to CO and OH. Further, the CO is hydrogenated subsequently forming formyl (CHO), formaldehyde (CH_2O), and methoxy (CH_3O) intermediates. The increasing order of MeOH yield was Au/Cu(111), Cu(111), Pd/Cu(111), Ph/Cu(111), Pt/Cu(111) and Ni/Cu(111) [98]. Besides, Graciani et al. proposed similar mechanism over Cu/CeO_2 catalyst with copper-ceria interface active site [99]. Moreover, Kattel et al. also predicted RWGS + CO hydrogenation pathway over Cu/ZrO_2 and Cu/TiO_2 catalysts. The synergistic effects between Cu^0 and reduced Zr^{3+} facilitate MeOH synthesis reaction, improving selective MeOH synthesis over Cu/ZrO_2 catalyst. The in situ DRIFTS confirms presence of formate species on the catalyst surface. However, the DFT calculations suggested HCOO as poison to the catalyst surface and is only spectator during the MeOH synthesis reaction [100]. Gaikwad et al. reported similar RWGS + CO hydrogenation pathways for MeOH synthesis reaction

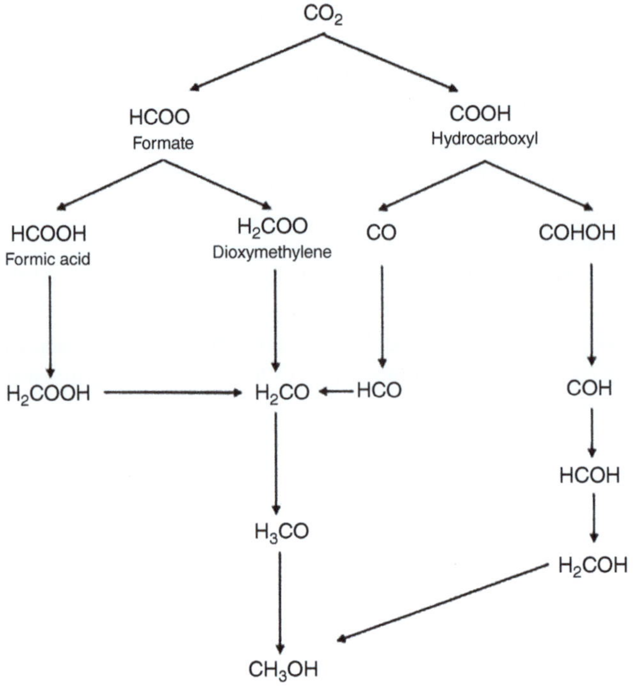

Figure 11.8 Reaction pathways proposed for CO_2 hydrogenation to methanol.

over $Cu/ZnO/Al_2O_3$ catalyst at high pressure (360 bar) [101]. DFT study over Pd/TiO_2 reveals that CO_2 activates and transforms to $CO_2^{\delta-}$ at the Pd–Ti interfacial sites. The reaction intermediates adsorbed on the interfacial sites were subsequently hydrogenated following $CO_2 \rightarrow COOH \rightarrow CO \rightarrow HCO \rightarrow H_2CO \rightarrow H_3CO \rightarrow CH_3OH$ route (RWGS + CO hydrogenation pathways) with CO hydrogenation to HCO as the RDS [102].

11.4 CO_2 Hydrogenation to Dimethyl Ether

11.4.1 Thermodynamics of Single-Step DME Synthesis

The direct CO_2 to DME synthesis process involves MeOH synthesis reaction (reaction (11.11)), RWGS reaction (reaction (11.7)), and MeOH dehydration reaction (reaction (11.9))

$$2CH_3OH \rightleftharpoons CH_3OCH_3 + H_2O \ \Delta H^0 = -23.4\ \text{kJ mol}^{-1}\ \Delta G^0 = -16.8\ \text{kJ mol}^{-1} \tag{11.9}$$

and a side reaction of paraffin formation, which mainly forms methane (reaction (11.10)):

$$nCO + (2n+1)H_2 \rightleftharpoons C_nH_{2n+2} + nH_2O\ (n = 1-3) \tag{11.10}$$

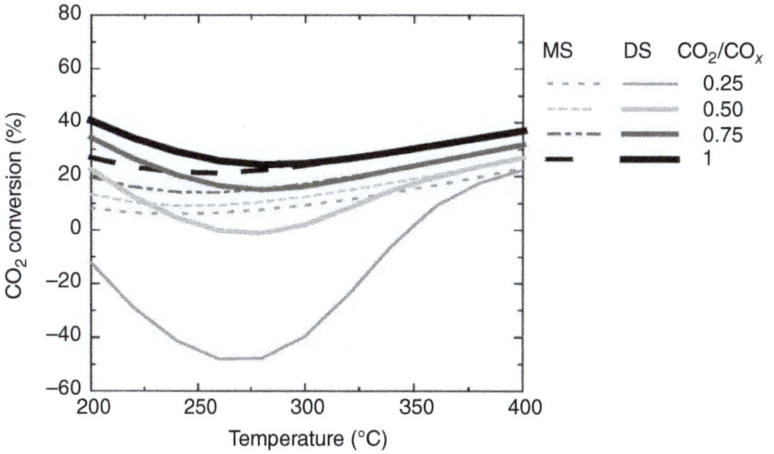

Figure 11.9 CO_2 conversion with temperature in methanol synthesis (MS) and direct DME synthesis (DS) with different $CO + CO_2$ feed. Source: Ateka et al. [106]/with permission of Elsevier.

The advantages of single-step DME synthesis include the following factors. (i) Thermodynamic factors i.e. performing MeOH dehydration reaction in the same reactor, shift the equilibrium of the MeOH synthesis reaction. (ii) direct DME synthesis is cheaper compared to the conventional two-step process [103]. First, the operating cost and energy requirement are lowered by 5% as a single reactor is to be operated. Second, the capital cost also decreases by 8% [104, 105]. Literature reporting thermodynamics of single-step DME synthesis from CO_2 feedstock is limited; major attention is given to DME synthesis by CO hydrogenation.

Bilbao and coworkers compared the thermodynamics of CO_2 to MeOH reaction and single-step DME synthesis reaction. The effect of process parameters was also determined. As CO_2 and CO hydrogenation is an exothermic reaction with a reduction in the number of moles, the high pressure is suitable for moving the reaction forward. At the same time, the high temperature lowers the DME yield. The authors report that the preferred temperature range for valorizing CO_2 mixed feed is 250–300 °C. Moreover, higher CO_2 conversion was observed for single-step DME synthesis reaction than for the MeOH synthesis reaction [106] (Figure 11.9).

The aforesaid discussion indicates the thermodynamic limitations of using CO_2 as feed for MeOH and DME synthesis. The single-pass CO_2 conversion is low compared to CO, and to achieve the same MeOH yield on an industrial scale, the plant must be operated with higher recycle ratio. Therefore, a suitable catalytic system is required to improve the MeOH yield, and the typical operating condition for MeOH and DME synthesis lies between 220–300 °C temperature and 20–50 bar pressure.

11.4.2 Catalysts for Single-Step DME Synthesis by CO_2 Hydrogenation

Solid-acid catalysts are used for MeOH dehydration to DME. The catalyst must be stable, hydrophobic selective, and active during the dehydration reaction. γ-Al_2O_3

is the most reported solid-acid catalyst due to its low cost and high selectivity [107, 108]. The mechanism of MeOH dehydration depends on the reaction conditions i.e. temperature and pressure. The MeOH dehydration reaction can follow (i) associative pathways, i.e. two MeOH molecules co-adsorb on an acidic site to produce DME or (ii) dissociative pathways where MeOH adsorbs over acidic site and decompose, eliminating water molecules and transfer into a surface methoxy group to react to another MeOH molecules forming DME. The activity of γ-Al_2O_3 decreases due to its hydrophilic nature. The adsorbed water facilitates the dealumination leading to a decrease in acidity, causing deactivation. Further, water plays multiple roles in MeOH dehydration (i) shifts the thermodynamic equilibrium of the reaction. (ii) diminishes the catalyst acidity by adsorbing over the acidic sites [109, 110] (iii) reduces the formation of methoxy species by shifting the equilibrium [111].

$$\mathrm{Al-OH+CH_3OH \rightleftharpoons Al-O-CH_3+H_2O} \tag{11.11}$$

For single-step DME synthesis via CO_2 hydrogenation, the abovementioned feature is more prevalent due to the presence of excess water (both MeOH synthesis and DME synthesis). Therefore, the research paradigm on acid catalysts has centered on developing and modifying the novel catalytic systems to alleviate the activity decay due to H_2O adsorption. This can be achieved by reducing hydrophobicity and improving the desorption of water from the acidic sites, considering that the DME synthesis process requires a relatively weaker acidic site to prevent the formation of hydrocarbons [112]. Uzun and coworkers have used MCM-41-supported tungstophosphoric acid as the dehydration component physically mixed with CuO–ZnO–Al_2O_3 catalyst [113]. Besides modification to γ-Al_2O_3 zeolites of different frameworks such as EUO (epsilon uniform one), BEA (beta), FER (ferrierite), TON (theta one nought), MTW (mobil twelve), MOR (modernite) [114, 115], MFI (mobil five) [116, 117] and silicoaluminophosphates as SAPO-18, SAPO-11 [118] have been used as MeOH dehydration catalyst.

Catizzone and coworkers screened different zeolite frameworks and studied the effect of crystal size, acidity, and Si/Al ratio on their activity for MeOH dehydration and reported that the FER and MFI-type zeolite outperformed the others in terms of stability, selectivity, and deactivation (less coke deposition) [119, 120]. HZSM-5 (H form of zeolite Socony Mobil-5) is the most studied zeolite because of its hydrothermal stability and activity due to its topology and acidic properties. But the major drawback with the use of HZSM-5 is the deactivation due to coke deposition because of the Brønsted acidic sites [121]. To overcome this application of modified HZSM-5 via various techniques has been widely documented [122–124]. Majority of them focus on tuning the acidity of zeolite to reduce the coke formation [125, 126]. Zeng and coworkers reduced the strength of acidic sites and generated mesopores in the zeolite structure by partial desilication and dealumination. The modified HZSM-5 exhibited superior catalytic performance and was hydrothermally stable and resistant to sintering. Moreover, Ordomsky reported that silication of HZSM-5 resulted in improved catalyst stability by minimizing the side reaction i.e. hydrocarbon synthesis. Aboul–Fotouh and coworkers altered the zeolite acidity by chlorination/fluorination of the zeolite. Aloise et al. documented an increase in pore diameter after desilication, which resulted in improved accessibility of the acidic sites and subsequently in DME yield. Hocine and coworkers prepared a bifunctional catalyst by physical mixing of Cu/ZnO/Al_2O_3 catalyst with hollow nano-HZSM-5 and reported 74% DME selectivity.

Bilbao and coworkers compared the activity of SAPO-18 and SAPO-11 with thermally and dry streaming treated HZSM-5 zeolite with SiO_2/Al_2O_3 ratios 80 and 280. They reported that the performance of the SAPO-11 catalyst was better compared to the other modified HZSM-5 and SAPO-18. Around 80% DME selectivity was observed for Cu–ZnO–ZrO_2/SAPO-11 at 275 °C and 30 bar. The superior catalytic performance of SAPO-11 is ascribed to the presence of AEL (aluminophosphate epsilon-L) topology in the pores and the high density of weak-strength acidic sites; this reduces the hydrocarbon adsorption and its retention time over the active site. Therefore, the coke formation was less with the SAPO-11. Further, Yi and coworkers synthesized SAPO-11 with a weaker acidic strength by the seed-induced quasi-solvothermal method. The increase in specific surface area and mesoporosity improved the activity of the MeOH dehydration reaction. The SAPO-34 was more active in MeOH conversion and DME selectivity, but due to large channels and narrow pore opening, was prone to deactivation due to pore blockage by higher hydrocarbon. The bifunctional catalyst for single-step DME synthesis comprises two different individual catalytic systems MeOH synthesis catalyst (discussed in section) and MeOH dehydration catalyst. These two catalyst constituents are combined, and the general strategy is to increase the composition of the acidic component to displace the MeOH synthesis equilibrium. As the primary step in the DME step is the MeOH synthesis reaction therefore, it is the rate-limiting step of DME synthesis and controls the overall reaction rate. Next, for MeOH dehydration reaction, an intimacy of contact between the MeOH dehydration catalyst and MeOH synthesis catalyst is very crucial; therefore, studies addressing the issues of catalytic bed configuration have been extensively documented. Peng and coworkers proposed that the proximity between the acidic and MeOH synthesis components could alter the reaction pathways for the DME synthesis. The in situ DRIFTS results revealed that for closely contacted MSC (methanol synthesis catalyst) and MDC (methanol dehydration catalyst) the DME synthesis reaction followed a shortcut methoxy-DME route instead of the previously reported methoxy-MeOH-DME pathway [127].

The literature documents following reactor configuration for single-step DME synthesis (a) dual bed configuration i.e. in the fixed reactor, the first bed is composed of metallic component for MeOH synthesis followed by MDC for its dehydration to DME; (b) physical mixture of both MSC and MDC component; (c) the most common hybrid configuration, where both MSC and MDC are mixed conforming bifunctional catalyst; (d) core–shell catalyst; (e) structure catalyst.

Ateka et al. compared (a), (b), and (c) reactor configurations with CuO–ZnO–MnO and SAPO-18 as MSC and MDC components, respectively. The constituents of the hybrid catalyst i.e. the MSC and MDC were mixed at an optimal 2 : 1 mass ratio and kept constant. The DME selectivity was 85%, implying that the dual bed configuration results in considerable catalyst activity. However, the total carbon conversion for mixed CO_2 and CO feed (1 : 1) was less than the mixed MDC and MSC catalyst (strategy b). Further, the physically mixed single bifunctional catalyst, 95% MeOH selectivity is obtained with around 22% total carbon conversion due to close contact between MSC and MDC catalyst [106, 128, 129]. These results were like the report published by Peng and coworkers with Cu–In–Zr (MSC) and SAPO-34 (MDC) catalysts [127].

In contradiction to above mentioned experimental observation, Frusteri and coworkers reported that the performance of the dual functional catalytic bed setup was superior to the single catalytic bed composed of pre-pelletized individual MSC and MDC catalyst for Cu–ZnO–ZrO_2 and HZSM-5 catalyst [130]. The deterioration in the catalyst activity can be ascribed to the pore blockage of the zeolite by the MeOH synthesis component during mortar treatment and palletization. Later the same group documented the effect of different precipitating agents on the synthesis of MSC catalyst in the zeolite slurry. The catalytic activity of the synthesized catalyst improved significantly due to enhanced hydrogenation activity over the hybrid catalytic system [131].

The synthesis of bifunctional catalyst core–shell structure of the MeOH dehydration component over the MeOH synthesis component has also been explored [132, 133]. The core-shell structure consists of depositing either MSC or MDC over the nucleus core of the other. The core–shell catalyst has been prepared by either, single crystal crystallization, hydrothermal synthesis, dual-layer method, or physical adhesive method. The core–shell structure prevents catalyst deactivation caused due to sintering of MSC catalyst and coke deposition. Moreover, the DME selectivity is also improved in the confined pore structure of the shell. Many core–shell catalytic systems have been documented in the literature with Cu–ZnO–Al_2O_3 catalyst as MSC catalyst along with HZSM-5, γ-Al_2O_3, SAPO-11, and SiO_2–Al_2O_3 as MDC catalyst [134]. Groppi and coworkers model analysis to understand the effect of kinetics, adsorption capacity, and heat and mass transfer on the performance of the bifunctional catalysts. The study demonstrated that MSC@MDC is the most suitable catalyst design configuration with minimum influence of internal diffusion on the catalyst activity. The catalyst structure favored DME formation and small size particle diameter with a lesser number of contact points between the MDC and MSC catalysts limited the hot spot formation [135, 136].

Bilbao and coworkers synthesized CuO–ZnO–ZrO_2@SAPO-11 core-shell catalyst [132] by the physical adhesive method (Figure 11.10). The CuO–ZnO–ZrO_2 and SAPO-11 were mixed in 1 : 2 mass ratio with SiO_2 as adhesive. In the synthesized catalyst configuration, DME was produced by dehydration of diffused MeOH from the MSC catalyst to DME over the acidic shell of SAPO-11. The results reveal that the adopted catalyst synthesis methodology resisted the pore blockage of the mesopores SAPO-11 that occurs during the catalyst pelletization. For mixed feed ($CO + CO_2$, 1 : 1) at 323 °C, 30 bar, and 7.6 g_{cat} h^{-1} mol^{-1} 8.7% DME yield with 81% selectivity was observed for the core-shell catalyst. Authors also proposed that due to separate MSC and MDC catalysts at different regions in the catalysts, the deactivation caused by sintering by water was limited [132]. Nie et al. also reported

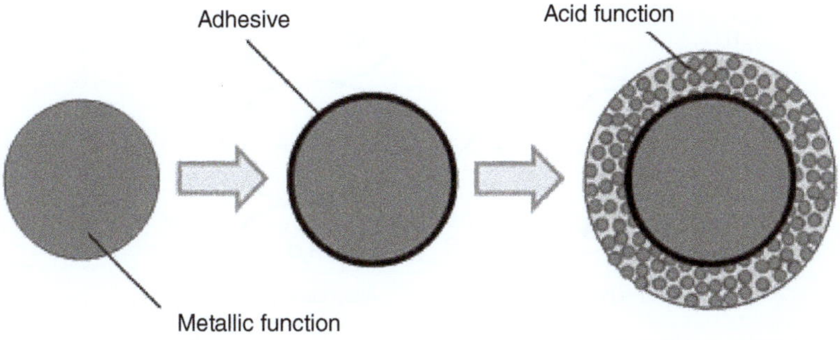

Figure 11.10 Schematic of core-shell zeolite synthesized by physical adhesive method. Source: Reproduced with permission.

the migration of metallic Cu toward zeolite was limited due to confined metallic Cu [137]. The deactivation of the core-shell Cu–ZnO–Al_2O_3 catalyst due to the migration of Al^{3+} species toward metallic Cu has been documented [138].

11.4.3 Mechanism of CO_2 Hydrogenation to DME

MeOH dehydration to DME is an industrially mature process which occurs at moderate pressure and temperature i.e. 20 bar and 150–300 °C [139]. Studies report MeOH dehydration by dissociative and associative pathways (Figure 11.11). The scientific community has not reached to a consensus about the prevailing reaction pathways during MeOH dehydration. In the case of protonated zeolites, both pathways are catalyzed by Brønsted acid sites. In the associative MeOH dehydration mechanism, two MeOH molecules adsorb over the acidic sites of the catalyst and produce DME and by-product water. The protonated MeOH dimer splits into either methyl carboxonium and carbenium ions or two methyl carboxonium ions, which combine to form DME [140]. Whereas, for dissociative DME synthesis, MeOH is adsorbed at the acidic site and converted to methoxy species after losing water. Subsequently, after the nucleophilic attack of the other MeOH molecule, the methoxy species is transformed into DME. Lee and coworkers summarized the disagreements in the literature describing the dominant reaction route, based on catalyst and reaction pathways. The authors reported that over the H-ZSM-5 dissociative route for DME synthesis was more prevalent and DME formation was the slowest kinetic step in the reaction [141]. In contrast, Strizhak and coworkers changed the reaction condition for MeOH dehydration and reported that MeOH adsorption was the rate-determining step. This was valid for both the associative and dissociative pathways [142].

11.5 CO_2 Hydrogenation to Light Olefins

Olefins, particularly ethylene, propylene, and isobutene, are central to modern industry and everyday life due to their extensive applications in producing polymers and derivatives. Ethylene, as the most commercially produced olefin, is

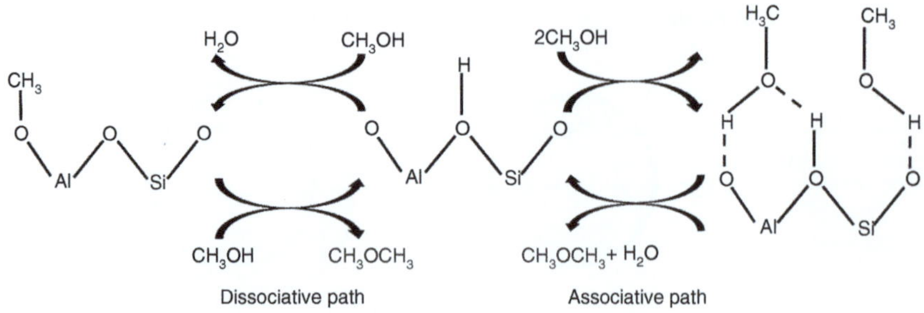

Figure 11.11 Methanol dehydration reaction pathways. Source: Reproduced with permission.

fundamental in creating polyethylene variants [LDPE (low density polyethylene) and HDPE (high density polyethylene)] for packaging, polyvinyl chloride (PVC), and ethylene oxide for textiles. Propylene contributes to polypropylene for packaging and daily use items such as furniture and medical containers, alongside other derivatives such as acrylonitrile and propylene oxide. Isobutene's role in manufacturing plastics, rubbers, and pharmaceuticals, as well as serving as a precursor for solvents and fuel additives such as MTBE (methyl tertiary-butyl ether) and ETBE (ethyl tertiary-butyl ether), underscores its importance [143]. The passage also touches on the value of higher olefins, such as linear α-olefins (LAOs), used in lubricants and detergents, highlighting the challenges in their production from CO_2, including the difficulty of controlling carbon chain growth and achieving high selectivity for desired hydrocarbon properties remains a formidable challenge in this field. This section provides a concise overview of the catalytic system and its underlying reaction mechanism [144, 145].

11.5.1 Catalysts for Light Olefin Synthesis by CO_2 Hydrogenation

The production of olefins from CO_2 through hydrogenation is achieved via two main catalytic routes: the pathways mediated by the RWGS reaction and those facilitated by MeOH intermediates. Each of these routes employs distinct catalysts and processes to transform CO_2 into olefins.

11.5.1.1 Reverse Water–Gas Shift (RWGS)-Mediated Pathway

In the RWGS-mediated pathway, CO_2 is first converted into CO through the RWGS reaction. This CO then undergoes further reactions (such as Fischer–Tropsch synthesis) to form olefins. Catalysts used in this pathway typically include:

Transition Metal Catalysts: Metals such as iron, cobalt, and ruthenium are common catalysts for the RWGS reaction and subsequent Fischer–Tropsch synthesis. These metals facilitate the reduction of CO_2 to CO and then aid in the polymerization of CO into longer-chain hydrocarbons, including olefins. Sodium and potassium play a pivotal role in enhancing the efficiency of catalysts used in olefin synthesis by altering the electronic properties of the catalyst surface and boosting the dispersion of iron and cobalt on the support material. This improvement increases the active surface area available for reactions and optimizes reactant adsorption, streamlining the catalytic process toward olefin production. Additionally, these alkali metals adjust adsorption sites and suppress the hydrogenation of olefins to saturated hydrocarbons, thereby shifting the production preference toward olefins rather than alkanes.

Wang et al. explored the impact of alkali metal promotion (Li, K, Na, Rb, and Cs) on iron oxide catalysts supported on various materials for CO_2 hydrogenation. Among these supports, Fe/ZrO_2 was noted for favoring the production of methane (CH_4) and C_2–C_4 saturated hydrocarbons as primary products [146]. However, the introduction of alkali metals, excluding lithium, shifted the product distribution toward higher olefins (C_2–C_4) and C5+ hydrocarbons, while significantly reducing the formation of CH_4 and saturated C_2–C_4 hydrocarbons. Notably, K and sodium (Na) exhibited comparable catalytic activities, but potassium-promoted samples stood out with a 43% CO_2 conversion and a 44% selectivity for C_2–C_4 olefins. This superior performance is linked to the enhanced formation of Fe_5C_2, an active phase for olefin production, as evidenced by in situ powder X-ray diffraction (XRD) analysis [147]. Conversely, alternative supports such as carbon nanotubes (CNTs) and silica (SiO_2) did not favor the carbon-to-olefins (CTO) pathway. CNTs showed a higher preference for CH_4 and saturated hydrocarbons due to their exceptional hydrogenation capabilities, attributed to their unique hydrogen adsorption and spillover properties. On the other hand, SiO_2 support resulted in lower olefin selectivity and catalytic activity, partly because it did not promote the formation of Fe_5C_2. The study also suggests that the formation of iron silicate, which resists reduction, might hinder the generation of the Fe_5C_2 phase, thereby affecting olefin production [148].

Supported Metal Catalysts: These catalysts involve metals supported on carriers such as zeolites, silica, or alumina, which can enhance the selectivity and stability of the catalyst. The support can also provide acid sites for the secondary reactions necessary for olefin formation.

Incorporating zeolites with iron-based catalysts shows promise for enhancing olefin selectivity, reducing CO by-products, and increasing the likelihood of chain growth. Ramirez and colleagues crafted a set of catalysts using KO_2-doped Fe_2O_3 paired with a highly acidic zeolite, mordenite (MOR), achieving direct CO_2 to light olefins conversion. MOR is particularly effective at selectively producing ethylene within its eight-membered rings, thus improving olefin selectivity. The base catalyst, Fe_2O_3 combined with KO_2, achieved a 48.1% conversion of CO_2, with a total olefin selectivity of 57.4%, 15.2% methane selectivity, and 15.9% CO selectivity, where light olefins comprised 49.3% of the total olefins, having an olefin/paraffin ratio of 8.5. Introducing MOR to this mix notably boosted ethylene (11.5%) and propylene (12.5%) formation, while CO_2 conversion rates remained consistent with those observed using the standalone Fe_2O_3@KO_2 catalyst. Altering the catalyst system's configuration to a dual bed setup with a 1 : 1 mass ratio dramatically shifted product distribution, whereas a mortar-mixed system resulted in poorer performance, primarily producing CO, likely due to mutual deactivation caused by the interaction between the zeolite's acidity and the alkalinity of potassium. Switching MOR to

ZSM-5 zeolite led to a preference for aromatic compounds [149]. Further modifications, including replacing the zeolite with SAPO-34 and applying a ZrS (zirconium sulfide) coating, were found to enhance light olefin selectivity. The Zr precursor forms a bulky tetramer in solution, resulting in a thin layer that doesn't penetrate the SAPO-34 pores. Configuring the Fe_2O_3@KO_2/ZrS/SAPO-34 system in a dual bed arrangement, equally balanced by mass, bolstered the production of ethylene and propylene, increasing light olefin selectivity to 43% among hydrocarbons and decreasing CO selectivity to 13%. The ZrS coating on SAPO-34 pushed the light olefin fraction up to 48%, demonstrating that the positive impact of SAPO-34 was retained, with CO selectivity unchanged. The addition of ZrS to SAPO-34, an acidic modifier, enabled the cracking of C5+ hydrocarbons, thereby boosting light olefin selectivity by over 7% for the same level of conversion, overcoming the challenges posed by lower reaction temperatures and diffusion limitations within the SAPO-34 structure [150].

11.5.1.2 Methanol-Mediated Pathways

The MeOH-mediated pathway involves the conversion of CO_2 to MeOH as an intermediate step, followed by the conversion of MeOH to olefins through processes such as the MeOH-to-olefins (MTO) reaction.

Within the realm of catalyst development for olefin synthesis, silico–alumino–phosphate (SAPO) catalysts, especially SAPO-34 with its chabazite (CHA) structure, have garnered significant interest [151]. The CHA structure of SAPO-34 is distinguished by its small pore size, medium reactivity, and notable hydrothermal stability, maintaining 80% of its crystallinity even after being subjected to steam treatment for 45 h at 800 °C. The pores in CHA are cylindrical cages with eight-membered ring openings, making them adept at capturing olefin intermediates. The silicon content within the SAPO structure also plays a crucial role in the reaction dynamics. Lower silicon levels, where silicon is primarily surrounded by aluminum, enhance the catalyst's activity. This configuration leads to a low hydrogen transfer index and improves long-term stability. However, the narrow pore size of SAPO-34 can lead to rapid deactivation, a challenge that can be mitigated in industrial applications by transitioning from a stationary catalyst bed to a fluidized bed system, which allows for more efficient handling of deactivation issues [152].

Leveraging the known high activity of bare indium oxide (In_2O_3) toward MeOH production due to its abundant oxygen vacancies, Numpilai et al. devised a bifunctional catalyst by amalgamating In_2O_3 with SAPO-34, achieving a 34.6% CO_2 conversion rate and 7.31% selectivity for light olefins [153]. The creation of an optimal interface between the components of this bifunctional catalyst is critical for enhanced catalytic efficacy. Gao et al. further refined this catalyst by incorporating zirconium dioxide (ZrO_2), facilitating the diffusion of MeOH intermediates from the In_2O_3/ZrO_2 surface into SAPO-34 pores for olefin conversion. Pressure studies revealed that an increase from 10 to 30 bar boosts hydrocarbon selectivity while reducing CO formation. Adjustments in space velocity (SV) showed a decrease in CO_2 conversion but an improvement in hydrocarbon selectivity. Comparisons with established MeOH catalysts such as $Cu/ZnO/Al_2O_3$ and $CrZnO_x$, when mixed with SAPO-34, demonstrated that InZr/SAPO-34 outperformed these alternatives in terms of activity. Investigating the spatial arrangement impact, a dual bed setup with SAPO-34 positioned under an oxide layer and separated by quartz sand was tested, resulting in over 55% methane (CH_4) selectivity and 30% selectivity for C_2–C_4 olefins [154].

11.5.2 Mechanism of CO_2 Hydrogenation to Olefins

In examining the underlying mechanisms of light olefin production, it becomes evident that managing the ratio of active hydrogen to carbon plays a crucial role. An abundance of hydrogen atoms (H*) on the catalyst's surface leads to over-hydrogenation, resulting in the production of methane, whereas a scarcity of H* limits the catalyst's capacity to hydrogenate, thus diminishing its effectiveness in converting CO_2. Fundamentally, the essential processes in transforming CO_2 into light olefins involve breaking the carbon–oxygen (C—O) bonds and creating carbon-carbon (C—C) bonds [155].

Iron-based catalysts have garnered significant attention in the CO_2 Fischer–Tropsch synthesis (CO_2-FTS) due to their effectiveness and high catalytic performance in both the RWGS and FTS reactions. In the utilization of Fe-based catalysts for CO_2-FTS, the initial iron oxide phase (Fe_2O_3) undergoes reduction by hydrogen to form magnetite (Fe_3O_4) or a combination of magnetite and wüstite (FeO). This magnetite phase is crucial for the RWGS reaction, with the potential to be further reduced to metallic iron. The proposed mechanism for the CO_2-FTS process begins with CO_2 being adsorbed and activated on the RWGS active sites, such as Fe_3O_4, leading to the formation of a carboxylate intermediate ($*CO_2$, where * denotes the adsorbed state). This $*CO_2$ is then hydrogenated to an *HOCO intermediate, which subsequently splits into *OH and *CO. The *OH is further converted to $*H_2O$ through hydrogenation. The *CO may either be released as CO gas or engage in further FTS reactions, where it is partially hydrogenated to *HCO, followed by complete hydrogenation, dissociation, and dehydration to generate $*CH_x$ intermediates, which serve as the building blocks for olefin production. Alternatively, *CO could break down into *C and *O, with some C diffusing into the iron metal lattice to form metal carbides, such as χ-Fe_5C_2, which acts as the active site for FTS. The carbon species on the χ-Fe_5C_2 surface are then hydrogenated to form CH_x intermediates, leading to hydrocarbon chain growth through carbon-carbon coupling reactions [114].

During the Fischer–Tropsch synthesis (FTS) process, the carbon atoms (*C) produced from the decomposition of carbon monoxide (*CO) may migrate into the iron (α-Fe) metal structure, leading to the creation of various iron carbide phases such as Fe_7C_3, χ-Fe_5C_2, θ-Fe3C, ε'-Fe2.2C, and ε-Fe2C, the prevalence of which is influenced by the specific conditions of the reaction. Iron carbides are pivotal for the hydrogenation and dissociation of CO, as well as for facilitating carbon–carbon (C—C) bond formation. It has been posited by some researchers that χ-Fe_5C_2

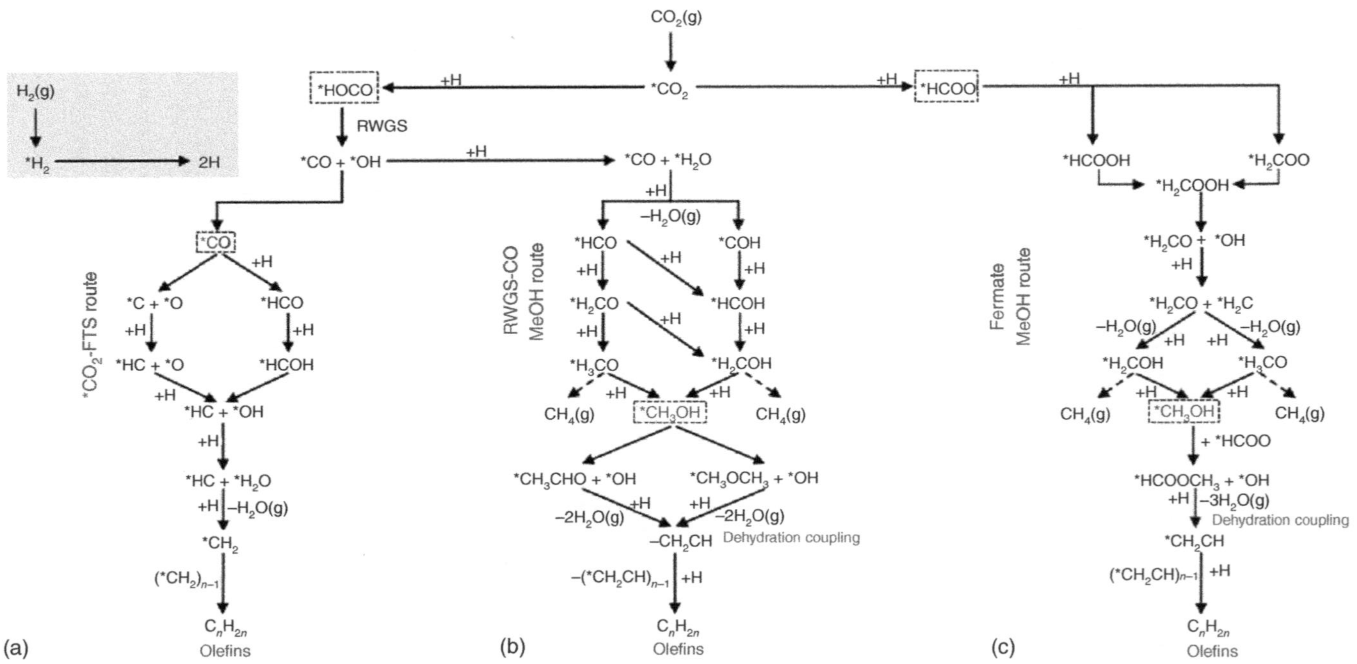

Figure 11.12 Mechanism of CO_2 hydrogenation to olefins. Source: Ref. [115]/MDPI/CC-BY-4.0.

represents the most active phase, whereas θ-Fe_3C shows lesser activity and might contribute to catalyst deactivation by generating graphite. This graphite formation is more stable under conventional FTS conditions and could inhibit the emergence of other active phases by covering them. On a different note, the mechanism for MeOH synthesis involves two distinct routes as illustrated in the proposed schemes (Figure 11.12c). The first pathway is driven by CO, where the *CO intermediate, generated from the RWGS reaction through the dissociation of a carboxyl (*HOCO) species, is converted into MeOH through intermediates such as *HCO and *COH. The second pathway involves formate (*HCOO) formation through the hydrogenation of the carboxylate intermediate (*CO_2), which then undergoes further reactions to become *H_2COOH, *H_2CO, *H_2COH, and finally *H_3CO. Dehydration coupling then transforms these intermediates into *CH_2CH, which can further hydrogenate to form olefins. This dual-pathway approach highlights the complexity and versatility of catalysts in facilitating different chemical conversions from CO_2 and CO to valuable chemical products [115].

11.6 Conclusions and Future Prospects

Despite the growing recognition of climate change and the push for clean energy adoption, fossil fuels continue to dominate as the primary energy source. The combustion of these fuels invariably results in carbon dioxide (CO_2) emissions. Addressing this issue fundamentally requires capturing and converting CO_2 to reintegrate the carbon back into the life cycle. CO_2 utilization technologies focus on transforming it into hydrocarbons, especially smaller olefins such as ethylene (C_2H_4) and alcohols such as MeOH. These conversion processes are hydrogen-intensive, with hydrogen itself being a valuable commodity predominantly sourced from fossil fuels. Consequently, the economic feasibility and process costs of CO_2 hydrogenation largely dictate the choice of end products.

The status of CO_2 hydrogenation to methane, MeOH, DME, and olefin synthesis catalysts and the underlying mechanism have been reviewed. Ni, Fe, and Co are the main active catalysts for methane synthesis, and noble metals such as Rh have also been explored. The Cu and Pd-based catalyst is used for MeOH synthesis, and formate reaction pathways are reported as the reaction mechanism. Bifunctional catalysts for single-step DME and olefin synthesis have been explored to lower the operating cost.

While significant progress has been made in developing catalysts for CO_2 hydrogenation to various chemicals, several challenges remain. These include improving catalyst stability, reducing the cost of noble metal catalysts, enhancing selectivity toward desired products, and understanding the detailed mechanisms to design more effective catalysts. Future research is likely to focus on the development of more efficient and selective catalysts, the optimization of reaction conditions to maximize yield and selectivity, and the integration of renewable energy sources to power the hydrogenation processes sustainably.

References

1 Zhong, J., Yang, X., Wu, Z. et al. (2020). *Chemical Society Reviews* 49: 1385–1413.

2 Ye, R.P., Ding, J., Gong, W. et al. *Nature Communications* 10 (1): 5698. https://doi.org/10.1038/s41467-019-13638-9.

3 Kanuri, S., Roy, S., Chakraborty, C. et al. (2022). *International Journal of Energy Research* 46: 5503–5522.

4 Saeidi, S., Amin, N.A.S., and Rahimpour, M.R. (2014). *Journal of CO_2 Utilization* 5: 66–81.

5 Wang, W., Wang, S., Ma, X., and Gong, J. (2011). *Chemical Society Reviews* 40: 3703.

6 Cao, L., Zhang, H., Zheng, M., and Wang, S. *Environmental Research Letters* 9 (2): 024012. https://doi.org/10.1088/1748-9326/9/2/024012.

7 Olah, G.A., Goeppert, A., and Prakash, G.K.S. (2009). *The Journal of Organic Chemistry* 74 (2): 487–498.

8 Goeppert, A., Czaun, M., Jones, J.-P. et al. (2014). *Chemical Society Reviews* 43: 7995–8048.

9 Koytsoumpa, E.I., Bergins, C., and Kakaras, E. (2018). *Journal of Supercritical Fluids* 132: 3–16.

10 Leung, D.Y.C., Caramanna, G., and Maroto-Valer, M.M. (2014). *Renewable and Sustainable Energy Reviews* 39: 426–443.

11 Vu, T.T.N., Desgagnés, A., and Iliuta, M.C. (2021). *Applied Catalysis A: General* 21 (1): 617.

12 Saeidi, S., Najari, S., Hessel, V. et al. (2021). *Progress in Energy and Combustion Science* 20: 85.

13 Fan, W.K. and Tahir, M. *Journal of Environmental Chemical Engineering* 9 (4): 105460. https://doi.org/10.1016/j.jece.2021.105460.

14 Ashok, J., Ang, M.L., and Kawi, S. (2017). *Catalysis Today* 281: 304–311.

15 Ashok, J., Pati, S., Hongmanorom, P. et al. (2020). *Catalysis Today* 356: 471–489.

16 Aldana, P.A.U., Ocampo, F., Kobl, K. et al. (2013). *Catalysis Today* 215: 201–207.

17 Vogt, C., Monai, M., Kramer, G.J., and Weckhuysen, B.M. (2019). *Nature Catalysis* 2: 188–197.

18 Pan, Q., Peng, J., Sun, T. et al. (2014). *Catalysis Communications* 45: 74–78.

19 Jia, X., Zhang, X., Rui, N. et al. (2019). *Applied Catalysis B: Environmental* 244: 159–169.

20 Heyl, D., Rodemerck, U., and Bentrup, U. (2016). *ACS Catalysis* 6: 6275–6284.

21 Khan, M.U., Wang, L., Liu, Z. et al. (2016). *Angewandte Chemie, International Edition* 55: 9548–9552.

22 Ganesh, I. (2014). *Renewable and Sustainable Energy Reviews* 31: 221–257.

23 Stangeland, K., Li, H., and Yu, Z. (2018). *Industrial and Engineering Chemistry Research* 57: 4081–4094.

24 Bowker, M. (2019). *ChemCatChem* 11: 4238–4246.

25 Cerveny, L. (1989). *Chemical Engineering Communications* 83 (1): 31–63.

26 Teschner, D., Révay, Z., Borsodi, J. et al. (2008). *Angewandte Chemie, International Edition* 47: 9274–9278.

27 Tungler, A., Tarnai, T., and Hegedus, L. (1998). *Platinum Metals Review* 42: 108–115.

28 Jiang, X., Koizumi, N., Guo, X., and Song, C. (2015). *Applied Catalysis B: Environmental* 171: 173–185.

29 Lin, F., Jiang, X., Boreriboon, N. et al. (2019). *Applied Catalysis A: General* 585: 117210.

30 Nie, X., Jiang, X., Wang, H. et al. (2018). *ACS Catalysis* 8: 4873–4892.

31 Jiang, X., Wang, X., Nie, X. et al. (2018). *Catalysis Today* 316: 62–70.

32 Jiang, X., Nie, X., Wang, X. et al. (2019). *Journal of Catalysis* 369: 21–32.

33 Rodriguez, J.A. and Goodman, D.W. (1992). *Science (1979)* 257: 897–903.

34 Vizdal, J., Kroupa, A., Popovic, J., and Zemanova, A. (2006). *Advanced Engineering Materials* 8: 164–176.

35 Liang, X., Dong, X., Lin, G., and Zhang, H. (2009). *Applied Catalysis B: Environmental* 88: 315–322.

36 Ota, A., Kunkes, E.L., Kasatkin, I. et al. (2012). *Journal of Catalysis* 293: 27–38.

37 Liao, F., Wu, X., Zheng, J. et al. (2017). *Green Chemistry* 19: 270–280.

38 Yin, Y., Hu, B., Li, X. et al. (2018). *Applied Catalysis B: Environmental* 234: 143–152.

39 Bahruji, H., Bowker, M., Hutchings, G. et al. (2016). *Journal of Catalysis* 343: 133–146.

40 T. Fujitani, M. Saito, Y. Kanai, T. Watanabe, J. Nakamura and T. Uchijima, 1995, 125, 0–3.

41 Bonivardi, A.L., Chiavassa, D.L., Querini, C.A., and Baltan, M.A. (2000). *Studies in Surface Science and Catalysis* 130: 3747–3752.

42 Oyola-rivera, O., Baltana, M.A., and Cardona-martı, N. (2015). *Journal of CO_2 Utilization* 9: 8–15.

43 Ye, J., Liu, C.J., Mei, D., and Ge, Q. (2014). *Journal of Catalysis* 317: 44–53.

44 Ye, J., Ge, Q., and Liu, C. (2015). *Chemical Engineering Science* 135: 193–201.

45 Rui, N., Wang, Z., Sun, K. et al. (2017). *Applied Catalysis B: Environmental* 218: 488–497.

46 Jiang, H., Lin, J., Wu, X. et al. (2020). *Journal of CO_2 Utilization* 26: 33–39.

47 Abdel-Mageed, A.M., Klyushin, A., Knop-Gericke, A. et al. (2019). *Journal of Physical Chemistry Letters* 10: 3645–3653.

48 Vourros, A., Garagounis, I., Kyriakou, V. et al. (2017). *Journal of CO_2 Utilization* 19: 247–256.

49 Rezvani, A., Abdel-Mageed, A.M., Ishida, T. et al. (2020). *ACS Catalysis* 10: 3580–3594.

50 Yang, X., Kattel, S., Senanayake, S.D. et al. (2015). *Journal of the American Chemical Society* 137: 10104–10107.

51 Rui, N., Zhang, F., Sun, K. et al. (2020). *ACS Catalysis* 10 (19): 11307–11317. https://doi.org/10.1021/acscatal.0c02120.

52 Men, Y.L., Liu, Y., Wang, Q. et al. (2019). *Chemical Engineering Science* 200: 167–175.

53 Bai, S., Shao, Q., Feng, Y. et al. (2017). *Small* 13: 1–6.
54 Toyao, T., Kayamori, S., Maeno, Z. et al. (2019). *ACS Catalysis* 9: 8187–8196.
55 Słoczyński, J., Grabowski, R., Kozłowska, A. et al. (2004). *Applied Catalysis A: General* 278: 11–23.
56 Köppel, R.A., Stöcker, C., and Baiker, A. (1998). *Journal of Catalysis* 179: 515–527.
57 Grabowski, R., Słoczyński, J., Śliwa, M. et al. (2011). *ACS Catalysis* 1: 266–278.
58 Xiao, J. and Frauenheim, T. (2013). *Journal of Physical Chemistry C* 117: 1804–1808.
59 Dostagir, N.H.M., Thompson, C.B., Kobayashi, H. et al. (2020). *Catalysis Science & Technology* 10 (24): 8196–8202. https://doi.org/10.1039/d0cy01789b.
60 Li, M.M.J., Zou, H., Zheng, J. et al. (2020). *Angewandte Chemie, International Edition* 59: 16039–16046.
61 Wang, J., Sun, K., Jia, X., and jun Liu, C. (2020). *Catalysis Today* 20: 0–1.
62 Ye, J., Liu, C., Mei, D., and Ge, Q. (2013). *ACS Catalysis* 3: 1296–1306.
63 Ye, J., Liu, C., and Ge, Q. (2012). *The Journal of Physical Chemistry C* 116: 7817–7825.
64 Sun, K., Fan, Z., Ye, J. et al. (2020). *Journal of CO_2 Utilization* 12: 1–6.
65 Dang, S., Qin, B., Yang, Y. et al. *Science Advances*, https://doi.org/10.1126/sciadv.aaz2060 6 (25): eaaz2060.
66 Dou, M., Zhang, M., Chen, Y., and Yu, Y. (2018). *Surface Science* 672: 7–12.
67 Martin, O., Martín, A.J., Mondelli, C. et al. (2016). *Angewandte Chemie International Edition* 55: 6261–6265.
68 Jiang, X., Nie, X., Gong, Y. et al. (2020). *Journal of Catalysis* 383: 283–296.
69 Chen, P., Tao, L., Zhu, J. et al. (2019). *Energy Technology* 7 (3): 1800747. https://doi.org/10.1002/ente.201800747.
70 Akkharaphatthawon, N., Chanlek, N., Cheng, C.K. et al. (2019). *Applied Surface Science* 489: 278–286.
71 Chou, C.Y. and Lobo, R.F. (2019). *Applied Catalysis A: General* 583: 117144.
72 Ralston, W.T., An, K., Brooks, C. et al. (2015). *Nature Communications* 34: 1–5.
73 Stangeland, K., Kalai, D.Y., Ding, Y., and Yu, Z. (2019). *Journal of CO_2 Utilization* 32: 146–154.
74 Wang, J., Li, G., Li, Z. et al. (2017). *Science Advances* 3: e1701290.
75 Wang, J., Tang, C., Li, G. et al. (2019). *ACS Catalysis* 9: 10253–10259.
76 Kim, Y., Si, T., Trung, B. et al. *ACS Catalysis* 6 (2): 1037–1044. https://doi.org/10.1021/acscatal.5b02083.
77 Lunkenbein, T., Schumann, J., Behrens, M. et al. (2015). *Angewandte Chemie, International Edition* 54: 4544–4548.
78 Medina, J.C., Figueroa, M., Manrique, R. et al. (2017). *Catalysis Science & Technology* 7: 3375–3387.
79 Volnina, E.A. and Kipnis, M.A. (2020). *Kinetics and Catalysis* 61: 119–129.
80 Kuld, S., Thorhauge, M., Falsig, H. et al. (2016). *Science (1979)* 352: 969–974.
81 Behrens, M., Studt, F., Kasatkin, I. et al. (2012). *Science (1979)* 759: 893–898.
82 Kattel, S., Chen, J.G., and Liu, P. (2017). *Science (1979)* 355: 1296–1299.
83 Wu, X.K., Xia, G.J., Huang, Z. et al. (2020). *Applied Surface Science* 525: 146481.

84 Eren, B., Weatherup, R.S., Liakakos, N. et al. (2016). *Journal of the American Chemical Society* 138 (26): 8207–8211. https://doi.org/10.1021/jacs.6b04039.

85 Larmier, K., Liao, W., Tada, S. et al. (2017). *Angewandte Chemie - International Edition* 25: 2318–2323.

86 Yang, Y., Evans, J., Rodriguez, J.A., and White, G. (2010). *Physical Chemistry Chemical Physics* 12 (33): 9909–9917.

87 Wang, Y.H., Gao, G., Wang, H., and Zheng, Y.E. (2017). *RSC Advances* 7: 8709–8717.

88 Ezeh, C.I., Yang, X., He, J. et al. (2018). *Ultrasonics Sonochemistry* 42: 48–56.

89 Grabow, L.C. and Mavrikakis, M. (2011). *ACS Catalysis* 21: 365–384.

90 Behrens, M., Studt, F., Kasatkin, I. et al. (2012). *Science (1979)* 336: 893–897.

91 Dou, M., Zhang, M., Chen, Y., and Yu, Y. (2018). *Computational & Theoretical Chemistry* 1126: 7–15.

92 Frei, M.S., Capdevila-Cortada, M., García-Muelas, R. et al. (2018). *Journal of Catalysis* 361: 313–321.

93 Wu, P. and Yang, B. (2019). *Catalysis Science & Technology* 9: 6102–6113.

94 Zhao, Y., Yang, Y., Mims, C. et al. (2011). *Journal of Catalysis* 281: 199–211.

95 Yang, Y., Mims, C.A., Mei, D.H. et al. (2013). *Journal of Catalysis* 298: 10–17.

96 Yang, Y., Mims, C.A., Disselkamp, R.S. et al. (2010). *Journal of Physical Chemistry C* 114: 17205–17211.

97 Liu, L., Yao, H., Jiang, Z., and Fang, T. (2018). *Applied Surface Science* 451: 333–345.

98 Yang, Y., White, M.G., and Liu, P. (2012). *The Journal of Physical Chemistry C* 248–256.

99 Graciani, J., Mudiyanselage, K., Xu, F. et al. (2014). *Science (1979)* 345: 546–550.

100 Kattel, S., Yan, B., Yang, Y. et al. (2016). *Journal of the American Chemical Society* 138: 12440–12450.

101 Gaikwad, R., Bansode, A., and Urakawa, A. (2016). *Journal of Catalysis* 343: 127–132.

102 Ou, Z., Ran, J., Niu, J. et al. (2020). *International Journal of Hydrogen Energy* 45: 6328–6340.

103 Trippe, F., Fröhling, M., Schultmann, F. et al. (2013). *Fuel Processing Technology* 106: 577–586.

104 Kabir, K.B., Hein, K., and Bhattacharya, S. (2013). *Computers and Chemical Engineering* 48: 96–104.

105 Pontzen, F., Liebner, W., Gronemann, V. et al. (2011). *Catalysis Today* 171: 242–250.

106 Ateka, A., Pérez-Uriarte, P., Gamero, M. et al. (2017). *Energy* 120: 796–804.

107 Hosseini, Z., Taghizadeh, M., and Yaripour, F. (2011). *Journal of Natural Gas Chemistry* 20: 128–134.

108 Liu, D., Yao, C., Zhang, J. et al. (2011). *Fuel* 90: 1738–1742.

109 Akarmazyan, S.S., Panagiotopoulou, P., Kambolis, A. et al. (2014). *Applied Catalysis B: Environmental* 145: 136–148.

110 Xu, M., Lunsford, J.H., Goodman, D.W., and Bhattacharyya, A. (1997). *Applied Catalysis A: General* 149: 289–301.

111 Sierra, I., Ereña, J., Aguayo, A.T. et al. (2011). *Applied Catalysis B: Environmental* 106: 167–173.
112 Sun, X., Mueller, S., Liu, Y. et al. (2014). *Journal of Catalysis* 317: 185–197.
113 Şeker, B., Dizaji, A.K., Balci, V., and Uzun, A. (2021). *Renewable Energy* 171: 47–57.
114 Wang, D., Xie, Z., Porosoff, M.D., and Chen, J.G. (2021). *Chem* 7: 2277–2311.
115 Weber, D., He, T., Wong, M. et al. (2021). *Catalysts* 11 (12): 1447.
116 Vanoye, L., Favre-Réguillon, A., Munno, P. et al. (2013). *Catalysis Today* 215: 239–242.
117 Ereña, J., Garoña, R., Arandes, J.M. et al. (2005). *Catalysis Today* 107–108: 467–473.
118 Sánchez-Contador, M., Ateka, A., Aguayo, A.T., and Bilbao, J. (2018). *Journal of Industrial and Engineering Chemistry* 63: 245–254.
119 Catizzone, E., Giglio, E., Migliori, M. et al. (2020). *Materials* 13: 1–15.
120 Catizzone, E., Aloise, A., Migliori, M., and Giordano, G. (2015). *Applied Catalysis A: General* 502: 215–220.
121 Abu-Dahrieh, J., Rooney, D., Goguet, A., and Saih, Y. (2012). *Chemical Engineering Journal* 203: 201–211.
122 Zhang, M.H., Liu, Z.M., Lin, G.D., and Bin Zhang, H. (2013). *Applied Catalysis A: General* 451: 28–35.
123 García-Trenco, A. and Martínez, A. (2012). *Applied Catalysis A: General* 411–412: 170–179.
124 Aboul-Fotouh, S.M., Ali, L.I., Naghmash, M.A., and Aboul-Gheit, N.A.K. (2017). *Journal of Fuel Chemistry and Technology* 45 (5): 581–588.
125 Catizzone, E., Aloise, A., Migliori, M., and Giordano, G. (2017). *Journal of Energy Chemistry* 26: 406–415.
126 Müller, J.M., Mesquita, G.C., Franco, S.M. et al. (2015). *Microporous and Mesoporous Materials* 204: 50–57.
127 Yao, L., Shen, X., Pan, Y., and Peng, Z. (2020). *Energy and Fuels* 34: 8635–8643.
128 Ateka, A., Sierra, I., Ereña, J. et al. (2016). *Fuel Processing Technology* 152: 34–45.
129 Ateka, A., Sánchez-Contador, M., Ereña, J. et al. (2018). *Reaction Kinetics, Mechanisms and Catalysis* 124: 401–418.
130 Bonura, G., Cordaro, M., Spadaro, L. et al. (2013). *Applied Catalysis B: Environmental* 140–141: 16–24.
131 Bonura, G., Cordaro, M., Cannilla, C. et al. (2014). *Catalysis Today* 228: 51–57.
132 Sánchez-Contador, M., Ateka, A., Aguayo, A.T., and Bilbao, J. (2018). *Fuel Processing Technology* 179: 258–268.
133 Chen, Z., Li, X., Xu, Y. et al. (2018). *Catalysis Communications* 103: 1–4.
134 Wang, Y., Wang, W., Chen, Y. et al. (2014). *Chemical Engineering Journal* 250: 248–256.
135 Guffanti, S., Visconti, C.G., and Groppi, G. (2020). *Industrial and Engineering Chemistry Research* 59: 14252–14266.
136 Guffanti, S., Visconti, C.G., and Groppi, G. (2021). *Industrial and Engineering Chemistry Research* 60: 6767–6783.

137 Nie, R., Lei, H., Pan, S. et al. (2012). *Fuel* 96: 419–425.

138 García-Trenco, A. and Martínez, A. (2014). *Catalysis Today* 227: 144–153.

139 Ateka, A., Rodriguez-Vega, P., Ereña, J. et al. (2022). *Fuel Processing Technology* 233.

140 Ha, K.S., Lee, Y.J., Bae, J.W. et al. (2011). *Applied Catalysis A: General* 395: 95–106.

141 Park, J., Cho, J., Park, M.J., and Lee, W.B. (2021). *Catalysis Today* 375: 314–323.

142 Trypolskyi, A., Zhokh, A., Gritsenko, V. et al. (2021). *Chemical Papers* 75: 3429–3442.

143 Goud, D., Gupta, R., Maligal-Ganesh, R., and Peter, S.C. (2020). *ACS Catalysis* 10: 14258–14282.

144 Chen, S., Wang, J., Feng, Z. et al. (2024). *Angewandte Chemie, International Edition* 63 (8): e202316874. https://doi.org/10.1002/anie.202316874.

145 Liu, Y., He, C., Jiang, X. et al. *Chemical Engineering Journal* 479: 147927. https://doi.org/10.1016/j.cej.2023.147927.

146 Wang, J., You, Z., Zhang, Q. et al. (2013). *Catalysis Today* 215: 186–193.

147 Ngantsoue-Hoc, W., Zhang, Y., O'brien, R.J. et al. (2002). *Applied Catalysis A: General* 236 (1–2): 77–89.

148 Kang, J., Zhang, S., Zhang, Q., and Wang, Y. (2009). *Angewandte Chemie, International Edition* 48: 2565–2568.

149 Ramirez, A., Dutta Chowdhury, A., Dokania, A. et al. (2019). *ACS Catalysis* 9: 6320–6334.

150 Ramirez, A., Dutta Chowdhury, A., Caglayan, M. et al. (2020). *Catalysis Science & Technology* 10: 1507–1517.

151 Bleken, F., Bjørgen, M., Palumbo, L. et al. (2009). *Topics in Catalysis* 52: 218–228.

152 Tian, P., Wei, Y., Ye, M., and Liu, Z. (2015). *ACS Catalysis* 5: 1922–1938.

153 Numpilai, T., Wattanakit, C., Chareonpanich, M. et al. (2019). *Energy Conversion and Management* 180: 511–523.

154 Gao, P., Dang, S., Li, S. et al. (2018). *ACS Catalysis* 8: 571–578.

155 Li, W., Wang, H., Jiang, X. et al. (2018). *RSC Advances* 8: 7651–7669.

146 [illegible]
147 [illegible]
148 [illegible]
149 [illegible]

Index

Solid Base Catalysts: Synthesis, Characterization, and Applications, First Edition.
Edited by Ravi Tomar, K.K. Pant, and Ramesh Chandra.

t

Z